THE TOPOLOGY PROBLEM SOLVER ®

REGISTERED TRADEMARK

A Complete Solution Guide to Any Textbook

P. 17
P. 19

X P. 22
P. 48
P. 54
P. 56
P. 64
P. 65
P. 77

P. 91
P. 79, 80
P 97

P. 160, 161
P. 164

Emil G. Milewski, Ph.D.

Research and Education Association
61 Ethel Road West
Piscataway, New Jersey 08854

THE TOPOLOGY PROBLEM SOLVER ®

Printed in the United States of America

Library of Congress Catalog Card Number 94-65507

International Standard Book Number 0-87891-925-2

PROBLEM SOLVER is a registered trademark of
Research and Education Association, Piscataway, New Jersey

WHAT THIS BOOK IS FOR

Students have generally found topology a difficult subject to understand and learn. Despite the publication of hundreds of textbooks in this field, each one intended to provide an improvement over previous textbooks, students continue to remain perplexed as a result of the numerous conditions that must often be remembered and correlated in solving a problem. Various possible interpretations of terms used in topology have also contributed to much of the difficulties experienced by students.

In a study of the problem, REA found the following basic reasons underlying students' difficulties with topology taught in schools:

(a) No systematic rules of analysis have been developed which students may follow in a step-by-step manner to solve the usual problems encountered. This results from the fact that the numerous different conditions and principles which may be involved in a problem, lead to many possible different methods of solution. To prescribe a set of rules to be followed for each of the possible variations, would involve an enormous number of rules and steps to be searched through by students, and this task would perhaps be more burdensome than solving the problem directly with some accompanying trial and error to find the correct solution route.

(b) Textbooks currently available will usually explain a given principle in a few pages written by a professional who has an insight in the subject matter that is not shared by students. The explanations are often written in an abstract manner which leaves the students confused as to the application of the principle. The explanations given are not sufficiently detailed and extensive to make the student aware of the wide range of applications and different aspects of the principle being studied. The numerous possible variations of principles and their applications are usually not discussed, and it is left for the students to discover these for themselves while doing exercises. Accordingly, the average student is expected to rediscover that which has been long known and practiced, but not published or explained extensively.

(c) The examples usually following the explanation of a topic are too few in number and too simple to enable the student to obtain a thorough grasp of the principles involved. The explanations do not provide sufficient basis to enable a student to solve problems that may be subsequently assigned for homework or given on examinations.

The examples are presented in abbreviated form which leaves out much material between steps, and requires that students derive the omitted material themselves. As a result, students find the examples difficult to understand—contrary to the purpose of the examples.

Examples are, furthermore, often worded in a confusing manner. They do not state the problem and then present the solution. Instead, they pass through a general discussion, never revealing what is to be solved for.

Examples, also, do not always include diagrams/graphs, wherever appropriate, and students do not obtain the training to draw diagrams or graphs to simplify and organize their thinking.

(d) Students can learn the subject only by doing the exercises themselves and reviewing them in class, to obtain experience in applying the principles with their different ramifications.

In doing the exercises by themselves, students find that they are required to devote considerably more time to topology than to other subjects of comparable credits, because they are uncertain with regard to the selection and application of the theorems and principles involved. It is also often necessary for students to discover those "tricks" not revealed in their texts (or review books), that make it possible to solve problems easily. Students must usually resort to methods of trial-and-error to discover these "tricks," and as a result they find that they may sometimes spend several hours to solve a single problem.

(e) When reviewing the exercises in classrooms, instructors usually request students to take turns in writing solutions on the boards and explaining them to the class. Students often find it difficult to explain in a manner that holds the interest of the class, and enables the remaining students to follow the material written on the boards. The remaining students seated in the class are, furthermore, too occupied with copying the material from the boards, to listen to the oral explanations and concentrate on the methods of solution.

This book is intended to aid students in topology overcoming the difficulties described, by supplying detailed illustrations of the solution methods which are usually not apparent to students. The solution methods are illustrated by problems selected from those that are most often assigned for class work and given on examinations. The problems are arranged in order of complexity to enable students to learn and understand a particular topic by reviewing the problems in sequence. The problems are illustrated with detailed step-by-step explanations, to save the students the large amount of time that is often needed to fill in the gaps that are usually found between steps of illustrations in textbooks or review/outline books.

The staff of REA considers topology a subject that is best learned by allowing students to view the methods of analysis and solution techniques themselves. This approach to learning the subject matter is similar to that practiced in various scientific laboratories, particularly in the medical fields.

In using this book, students may review and study the illustrated problems at their own pace; they are not limited to the time allowed for explaining problems on the board in class.

When students want to look up a particular type of problem and solution, they can readily locate it in the book by referring to the index which has been extensively prepared. It is also possible to locate a particular type of problem by glancing at just the material within the boxed portions. To facilitate rapid scanning of the problems, each problem has a heavy border around it. Furthermore, each problem is identified with a number immediately above the problem at the right-hand margin.

To obtain maximum benefit from the book, students should familiarize themselves with the section, "How To Use This Book," located in the front pages.

Special thanks are due to Dr. Nathan Busch for his technical editing of the book.

MAX FOGIEL, Ph.D.
Program Director

HOW TO USE THIS BOOK

This book can be an invaluable aid to students in topology as a supplement to their textbooks. The book is subdivided into 19 chapters, each dealing with a separate topic. The subject matter is developed beginning with the fundamental concepts of sentence calculus, sets, mappings, and extending through metric and topological spaces, continuity, homeomorphisms, axioms, compactness, connectedness, and homotopy theory.

TO LEARN AND UNDERSTAND
A TOPIC THOROUGHLY

1. Refer to your class text and read the section pertaining to the topic. You should become acquainted with the principles discussed there. These principles, however, may not be clear to you at that time.

2. Then locate the topic you are looking for by referring to the "Table of Contents" in front of this book, "The Topology Problem Solver."

3. Turn to the page where the topic begins and review the problems under each topic, in the order given. For each topic, the problems are arranged in order of complexity, from the simplest to the more difficult. Some problems may appear similar to others, but each problem has been selected to illustrate a different point or solution method.

To learn and understand a topic thoroughly and retain its contents, it will be generally necessary for students to review the problems several times. Repeated review is essential in order to gain experience in recognizing the principles that should be applied, and in selecting the best solution technique.

TO FIND A PARTICULAR PROBLEM

To locate one or more problems related to a particular subject matter, refer to either the index, located at the back of the book or the indexes found at the beginning of each chapter. In using the indexes, be certain to note that the numbers given refer to problem numbers, not page numbers. This arrangement of the indexes is intended to facilitate finding a problem more rapidly, since two or more problems may appear on a page.

If a particular type of problem cannot be found readily, it is recommended that the student refer to the "Table of Contents" in the front pages, and then turn to the chapter which is applicable to the problem being sought. By scanning or glancing at the material that is boxed, it will generally be possible to find problems related to the one being sought, without consuming considerable time. After the problems have been located, the solutions can be reviewed and studied in detail. For this purpose of locating problems rapidly, students should acquaint themselves with the organization of the book as found in the "Table of Contents."

In preparing for an exam, locate the topics to be covered on the exam in the "Table of Contents," and then review the problems under those topics several times. This should equip the student with what might be needed for the exam.

CONTENTS

CHAPTER 1

INTRODUCTION TO SENTENCE CALCULUS

Let α, β, γ, ... denote sentences, each of which has one of two logical values 0 or 1. We assign the value 0 to a false sentence and 1 to a true sentence. To express the fact that α is a true sentence we write

$$\alpha \equiv 1 \text{ or } \beta \equiv 0 \qquad (1)$$

if β is false.

We consider here only sentences of mathematical nature, i.e., sentences which take values either 0 or 1. For example, the sentence: "two times two is four" is (no doubt) a true one. On the other hand, the statement: "Goethe is a great poet" is not a sentence even though most people would agree that he indeed was a great poet.

Define the sum, the product and negation of sentences and establish the truth tables for each case.

SOLUTION:

If α and β are two sentences, then the sentence "α or β," denoted by $\alpha \vee \beta$, is called the sum of α and β. The sentence "α and β," denoted by $\alpha \wedge \beta$, is called the product of α and β. Clearly, the sentence $\alpha \vee \beta$ is true if at least one of the components is a true proposition. It has $2 \times 2 = 4$ logical possibilities.

Table 1

α	β	$\alpha \vee \beta$
1	1	1
1	0	1
0	1	1
0	0	0

Sentence $\alpha \wedge \beta$ is true if both factors are true sentences.

Table 2

α	β	$\alpha \wedge \beta$
1	1	1
1	0	0
0	1	0
0	0	0

The sum and product of sentences are commutative and associative, i.e.

$$\alpha \vee \beta \equiv \beta \vee \alpha, \quad \alpha \wedge \beta \equiv \beta \wedge \alpha \qquad (2)$$

$$\alpha \vee (\beta \vee \gamma) \equiv (\alpha \vee \beta) \vee \gamma, \quad \alpha \wedge (\beta \wedge \gamma) \equiv (\alpha \wedge \beta) \wedge \gamma. \qquad (3)$$

The distributive law holds

$$\alpha \wedge (\beta \vee \gamma) \equiv (\alpha \wedge \beta) \vee (\alpha \wedge \gamma). \qquad (4)$$

In the above formulas we used the equivalence sign $\equiv$. The equivalence $\alpha \equiv \beta$ holds, if and only if α and β have the same logical value. The operation of negation of a sentence α denoted by α' (or by $\rceil \alpha$ or $\sim \alpha$) is defined in such a way that if α is true, then α' is false, and if α is false, then α' is true.

Table 3

α	α'
0	1
1	0

From this we obtain the law of double negation

$$\alpha'' \equiv \alpha. \qquad (5)$$

● **PROBLEM 1–2**

Using the logical notation introduced in Problem 1-1, write down two fundamental theorems of Aristotelian logic. Prove that $\alpha \vee 1 \equiv 1$, $\alpha \wedge 0 \equiv 0$.

SOLUTION:

The first theorem is called the law of the excluded middle (principium tertii exclusi) and, in classical logic, is formulated in the following manner:

From two contradictory sentences, one is true.

Let α represent a sentence, then α' is its contradiction. The law can be written as

$$\alpha \vee \alpha' \equiv 1. \qquad (1)$$

The second theorem, called the law of contradiction, states: no sentence can

3

be true simultaneously with its negation. It can be written briefly as follows:

$$\alpha \wedge \alpha' \equiv 0. \tag{2}$$

Let us examine the truth table for the statement $\alpha \vee 1$:

α	$\alpha \vee 1$
1	1
0	1

Hence, $\alpha \vee 1$ is a statement which is always true.

$$\alpha \vee 1 \equiv 1. \tag{3}$$

Similarly, the truth table for $\alpha \wedge 0$ is

α	$\alpha \wedge 0$
1	0
0	0

and

$$\alpha \wedge 0 \equiv 0. \tag{4}$$

● **PROBLEM 1-3**

So far, we defined three operations on sentences: product, sum and negation. By applying DeMorgan's laws we see that the number of these fundamental operations can be reduced to two.

SOLUTION:

DeMorgan's laws play an important role in mathematical logic. They can be formulated as follows:

$$(\alpha \vee \beta)' \equiv \alpha' \wedge \beta' \tag{1}$$

$$(\alpha \wedge \beta)' \equiv \alpha' \vee \beta'. \tag{2}$$

The first law (1) asserts that, if it is not true that one of the sentences α and β is true, then both of these sentences are false and conversely.

Taking the negation of (1) we obtain:

$$(\alpha \vee \beta)'' \equiv \alpha \vee \beta \equiv (\alpha' \wedge \beta')'. \tag{3}$$

4

Equation (3) can be treated as a definition of the sum of sentences.

Here, the sum is defined in terms of two operations: product and negation. Hence, we limit the number of fundamental operations to two, product and negation. Similarly, taking the negation of (2) we obtain

$$\alpha \wedge \beta \equiv (\alpha' \vee \beta')'. \tag{4}$$

Now, the product is defined in terms of sum and negation. Again the number of fundamental operations is reduced to two. Product is defined with the aid of sum and negation. It is easy to verify DeMorgan's laws using the truth tables.

Table 1
for $(\alpha \vee \beta)' \equiv \alpha' \wedge \beta'$

α	β	$(\alpha \vee \beta)'$	$\alpha' \wedge \beta'$
1	1	0	0
1	0	0	0
0	1	0	0
0	0	1	1

Table 2
for $(\alpha \wedge \beta)' \equiv \alpha' \vee \beta'$

α	β	$(\alpha \wedge \beta)'$	$\alpha' \vee \beta'$
1	1	0	0
1	0	1	1
0	1	1	1
0	0	1	1

● PROBLEM 1-4

Generalize DeMorgan's laws for n components.

SOLUTION:

DeMorgan's laws state that

$$(\alpha_1 \vee \alpha_2)' \equiv \alpha_1' \wedge \alpha_2' \tag{1}$$

$$(\alpha_1 \wedge \alpha_2)' \equiv \alpha_1' \vee \alpha_2'. \tag{2}$$

We shall generalize (1) for the system of n sentences: $\alpha_1, \alpha_2, \ldots, \alpha_n$.

$$(\alpha_1 \vee \alpha_2 \vee \ldots \vee \alpha_{n-1} \vee \alpha_n)' \equiv [(\alpha_1 \vee \alpha_2 \ldots \vee \alpha_{n-1}) \vee \alpha_n]' \equiv \qquad (3)$$

The sum of the sentences is associative. Applying (1) to (3) we find

$$\equiv (\alpha_1 \vee \alpha_2 \vee \ldots \vee \alpha_{n-1})' \wedge \alpha_n' \equiv$$

$$\equiv [(\alpha_1 \vee \alpha_2 \vee \ldots \vee \alpha_{n-2}) \vee \alpha_{n-1}]' \wedge \alpha_n' \equiv \qquad (4)$$

Applying (1) again we obtain

$$(\alpha_1 \vee \alpha_2 \vee \ldots \vee \alpha_{n-1})' \wedge \alpha_n' \equiv (\alpha_1 \vee \alpha_2 \vee \ldots \vee \alpha_{n-2})' \wedge \alpha_{n-1}' \wedge \alpha_n'. \qquad (5)$$

Observe that applying the above procedure to the sentence $(\alpha_1 \vee \alpha_2 \vee \ldots \vee \alpha_n)'$, $(n-1)$ times we find

$$(\alpha_1 \vee \alpha_2 \vee \ldots \vee \alpha_n)' \equiv \alpha_1' \wedge \alpha_2' \wedge \ldots \wedge \alpha_n'. \qquad (6)$$

Now, take the negation of the product, $(\alpha_1 \wedge \alpha_2 \wedge \ldots \wedge \alpha_n)'$. Since the product is associative, we have

$$(\alpha_1 \wedge \alpha_2 \wedge \ldots \wedge \alpha_n)' \equiv [(\alpha_1 \wedge \alpha_2 \wedge \ldots \wedge \alpha_{n-1}) \wedge \alpha_n]' \equiv \qquad (7)$$

Applying (2) to (7) we obtain

$$\equiv (\alpha_1 \wedge \alpha_2 \wedge \ldots \wedge \alpha_{n-1})' \vee \alpha_n' \equiv \qquad (8)$$

This procedure can be repeated again:

$$\equiv [(\alpha_1 \wedge \alpha_2 \wedge \ldots \wedge \alpha_{n-2}) \wedge \alpha_{n-1}]' \vee \alpha_n' \equiv$$

$$\equiv (\alpha_1 \wedge \ldots \wedge \alpha_{n-2})' \vee \alpha_{n-1} \vee \alpha_n' \equiv \qquad (9)$$

After $(n-1)$ steps we obtain

$$(\alpha_1 \wedge \alpha_2 \wedge \ldots \wedge \alpha_n)' \equiv \alpha_1' \vee \alpha_2' \vee \ldots \vee \alpha_n'. \qquad (10)$$

● **PROBLEM 1-5**

Another important operation on sentences is implication. It is defined by

$$(\alpha \Rightarrow \beta) \equiv (\alpha' \vee \beta) \qquad (1)$$

The sign "$\Rightarrow$" is read "implies" or "if ..., then ...".

1. Write the truth table for the implication $\alpha \Rightarrow \beta$.

2. Explain the difference between "if ..., then ..." as defined in a strictly mathematical way by (1) and everyday "if ..., then ...."

6

3. Show that

$$\text{if } \alpha \Rightarrow \beta \text{ and } \beta \Rightarrow \alpha, \text{ then } \alpha \equiv \beta. \tag{2}$$

SOLUTION:

Let us start with the truth table for $\alpha \Rightarrow \beta$.

Table 1

α	β	$(\alpha \Rightarrow \beta) \equiv (\alpha' \vee \beta)$
1	1	1
1	0	0
0	1	1
0	0	1

1. The implication $\alpha \Rightarrow \beta$ is always true, except when α is true ($\alpha \equiv 1$) and β is false ($\beta \equiv 0$).

2. Implication has properties similar to deduction. There is, however, a small but significant difference. Let α denote the statement "The sun is shining" and let β denote the statement "I am swimming." The statement "If the sun is shining, then I am swimming" makes sense. When is such a statement false?

When the sun is shining, I am not swimming, i.e., $\alpha \wedge \beta' \equiv 1$. But, that is equivalent to

$$1' \equiv 0 \equiv (\alpha \wedge \beta')' \equiv \alpha' \vee \beta \tag{3}$$

in agreement with Table 1. In everyday language a sentence of the form "If α, then β" means that β is true whenever α is true. In such a case, the statement "If two times two is five, then Paris is the capital of Germany" is regarded as nonsense, since both components are false.

The same statement, in terms of formal language, not only makes sense but is also true. To each of the four possibilities shown in Table 1 (for $\alpha \Rightarrow \beta$), the system of formal language assigns either 0 or 1, even though two of the cases ($\alpha \equiv 0$ and $\beta \equiv 0$, $\alpha \equiv 0$ and $\beta \equiv 1$) appear to be pure nonsense in ordinary language.

3. To prove (2), note that

$$(\alpha \Rightarrow \beta) \text{ and } (\beta \Rightarrow \alpha) \equiv (\alpha' \vee \beta) \wedge (\beta' \vee \alpha) \equiv$$

$$\equiv (\alpha' \wedge \beta') \vee (\alpha' \wedge \alpha) \vee (\beta \wedge \beta') \vee (\beta \wedge \alpha) \equiv$$

$$\equiv (\alpha' \wedge \beta') \vee (\alpha \wedge \beta). \tag{4}$$

7

The sentence $(\alpha' \wedge \beta') \vee (\alpha \wedge \beta)$ is true when both components α and β have the same logical value; i.e., they are both true or both false.

Hence,

$$\text{if } \alpha \Rightarrow \beta \text{ and } \beta \Rightarrow \alpha, \text{ then } \alpha \equiv \beta. \tag{5}$$

Symbols $\Leftrightarrow$ and $\equiv$ may be used interchangeably.

The symbol $\Leftrightarrow$ is read as: "if, and only if," and indicates implication in both directions. Sometimes, instead of if, and only if, we write iff.

Table 2 lists the symbols used so far.

Table 2

Symbol	Meaning
$\alpha, \beta, \gamma, \delta, \ldots$	logical sentences
$'$ or $\sim$ or $\rceil$	negation (not)
$\vee$	sum (or)
$\wedge$	product (and)
$\Rightarrow$	implication (if ..., then ...)
$\Leftrightarrow$	iff (if, and only if)

● **PROBLEM 1–6**

1. Prove the syllogism law (sometimes called the law of implication):

$$\text{If } \alpha \Rightarrow \beta \text{ and } \beta \Rightarrow \gamma, \text{ then } \alpha \Rightarrow \gamma. \tag{1}$$

2. Prove the law of contraposition, which is the basis of indirect proof (proof by contradiction, reductio ad absurdum):

$$(\alpha \Rightarrow \beta) \Leftrightarrow (\beta' \Rightarrow \alpha'). \tag{2}$$

3. Prove the law of Duns Scotus:

$$\text{If } \alpha \equiv 0, \text{ then } \alpha \Rightarrow \beta. \tag{3}$$

4. Prove the law of Clausius:

$$\text{If } \alpha' \Rightarrow \beta \text{ for each } \beta, \text{ then } \alpha \equiv 1. \tag{4}$$

SOLUTION:

1. Let us write (1) in the form

$$[(\alpha' \vee \beta) \wedge (\beta' \vee \gamma)] \Rightarrow (\alpha' \vee \gamma). \tag{5}$$

Note that the implication $\alpha \Rightarrow \beta$ is always true, except in the case when $\alpha \equiv 1$ and $\beta \equiv 0$. Suppose

$$(\alpha' \vee \beta) \wedge (\beta' \vee \gamma) \equiv 1 \quad \text{and} \quad \alpha' \vee \gamma \equiv 0. \tag{6}$$

From $\alpha' \vee \gamma \equiv 0$, we conclude that $\alpha \equiv 1$ and $\gamma \equiv 0$. Substituting into (6) we obtain

$$(0 \vee \beta) \wedge (\beta' \vee 0) \equiv 1 \quad \text{or} \quad \beta \wedge \beta' \equiv 1 \tag{7}$$

which is a contradiction.

2. $\Rightarrow$ We shall prove that

$$(\alpha \Rightarrow \beta) \Rightarrow (\beta' \Rightarrow \alpha') \tag{8}$$

which is equivalent to

$$(\alpha \Rightarrow \beta)' \vee (\beta' \Rightarrow \alpha') \tag{9}$$

and to

$$(\alpha' \vee \beta)' \vee (\beta \vee \alpha') \tag{10}$$

and to

$$(\alpha \wedge \beta') \vee (\alpha' \vee \beta). \tag{11}$$

The last statement is true for any combintation of α and β.

$\Leftarrow$
$$(\beta' \Rightarrow \alpha') \Rightarrow (\alpha \Rightarrow \beta) \tag{12}$$

or

$$(\beta \vee \alpha')' \vee (\alpha' \vee \beta) \tag{13}$$

or

$$(\beta' \wedge \alpha) \vee (\alpha' \vee \beta) \equiv 1. \tag{14}$$

■

3. Suppose $\alpha \equiv 0$, then $\alpha' \equiv 1$. Therefore

$$(\alpha \Rightarrow \beta) \equiv (\alpha' \vee \beta) \equiv 1. \tag{15}$$

4. If for $\beta \equiv 0$ and $\beta \equiv 1$ we have

$$(\alpha' \Rightarrow \beta) \equiv (\alpha \vee \beta) \equiv 1 \tag{16}$$

then

$$\alpha \equiv 1. \tag{17}$$

Throughout this book we will be using a box (■) to indicate the end of a proof, theorem, or definition.

Note that there is a shorter way to prove (2)

$$(\alpha \Rightarrow \beta) \equiv (\beta' \vee \alpha').$$ (18)

Indeed,

$$(\alpha' \vee \beta) \equiv (\beta'' \vee \alpha').$$ (19)

● **PROBLEM 1-7**

Prove that:

1. $\alpha \Rightarrow \alpha \vee \beta$ (1)

2. $\alpha \Rightarrow \alpha$ (2)

3. $\alpha \wedge \beta \Rightarrow \alpha \wedge \beta$ (3)

4. $\alpha \wedge \beta \Rightarrow \beta$ (4)

SOLUTION:

1. We shall apply the definition of implication (see Problem 1–5, (1))

$$(\alpha \Rightarrow \beta) \equiv \alpha' \vee \beta .$$ (5)

$$(\alpha \Rightarrow \alpha \vee \beta) \equiv \alpha' \vee \alpha \vee \beta \equiv (\alpha' \vee \alpha) \vee \beta \equiv$$

$$\equiv 1 \vee \beta \equiv 1.$$ (6)

Here, we applied the law of excluded middle (see Problem 1–2, (1)).

2. $$(\alpha \Rightarrow \alpha) \equiv \alpha' \vee \alpha \equiv 1.$$ (7)

3. In (2) we can replace α by $\alpha \wedge \beta$ to obtain (3).

4. $$(\alpha \wedge \beta \Rightarrow \beta) \equiv (\alpha \wedge \beta)' \vee \beta \equiv$$

$$\equiv \alpha' \vee \beta' \vee \beta \equiv 1.$$ (8)

In (8) we applied DeMorgan's law (See Problem 1–3, (2)).

Note that each of the above problems can be solved "automatically" using the truth tables.

Table

α	β	$\alpha \vee \beta$	$\alpha \wedge \beta$	$\alpha \Rightarrow \alpha \vee \beta$	$\alpha \Rightarrow \alpha$	$\alpha \wedge \beta \Rightarrow \alpha \wedge \beta$	$\alpha \wedge \beta \Rightarrow \beta$
1	1	1	1	1	1	1	1
1	0	1	0	1	1	1	1
0	1	1	0	1	1	1	1
0	0	0	0	1	1	1	1

Now it should be clear why this system is called a formal language.

● **PROBLEM 1-8**

1. Prove that:

$$(\alpha \Rightarrow \beta) \Rightarrow (\alpha \wedge \gamma \Rightarrow \beta \wedge \gamma). \qquad (1)$$

2. Prove that:

$$(\alpha \equiv \beta) \equiv \quad (\alpha \wedge \beta) \vee (\alpha' \wedge \beta'). \qquad (2)$$

SOLUTION:

1. From the definition of implication, we conclude that statement (1) is equivalent to

$$(\alpha \Rightarrow \beta)' \vee (\alpha \wedge \gamma \Rightarrow \beta \wedge \gamma) \qquad (3)$$

which is equivalent to

$$(\alpha' \vee \beta)' \vee [(\alpha \wedge \gamma)' \vee (\beta \wedge \gamma)]. \qquad (4)$$

From DeMorgan's law, we find that (4) can be replaced by

$$(\alpha \wedge \beta') \vee (\alpha' \vee \gamma') \vee (\beta \wedge \gamma). \qquad (5)$$

Statement (5) is always true. Indeed, it is true when $\alpha' \equiv 1$ or $\gamma' \equiv 1$. Suppose $\alpha' \equiv 0$ and $\gamma' \equiv 0$, then $\alpha \equiv 1$ and $\gamma \equiv 1$ and for any β, $(1 \wedge \beta') \vee (\beta \wedge 1) \equiv 1$. Thus

$$(\alpha \wedge \beta') \vee (\alpha' \vee \gamma') \vee (\beta \wedge \gamma) \equiv 1. \qquad (6)$$

2. Remember that $\equiv$ can be replaced by $\Leftrightarrow$ (See Problem 1-5). Then

$$(\alpha \equiv \beta) \equiv (\alpha \Leftrightarrow \beta) \equiv (\alpha \Rightarrow \beta) \wedge (\beta \Rightarrow \alpha) \equiv$$

11

$$\equiv (\alpha' \lor \beta) \land (\beta' \lor \alpha) \equiv (\alpha' \land \beta') \lor (\alpha' \land \alpha) \lor (\beta \land \beta') \lor (\alpha \land \beta) \equiv$$

$$\equiv (\alpha' \land \beta') \lor (\alpha \land \beta). \tag{7}$$

As always, the proofs can be carried out by means of the truth tables.

● PROBLEM 1–9

1. Prove that if $\alpha \Rightarrow \beta$, then

$$\alpha \land \beta \equiv \alpha \tag{1}$$

and

$$\alpha \lor \beta \equiv \beta. \tag{2}$$

2. Prove that if $\alpha \Rightarrow \beta$ and $\gamma \Rightarrow \delta$, then

$$\alpha \land \gamma \Rightarrow \beta \land \delta \tag{3}$$

$$\alpha \lor \gamma \Rightarrow \beta \lor \delta. \tag{4}$$

SOLUTION:

1. Since $\alpha \Rightarrow \beta$, we have

$$\alpha' \lor \beta \equiv 1 \tag{5}$$

and

$$\alpha \equiv \alpha \land 1 \equiv \alpha \land (\alpha' \lor \beta) \equiv (\alpha \land \alpha') \lor (\alpha \land \beta) \equiv$$

$$\equiv \alpha \land \beta. \tag{6}$$

Similarly, since $\alpha' \lor \beta \equiv 1$ we obtain

$$(\alpha' \lor \beta)' \equiv \alpha \land \beta' \equiv 1' \equiv 0 \tag{7}$$

and

$$\beta \equiv \beta \lor 0 \equiv \beta \lor (\alpha \land \beta') \equiv \alpha \lor \beta. \tag{8}$$

2. Since $\alpha \Rightarrow \beta$, then for any sentence γ we have

$$\alpha \land \gamma \Rightarrow \beta \land \gamma. \tag{9}$$

See Problem 1–8, (1).
    Similarly, since $\gamma \Rightarrow \delta$, then

$$\beta \land \gamma \Rightarrow \beta \land \delta. \tag{10}$$

12

Combining (9) and (10) we obtain, by virture of Problem 1–6,

$$\alpha \wedge \gamma \Rightarrow \beta \wedge \delta. \tag{11}$$

To prove (4), note that

$$(\alpha \Rightarrow \beta) \Rightarrow (\beta' \Rightarrow \alpha') \tag{12}$$

$$(\gamma \Rightarrow \delta) \Rightarrow (\delta' \Rightarrow \gamma'). \tag{13}$$

Applying (11) to (12) and (13), we find

$$\beta' \wedge \delta' \Rightarrow \alpha' \wedge \gamma'. \tag{14}$$

Applying the law of contraposition (Problem 1–6), (2)) to (14) we obtain from (14)

$$(\alpha' \wedge \gamma')' \Rightarrow (\beta' \wedge \delta')' \tag{15}$$

or

$$(\alpha \vee \gamma) \Rightarrow (\beta \vee \delta). \tag{16}$$

● **PROBLEM 1–10**

Prove the law of absorption:

$$\alpha \equiv \alpha \vee (\alpha \wedge \beta) \equiv \alpha \wedge (\alpha \vee \beta) \tag{1}$$

and its more general version

$$\alpha \vee (\beta \wedge \gamma) \equiv (\alpha \vee \beta) \wedge (\alpha \vee \gamma). \tag{2}$$

SOLUTION:

We have

$$\alpha \wedge (\alpha \vee \beta) \equiv (\alpha \wedge \alpha) \vee (\alpha \wedge \beta) \equiv \alpha \vee (\alpha \wedge \beta). \tag{3}$$

Now we shall show that

$$\alpha \equiv \alpha \vee (\alpha \wedge \beta). \tag{4}$$

Indeed, if $\alpha \equiv 1$, then

$$\alpha \vee (\alpha \wedge \beta) \equiv 1 \vee (1 \wedge \beta) \equiv 1 \equiv \alpha. \tag{5}$$

If $\alpha \equiv 0$, then

$$\alpha \vee (\alpha \wedge \beta) \equiv 0 \vee (0 \wedge \beta) \equiv 0 \equiv \alpha. \tag{6}$$

That completes (1).
To prove (2), note that

$$(\alpha \vee \beta) \wedge (\alpha \vee \gamma) \equiv [\alpha \wedge (\alpha \vee \gamma)] \vee [\beta \wedge (\alpha \vee \gamma)] \equiv \qquad (7)$$

By applying (1) we obtain

$$\equiv \alpha \vee [\beta \wedge (\alpha \vee \gamma)] \equiv \alpha \vee [(\alpha \wedge \beta) \vee (\beta \wedge \gamma)] \equiv$$

$$\equiv [\alpha \vee (\alpha \wedge \beta)] \vee (\beta \wedge \gamma) \equiv \qquad (8)$$

Again, by applying (1), we obtain from (8)

$$(\alpha \vee \beta) \wedge (\alpha \vee \gamma) \equiv \alpha \vee (\beta \wedge \gamma). \qquad (9)$$

● **PROBLEM 1–11**

The symmetric difference of the sentences α and β is defined by

$$(\alpha \div \beta) \equiv [(\alpha \wedge \beta') \vee (\alpha' \wedge \beta)]. \qquad (1)$$

Prove that:

$$\alpha \vee \beta \equiv [(\alpha \div \beta) \div (\alpha \wedge \beta)]. \qquad (2)$$

SOLUTION:

The easiest way to prove (2) is to establish the truth table and to compare the corresponding values. The truth table for $\alpha \div \beta$ is:

Table 1

α	β	$\alpha \wedge \beta'$	$\alpha' \wedge \beta$	$\alpha \div \beta$
1	1	0	0	0
1	0	1	0	1
0	1	0	1	1
0	0	0	0	0

Utilizing Table 1, we obtain

Table 2

α	β	$\alpha \vee \beta$	$(\alpha \div \beta) \div (\alpha \wedge \beta)$
1	1	1	1
1	0	1	1
0	1	1	1
0	0	0	0

By comparing the results of the last two columns of Table 2, we find that (2) is true.

We can also carry out the proof of (2) without the use of the truth table.

$$(\alpha \div \beta) \div (\alpha \wedge \beta) \equiv [(\alpha \div \beta) \wedge (\alpha \wedge \beta)'] \vee [(\alpha \div \beta)' \wedge (\alpha \wedge \beta)] \equiv$$

$$\equiv \{[(\alpha \wedge \beta') \vee (\alpha' \wedge \beta)] \wedge (\alpha' \vee \beta')\} \vee$$

$$\vee \{[(\alpha \wedge \beta') \vee (\alpha' \wedge \beta)]' \wedge (\alpha \wedge \beta)\} \equiv$$

$$\equiv [(\alpha' \vee \beta') \wedge (\alpha \wedge \beta')] \vee [(\alpha' \vee \beta') \wedge (\alpha' \vee \beta)] \vee$$

$$\vee [(\alpha \wedge \beta')' \wedge (\alpha' \wedge \beta)' \wedge (\alpha \wedge \beta)] \equiv$$

$$\equiv (\alpha \wedge \beta') \vee (\alpha' \wedge \beta) \vee [(\alpha' \vee \beta) \wedge (\alpha \vee \beta') \wedge (\alpha \wedge \beta)] \equiv$$

$$\equiv (\alpha \wedge \beta') \vee (\alpha' \wedge \beta) \vee (\alpha \wedge \beta) \equiv$$

$$\equiv (\alpha \wedge \beta') \vee [\beta \wedge (\alpha \vee \alpha')] \equiv$$

$$\equiv \beta \vee (\alpha \wedge \beta') \equiv \alpha \vee \beta. \qquad (3)$$

● PROBLEM 1–12

The digital (logic) circuits operate in the binary mode where each input and output voltage is either 1 or 0. This enables us to use Boolean algebra, which differs from ordinary algebra in that Boolean variables and constants are only allowed to have two possible values, 1 or 0. The voltage values in the circuit are predetermined, in the sense that any voltage in the range of 0V to 0.5V corresponds to "0" and any voltage in the range of 2V to 6V corresponds to "1." The values between 0.5V and 2V should not occur in the circuit. Table 1 lists some commonly used terms.

Table 1

Logic 1	Logic 0
True	False
On	Off
High	Low
Yes	No
Closed switch	Open switch

1. Describe the OR gate.

2. Describe the AND gate.

3. Describe the NOT gate (inverter).

SOLUTION:

1. In a digital circuit, an OR gate has two or more inputs, and one output is equal to the sum of the inputs.

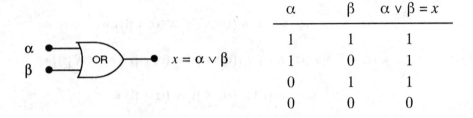

α	β	α ∨ β = x
1	1	1
1	0	1
0	1	1
0	0	0

FIGURE 1

Figure 1 shows the symbol of the OR gate with two inputs and its truth table.

An OR gate with 4 inputs is shown in Figure 2.

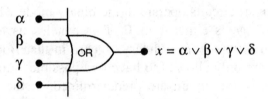

FIGURE 2

2. A two-input AND gate and its truth table is shown in Figure 3.

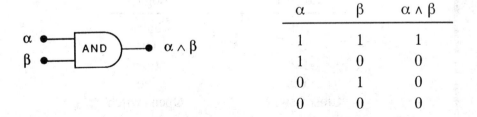

α	β	α ∧ β
1	1	1
1	0	0
0	1	0
0	0	0

FIGURE 3

16

3. The NOT gate (inverter) has one input and one output.

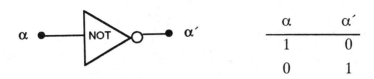

α	α'
1	0
0	1

FIGURE 4

● **PROBLEM 1–13**

1. Replace the following digital circuit with the simpler one:

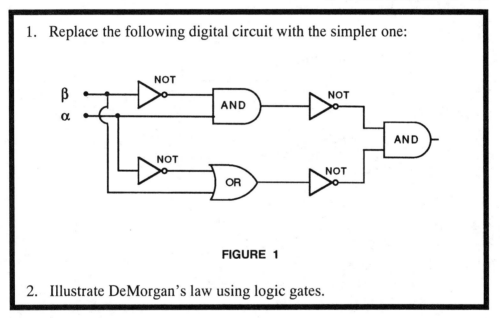

FIGURE 1

2. Illustrate DeMorgan's law using logic gates.

SOLUTION:

1. The circuit shown in Figure 1 has two inputs, one output, and it performs the function described by

$$(\beta' \wedge \alpha)' \wedge (\alpha' \vee \beta)'. \tag{1}$$

By applying DeMorgan's laws and some basic properties of sentence calculus, we obtain

$$(\beta' \wedge \alpha)' \wedge (\alpha' \vee \beta)' \equiv (\beta \vee \alpha') \wedge (\alpha \wedge \beta') \equiv$$

$$\equiv [(\alpha \wedge \beta') \wedge \beta] \vee [(\alpha \wedge \beta') \wedge \alpha'] \equiv$$

$$\equiv (\alpha \wedge \beta' \wedge \beta) \vee (\alpha \wedge \alpha' \wedge \beta') \equiv$$

17

$$\equiv (\alpha \wedge 0) \vee (0 \wedge \beta') \equiv 0 \vee 0 \equiv 0. \qquad (2)$$

No matter what the input, the output is always 0. To construct a circuit equivalent to the one shown in Figure 1, all that is needed is wire. See Figure 2.

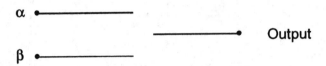

FIGURE 2

Luckily, not all of the computer logic circuits can be simplified to such an extent.

2. We will build the circuit for

$$(\alpha \wedge \beta)' \equiv \alpha' \vee \beta'. \qquad (3)$$

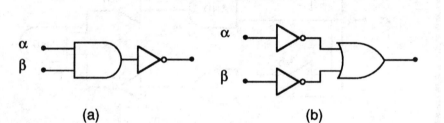

(a)   (b)

FIGURE 3

Circuits (a) and (b) shown in Figure 3 are equivalent.
   Now we will build the circuit for

$$(\alpha \vee \beta)' \equiv \alpha' \wedge \beta'. \qquad (4)$$

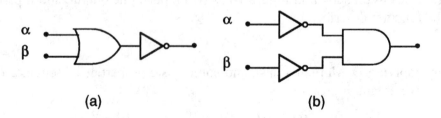

(a)   (b)

FIGURE 4

Circuits (a) and (b) of Figure 4 are equivalent.

Design the circuit that implements the expression

1. $$[(\alpha' \wedge \beta) \vee (\beta' \wedge \gamma)]'. \qquad (1)$$

2. $$[(\alpha \wedge \beta \wedge \gamma) \vee (\alpha' \wedge \beta')] \wedge \gamma'. \qquad (2)$$

SOLUTION:

1. Step by step we shall move α', which is realized by

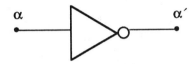

$\alpha' \wedge \beta$ is realized by

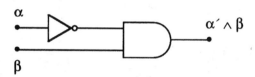

β' is realized by

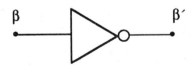

$\beta' \wedge \gamma$ is realized by

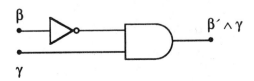

$(\alpha' \wedge \beta) \vee (\beta' \wedge \gamma)$ is realized by

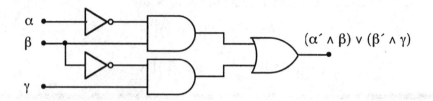

$[(\alpha' \wedge \beta) \vee (\beta' \wedge \gamma)]'$ is realized by

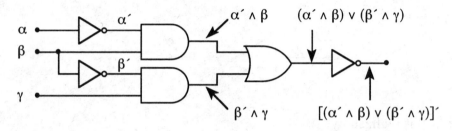

FIGURE 1

2. $\alpha \wedge \beta \wedge \gamma$ is implemented by

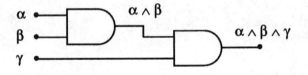

$\alpha' \wedge \beta'$ is implemented by

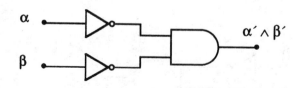

20

$[(\alpha \wedge \beta \wedge \gamma) \vee (\alpha' \wedge \beta')] \wedge \gamma'$ is implemented by

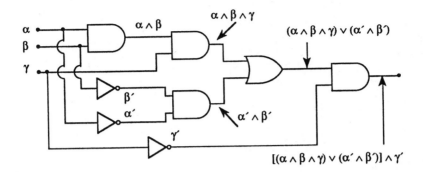

FIGURE 2

Note that since

$$[(\alpha \wedge \beta \wedge \gamma) \vee (\alpha' \wedge \beta')] \wedge \gamma' \equiv$$

$$\equiv (\alpha \wedge \beta \wedge \gamma \wedge \gamma') \vee (\alpha' \wedge \beta' \wedge \gamma') \equiv \alpha' \wedge \beta' \wedge \gamma' \qquad (3)$$

the circuit in the last part of Figure 2 can be replaced by

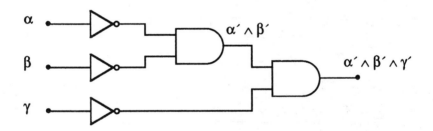

FIGURE 3

● **PROBLEM 1–15**

1. Define NOR and NAND gates.

2. Show that any logic circuit can be built exclusively with the NAND gates.

SOLUTION:

1. NOR gate consists of OR gate and an INVERTER.

21

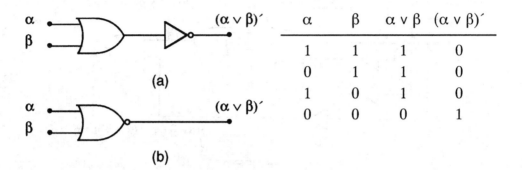

α	β	α ∨ β	(α ∨ β)´
1	1	1	0
0	1	1	0
1	0	1	0
0	0	0	1

(a)

(b)

FIGURE 1—NOR Gate

Both figures represent the same logic circuit. The symbol in Figure (b) is used for convenience.

NAND gate and its symbol are shown in Figure 2.

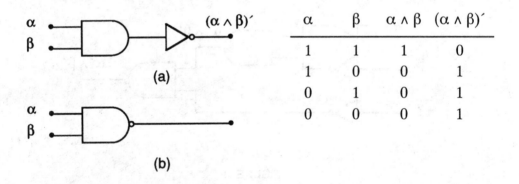

α	β	α ∧ β	(α ∧ β)´
1	1	1	0
1	0	0	1
0	1	0	1
0	0	0	1

(a)

(b)

FIGURE 2—NAND Gate

2. All Boolean expressions consist of various combinations of the basic operations of OR, AND, and INVERT. Thus, any circiut can be implemented using OR, AND, and INVERT gates. It is possible, however, to build any logic circuit using only NAND gates. Using NAND gates, we can design any logic operations OR, AND, and INVERT.

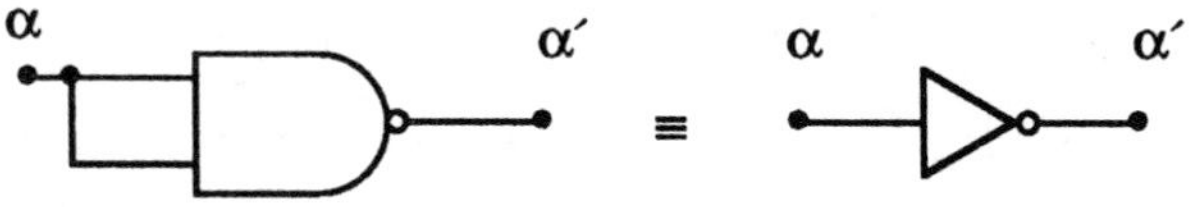

(a) INVERTER

Indeed $(\alpha \wedge \alpha)' \equiv \alpha'$

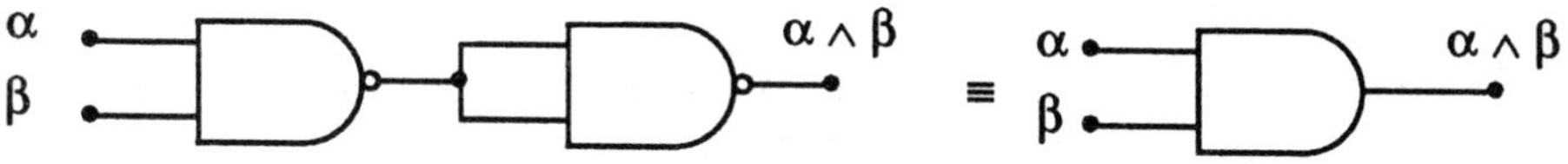

(b) AND Gate

Indeed $[(\alpha \wedge \beta)' \wedge (\alpha \wedge \beta)']' \equiv (\alpha \wedge \beta) \vee (\alpha \wedge \beta) \equiv \alpha \wedge \beta$.

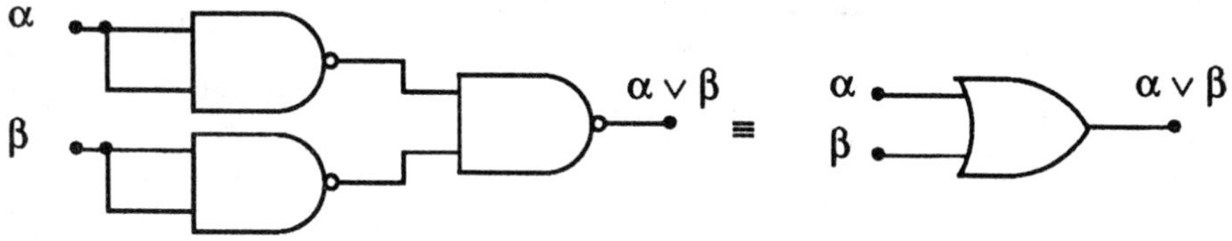

(c) OR Gate

Indeed $[(\alpha \wedge \alpha)' \wedge (\beta \wedge \beta)']' \equiv (\alpha' \wedge \beta')' \equiv \alpha \vee \beta$.

FIGURE 3

NAND gates can be used to perform any Boolean function.

23

1. Design a logic circuit that implements the function

$$(\alpha \wedge \beta) \vee (\gamma \wedge \delta). \tag{1}$$

2. Suppose you have the TTL IC's shown in Figure 1 at your disposal. Each IC is a quad; that is, each chip contains four identical two-input gates. Using the minimum number of IC's, design the circuit which performs function (1).

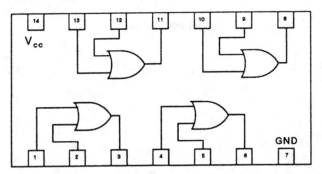

FIGURE 1a — IC 7432

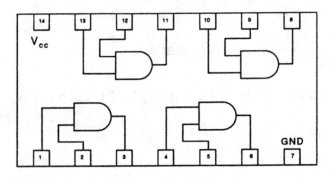

FIGURE 1b — IC 7408

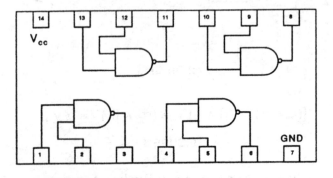

FIGURE 1c — IC 7400

SOLUTION:

1.

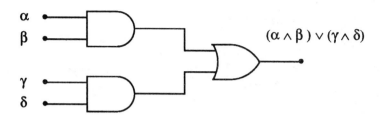

FIGURE 2

Figure 2 shows the circuit implementating operation $(\alpha \wedge \beta) \vee (\gamma \wedge \delta)$.

2. From Figure 2 we conclude that in order to build the circuit (1), we have to use two gates from IC 7408 and one gate from IC 7432, as shown in Figure 3.

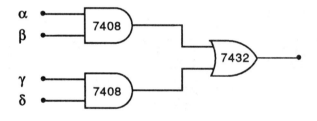

FIGURE 3

Each AND gate and OR gate can be replaced by a combination of NAND gates (See Problem 1–15).

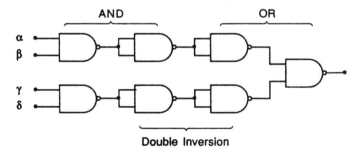

FIGURE 4

By eliminating double inversion, we obtain

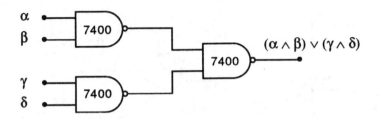

FIGURE 5

Now to implement the operation, we only need three NAND gates. There-fore, instead of using two IC's as in Figure 3, we can use one IC 7400, as shown in Figure 5 and Figure 6.

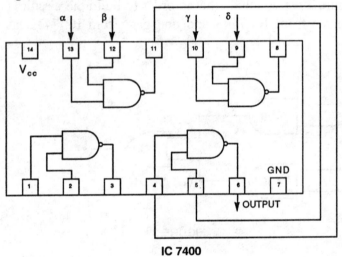

IC 7400

$$\text{OUTPUT} = (\alpha \wedge \beta) \vee (\gamma \wedge \delta)$$

FIGURE 6

CHAPTER 2

ALGEBRA OF SETS

Which of the following statements is true?

1. $7 \in A$, where

$$A = \{x : x \in N \text{ and } 5 < x < 8\}$$

and N is the set of natural numbers.

2. $B = 4$, where

$$B = \{x : x + 1 = 5\}.$$

SOLUTION:

The theory of sets was founded and developed into a mathematical system by G. Cantor (1845-1918). According to him, we are to understand by a set — a collection into a whole, of definite, well-distinguished objects (called the elements of A) of our perception or of our thought."

For example, the set of vertices of a square consists of four elements. The set can be defined by listing its elements.

For example:

$$A = \{\text{John, seven}\}$$

is the set consisting of two elements: John and seven. The statement "x is an element of A" is written

$$x \in A.$$

The negation of the statement $x \in A$ is

$$x \notin A.$$

The other way to define a set is to characterize the properties (or property) of its elements.

$$A = \{x : p(x)\}.$$

If $x \in A$, then $p(x)$ is true ($p(x) \equiv 1$).

For example:

$$N = \{n : n \text{ is a natural number}\}.$$

1. The set A consists of two elements

$$A = \{6, 7\}.$$

Thus

$$7 \in A.$$

2. The set B consists of a single element

$$B = \{4\}.$$

The number 4 belongs to the set B, $4 \in B$, but it does not equal B. The statement $B = 4$ is not true.

● **PROBLEM 2-2**

Show that equality of sets is reflexive, symmetric, and transitive, i.e., show that

1. $A = A$ for all sets A.

2. If $A = B$, then $B = A$ for all sets A, B.

3. If $A = B$ and $B = C$, then $A = C$ for all sets A, B, C.

SOLUTION:

Two sets, A and B, are said to be equal, $A = B$, provided that they contain the same elements. From this definition we conclude that:

1. $x \in A$ if and only if $x \in A$, thus

$$A = A.$$

Frequently, instead of writing if, and only if, we shall use iff.

$$\text{"iff"} = \text{if and only if.}$$

2. The statement $x \in A$ if and only if (iff) $x \in B$ is equivalent to $x \in B$, iff $x \in A$.

3. $(x \in A$ if and only if $x \in B)$ and $(x \in B$ if and only if $x \in C)$ imply $(x \in A$ if and only if $x \in C)$. Using the logic notation, we can write this statement in the form

$$[(x \in A \text{ iff } x \in B) \wedge (x \in B \text{ iff } x \in C)] \Rightarrow (x \in A \text{ iff } x \in C)$$

or

$$(A = B) \wedge (B = C) \Rightarrow (A = C).$$

1. Show that inclusion of sets is reflexive, anti-symmetric, and transitive; i.e.

    a. $A \subset A$ for all sets A.

    b. $A \subset B$ and $B \subset A$ imply $A = B$ for all A, B.

    c. If $A \subset B$ and $B \subset C$, then $A \subset C$ for all sets A, B, and C.

2. Prove

$$[(A \subset B) \wedge (B \subset C) \wedge (C \subset A)] \Rightarrow (A = B = C). \quad (1)$$

SOLUTION:

DEFINITION: INCLUSION

The set A is included in B; i.e., A is a subset of B, which is written as $A \subset B$ or $B \subset A$, when

$$x \in A \Rightarrow x \in B \quad (2)$$

■

1. Obviously

$$(x \in A) \Rightarrow (x \in A) \text{ for all } A. \quad (3)$$

Hence $$A \subset A.$$

$$(x \in A \Rightarrow x \in B) \wedge (x \in B \Rightarrow x \in A) \Rightarrow$$

$$(x \in A \text{ iff } x \in B). \quad (4)$$

Hence $$(A \subset B) \wedge (B \subset A) \Rightarrow A = B. \quad (5)$$

$$(x \in A \Rightarrow x \in B) \wedge (x \in B \Rightarrow x \in C) \Rightarrow$$

$$(x \in A \Rightarrow x \in C). \quad (6)$$

Thus

$$(A \subset B) \wedge (B \subset C) \Rightarrow A \subset C. \quad (7)$$

2. To prove that $A = B$, we must prove that $x \in A$ iff $x \in B$. Since $A \subset B$, it is

enough to show that $B \subset A$. But $B \subset C$ and $C \subset A$, thus $B \subset A$. And

$$A \subset B \text{ and } B \subset A \Rightarrow A = B. \qquad (8)$$

Similarly, we show that $B = C$.

$$(C \subset A \text{ and } A \subset B) \Rightarrow C \subset B. \qquad (9)$$

Thus

$$(B \subset C \text{ and } C \subset B) \Rightarrow C = B.$$

● **PROBLEM 2-4**

Using Venn diagrams, illustrate the union, intersection and difference of two sets A and B.

Show that intersection is associative; i.e., that

$$A \cap (B \cap C) = (A \cap B) \cap C \qquad (1)$$

for all sets A, B, and C.

SOLUTION:

The union of two sets A and B is the set whose elements are all the elements of the set A and all the elements of the set B and which does not contain any other elements. The union of A and B is denoted by $A \cup B$; see Figure 1.

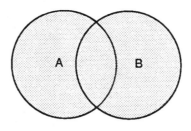

FIGURE 1. $A \cup B$ is shaded.

The common part of two sets is called the intersection. The intersection of A and B contains those, and only those, elements which belong to A and to B. The intersection is denoted by $A \cap B$.

31

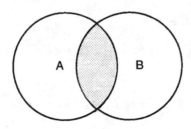

FIGURE 2. A ∩ B is shaded.

The difference of two sets A and B is the set consisting of those, and only those, elements which belong to A, but do not belong to B. The difference is denoted by $A - B$ (sometimes $A \setminus B$).

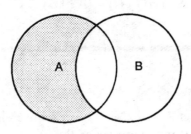

FIGURE 3. A − B is shaded.

Now we will use Venn diagrams to prove (1).

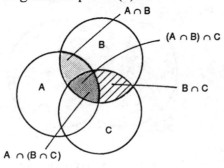

FIGURE 4.

From Figure 4 we conclude that

$$A \cap (B \cap C) = (A \cap B) \cap C.$$

By applying the sentence calculus, prove the following formula:

$$A \cup (A \cap B) = A = A \cap (A \cup B). \qquad (1)$$

SOLUTION:

Operations on sets are related to operations on sentences. To denote that x is an element of A we write

$$x \in A. \qquad (2)$$

Similarly

$$(x \notin A) \equiv (x \in A)' \qquad (3)$$

where $x \notin A$ means x is not an element of A.

The following equivalences hold for all x:

$$(x \in A \cup B) \equiv (x \in A) \vee (x \in B) \qquad (4)$$

$$(x \in A \cap B) \equiv (x \in A) \wedge (x \in B) \qquad (5)$$

$$(x \in A - B) \equiv (x \in A) \wedge (x \in B)' \qquad (6)$$

Using formulas (3) - (6) we can deduce theorems on the calculus of sets from analogous theorems of sentence calculus. Furthermore, let us note that

$$\text{if } x \in A \equiv x \in B \text{ holds for all } x, \text{ then } A = B. \qquad (7)$$

We shall now prove (1).

$$[x \in A \cup (A \cap B)] \equiv (x \in A) \vee (x \in A \cap B) \equiv$$

$$\equiv (x \in A) \vee [(x \in A) \wedge (x \in B)] \equiv (x \in A). \qquad (8)$$

Thus

$$A \cup (A \cap B) = A. \qquad (9)$$

Similarly

$$[x \in A \cap (A \cup B)] \equiv (x \in A) \wedge (x \in A \cup B) \equiv$$

$$\equiv (x \in A) \wedge [(x \in A) \vee (x \in B)] \equiv (x \in A). \qquad (10)$$

Hence

$$A \cap (A \cup B) = A. \qquad (11)$$

Note that in (8) and (10) we applied the law of absorption of sentence calculus.

$$\alpha \lor (\alpha \land \beta) \equiv \alpha \equiv \alpha \land (\alpha \lor \beta) \tag{12}$$

● **PROBLEM 2-6**

Prove these basic formulas for unions and intersections:

1. $A \cup A = A$

   } indempotency (1)

   $A \cap A = A$

2. $A \cup B = B \cup A$

   } commutativity (2)

   $A \cap B = B \cap A$

3. $A \cup (B \cup C) = (A \cup B) \cup C$

   } associativity (3)

   $A \cap (B \cap C) = (A \cap B) \cap C$

4. $A \cup (A \cap B) = A$

   } adjunction (4)

   $A \cap (A \cup B) = A$

5. $A \cup (B \cap C) = (A \cup B) \cap (A \cup C)$

   } distributivity (5)

   $A \cap (B \cup C) = (A \cap B) \cup (A \cap C)$

SOLUTION:

1.
$$(x \in A \cup A) \equiv (x \in A) \lor (x \in A) \equiv (x \in A) \tag{6}$$

Hence
$$A \cup A = A. \tag{7}$$

$$(x \in A \cap A) \equiv (x \in A) \land (x \in A) \equiv (x \in A) \tag{8}$$

Hence
$$A \cap A = A. \tag{9}$$

From (7) and (9) we see that, in contrast to arithmetic, neither multiples nor exponents arise in set theory.

2. $$(x \in A \cup B) \equiv (x \in A) \vee (x \in B) \equiv$$

$$\equiv (x \in B) \vee (x \in A) \equiv (x \in B \cup A). \tag{10}$$

Thus $$A \cup B = B \cup A. \tag{11}$$

$$(x \in A \cap B) \equiv (x \in A) \wedge (x \in B) \equiv$$

$$\equiv (x \in B) \wedge (x \in A) \equiv (x \in B \cap A). \tag{12}$$

Hence $$A \cap B = B \cap A. \tag{13}$$

3. $$[x \in A \cup (B \cup C)] \equiv [(x \in A) \vee (x \in B \cup C)] \equiv$$

$$\equiv [(x \in A) \vee (x \in B) \vee (x \in C)] \equiv [(x \in A \cup B) \vee (x \in C)] \equiv \tag{14}$$

$$\equiv [x \in (A \cup B) \cup C]$$

Thus $$A \cup (B \cup C) = (A \cup B) \cup C. \tag{15}$$

Replacing "$\vee$" by "$\wedge$" in (14) we prove

$$A \cap (B \cap C) = (A \cap B) \cap C. \tag{16}$$

4. See Problem 2–5.

5. $$[x \in A \cup (B \cap C)] \equiv [(x \in A) \vee (x \in B \cap C)] \equiv$$

$$\equiv \{(x \in A) \vee [(x \in B) \wedge (x \in C)]\} \equiv$$

$$\equiv \{[(x \in A) \vee (x \in B)] \wedge [(x \in A) \vee (x \in C)]\} \equiv$$

$$\equiv (x \in A \cup B) \wedge (x \in A \cup C) \equiv$$

$$\equiv [x \in (A \cup B) \cap (A \cup C)]. \tag{17}$$

Hence $$A \cup (B \cap C) = (A \cup B) \cap (A \cup C). \tag{18}$$

$$[x \in A \cap (B \cup C)] \equiv [(x \in A) \wedge (x \in B \cup C)] \equiv$$

$$\equiv \{(x \in A) \wedge [(x \in B) \vee (x \in C)]\} \equiv$$

$$\equiv \{[(x \in A) \wedge (x \in B)] \vee [(x \in A) \wedge (x \in C)]\} \equiv$$

$$\equiv [(x \in A \cap B) \vee (x \in A \cap C)] \equiv$$

$$\equiv [x \in (A \cap B) \cup (A \cap C)]. \tag{19}$$

Hence
$$A \cap (B \cup C) = (A \cap B) \cup (A \cap C). \tag{20}$$

● **PROBLEM 2-7**

1. Find the elements of the set
$$A = \{x : x + 1 = 3, x^2 = 1\}. \tag{1}$$

2. Prove that:

If A is a subset of the null set ϕ, then $A = \phi$.

3. Prove that:
$$(A - B) \cap B = \phi. \tag{2}$$

SOLUTION:

DEFINITION OF THE NULL SET

The null set (or the empty set or the void set) is the set which contains no elements. The null set is denoted by ϕ. Often the null set is defined as
$$\phi = \{x : x \neq x\}. \tag{3}$$

1. Number x, such that $x + 1 = 3$ and $x^2 = 1$ does not exist, hence
$$A = \phi. \tag{4}$$

2. The null set ϕ is a subset of every set, thus $\phi \subset A$. But, by hypothesis, $A \subset \phi$, therefore
$$A = \phi. \tag{5}$$

3.
$$[x \in (A - B) \cap B] \equiv$$

36

$$\equiv [(x \in A - B) \land (x \in B)] \equiv$$

$$\equiv [(x \in A) \land (x \in B)' \land (x \in B)] \equiv 0. \qquad (6)$$

Thus, the set $(A - B) \cap B$ contains no elements

$$(A - B) \cap B = \phi. \qquad (7)$$

● **PROBLEM 2-8**

1. Find the power set $P(A)$ of the set

$$A = \{1, 2, 3\}. \qquad (1)$$

2. Find the power set $P(B)$ of the set

$$B = \{1, \{2, 3\}\}. \qquad (2)$$

3. Set A consists of 10 elements. How many elements does the power set $P(A)$ have?

SOLUTION:

DEFINITION OF POWER SET

The power set $P(A)$ of A is the class of all subsets of A. ∎

1. The subsets of A are:

0 elements: ϕ
1 element: $\{1\}, \{2\}, \{3\}$ $\qquad\qquad (3)$
2 elements: $\{1, 2\}, \{1, 3\}, \{2, 3\}$
3 elements: $\{1, 2, 3\}$

Hence

$$P(A) = \{\phi, \{1\}, \{2\}, \{3\}, \{1,2\}, \{1,3\}, \{2,3\}, \{1, 2, 3\}\} \qquad (4)$$

$P(A)$ contains $2^3 = 8$ elements.

2. Note that B contains only two elements, 1 and $\{2, 3\}$. Thus, its power set $P(B)$ is

$$P(B) = \{\phi, \{1\}, \{2,3\}, \{1, \{2, 3\}\}\}. \qquad (5)$$

3. From the theory of permutations we have the equation for the number of different combinations that can be formed from n different elements, using m elements at a time.

$$\binom{n}{m} = \frac{n!}{m!(n-m)!} \tag{6}$$

where

$$n! = 1 \cdot 2 \cdot \ \ldots \ \cdot (n-1) \cdot n \tag{7}$$

and

$$0! = 1.$$

Suppose set A contains n elements; then there are

$$\binom{n}{0} = 1 \quad \text{subsets with 0 elements}$$

$$\binom{n}{1} = \frac{n!}{1!(n-1)!} = n \quad \text{subsets with 1 element}$$

$$\binom{n}{2} = \frac{n!}{2!(n-2)!} = \frac{n(n-1)}{2} \quad \text{subsets with 2 elements.}$$

The total number of subsets of a set A consisting of n elements is

$$\binom{n}{0} + \binom{n}{1} + \binom{n}{2} + \ldots + \binom{n}{n} = (1+1)^n = 2^n. \tag{8}$$

Set A consists of 10 elements, hence its power set $P(A)$ consists of $2^{10} = 1024$.

● **PROBLEM 2-9**

If A is any non-empty set, then $P(A)$ (or 2^A) is the power set of A. Prove that:

1. $2^A \cap 2^B = 2^{A \cap B}$ (1)

2. $2^A \cup 2^B \subset 2^{A \cup B}$ (2)

3. Show that $2^A \cup 2^B \neq 2^{A \cup B}$ (3)

SOLUTION:

1. Let α represent an element of $2^A \cap 2^B$. Recall that the power set 2^A of A is the class of all subsets of A. Suppose α consists of elements $p_1, p_2, \ldots$

$$\alpha = \{p_1, p_2, \ldots\} \tag{4}$$

Since $\alpha \in 2^A$ and $\alpha \in 2^B$, each of $p_1, p_2, \ldots$ belongs to $A \cap B$. Thus, α is a subset of $A \cap B$ and

$$\alpha \in 2^{A \cap B} \tag{5}$$

Hence,

$$2^A \cap 2^B \subset 2^{A \cap B}. \tag{6}$$

Now, suppose $\alpha \in 2^{A \cap B}$; i.e., α is a subset of $A \cap B$, and each element of $\alpha = \{p_1, p_2, \ldots\}$ belongs to $A \cap B$. Hence, each of $p_1, p_2, \ldots$ belongs to A and to B.

Therefore, α is a subset of A, $\alpha \in 2^A$; and α is a subset of B, $\alpha \in 2^B$.

$$\alpha \in 2^A \cap 2^B. \tag{7}$$

Hence,

$$2^{A \cap B} \subset 2^A \cap 2^B. \tag{8}$$

From (6) and (8) we conclude that

$$2^{A \cap B} = 2^A \cap 2^B. \tag{9}$$

2. Suppose $\alpha \in 2^A \cup 2^B$; that is, $\alpha \in 2^A$ or $\alpha \in 2^B$. Assume $\alpha \in 2^A$, then α is a subset of A. If α is a subset of A, then α is a subset of $A \cup B$ and

$$\alpha \in 2^{A \cup B}. \tag{10}$$

That proves

$$2^A \cup 2^B \subset 2^{A \cup B}. \tag{11}$$

3. Now we shall show that

$$2^A \cup 2^B \neq 2^{A \cup B}. \tag{12}$$

Let $A = \{a\}, B = \{b\}$, then

$$2^A = \{\phi, \{a\}\} \tag{13}$$

$$2^B = \{\phi, \{b\}\} \tag{14}$$

and $A \cup B = \{a, b\}$

$$2^{A \cup B} = \{\phi, \{a\}, \{b\}, \{a, b\}\}. \tag{14}$$

Hence,

$$2^A \cup 2^B = \{\phi, \{a\}, \{b\}\} \tag{15}$$

and

$$2^A \cup 2^B \neq 2^{A \cup B}. \tag{16}$$

● **PROBLEM 2–10**

Prove the following formulas:

1. $A \cap B \subset A \subset A \cup B$ (1)

2. $A - B \subset A$ (2)

3. If $A \subset B$ and $C \subset D$, then

$$A \cup C \subset B \cup D \tag{3}$$

and

$$A \cap C \subset B \cap D \tag{4}$$

4. $(A \subset B) \equiv (A \cup B = B) \equiv (A \cap B = A)$ (5)

5. $(A \cup B) \cap (A \cup C) = A \cup (B \cap C)$ (6)

SOLUTION:

1. $[x \in A \cap B] \equiv [(x \in A) \wedge (x \in B)] \Rightarrow (x \in A)$ (7)
Thus,

$$A \cap B \subset A \tag{8}$$

$$(x \in A)[(x \in A) \vee (x \in B)] \equiv (x \in A \cup B) \tag{9}$$

Hence,

$$A \subset A \cup B. \tag{10}$$

2. $[x \in A - B] \equiv [(x \in A) \vee (x \in B)'] \Rightarrow (x \in A)$ (11)
Hence,

$$A - B \subset A. \tag{12}$$

40

3. We have to show that

$$[(A \subset B) \land (C \subset D)] \Rightarrow \begin{pmatrix} A \cup C \subset B \cup D \\ and \\ A \cap C \subset B \cap D \end{pmatrix} \qquad (13)$$

$$[x \in A \cup C] \equiv [(x \in A) \lor (x \in C)] \Rightarrow$$

$$\Rightarrow [(x \in B) \lor (x \in D)] \equiv (x \in B \cup D) \qquad (14)$$

Hence,

$$A \cup C \subset B \cup D. \qquad (15)$$

$$(x \in A \cap C) \equiv [(x \in A) \land (x \in C)] \Rightarrow$$

$$\Rightarrow [(x \in B) \land (x \in D)] \equiv (x \in B \cap D) \qquad (16)$$

Hence,

$$A \cap C \subset B \cap D. \qquad (17)$$

4. Suppose $A \subset B$; then from $(A \subset B)$ and $(B \subset B)$ we conclude that

$$(A \cup B) \subset (B \cup B) = B \qquad (18)$$

by virtue of (13). But since $B \subset (A \cup B)$, we have

$$A \cup B = B. \qquad (19)$$

Conversely from $A \cup B = B$, it follows that $A \subset B$. Hence, the relations $A \subset B$ and $A \cup B = B$ are equivalent.

$$(A \subset B) \equiv (A \cup B = B). \qquad (20)$$

Similarly, combining $A \subset B$ with $A \subset A$, we obtain $A \subset (A \cap B)$. Since $A \cap B \subset A$,

$$A = A \cap B. \qquad (21)$$

Conversely from $A = A \cap B$ we conclude that $A \subset B$.

5. $(A \cup B) \cap (A \cup C) = (A \cap A) \cup (A \cap C) \cup (B \cap A) \cup (B \cap C) =$

$$= A \cup (A \cap C) \cup (A \cap B) \cup (B \cap C) \qquad (22)$$

But $A \cap B \subset A$, hence, $A \cup (A \cap B) = A$ and $A \cap C \subset A$, hence, $A \cup (A \cap C) = A$. Thus,

$$(A \cup B) \cap (A \cup C) = A \cup (B \cap C). \qquad (23)$$

41

Prove the following formulas:

1. $A \cap B = A - (A - B)$ (1)

2. $A \cup (B - A) = A \cup B$ (2)

3. $A - (A \cap B) = A - B$ (3)

4. $A \cap (B - C) = (A \cap B) - (A \cap C)$ (4)

5. $A \cup (B - C) \neq (A \cup B) - (A \cup C)$ (5)

SOLUTION:

1. $[x \in A - (A - B)] \equiv [(x \in A) \wedge (x \in A - B)'] \equiv$

           $\equiv \{(x \in A) \wedge [(x \in A) \wedge (x \in B)']'\} \equiv$

           $\equiv \{(x \in A) \wedge [(x \in A)' \vee (x \in B)]\} \equiv$

           $\equiv [(x \in A) \wedge (x \in B)] \equiv (x \in A \cap B).$ (6)

2. $[x \in A \cup (B - A)] \equiv [(x \in A) \vee (x \in B - A)] \equiv$

           $\equiv \{(x \in A) \vee [(x \in B) \wedge (x \in A)']\} \equiv$

      $\equiv \{[(x \in A) \vee (x \in B)] \wedge [(x \in A) \vee (x \in A)']\} \equiv$

           $\equiv [(x \in A) \vee (x \in B)] \equiv (x \in A \cup B).$ (7)

Hence,

$$A \cup (B - A) = A \cup B. \qquad (8)$$

3. $[x \in A - (A \cap B)] \equiv [(x \in A) \wedge (x \in A \cap B)'] \equiv$

           $\equiv \{(x \in A) \wedge [(x \in A) \wedge (x \in B)]'\} \equiv$

           $\equiv \{(x \in A) \wedge [(x \in A)' \vee (x \in B)']\} \equiv$

$$\equiv \{[(x \in A) \wedge (x \in A)'] \vee [(x \in A) \wedge (x \in B)']\} \equiv$$

$$\equiv [(x \in A) \wedge (x \in B)'] \equiv (x \in A - B). \qquad (9)$$

Hence,

$$A - (A \cap B) = A - B. \qquad (10)$$

4. $\qquad [x \in (A \cap B) - (A \cap C)] \equiv [(x \in A \cap B) \wedge (x \in A \cap C)'] \equiv$

$$\equiv \{(x \in A) \wedge (x \in B) \wedge [(x \in A) \wedge (x \in C)]'\} \equiv$$

$$\equiv \{(x \in A) \wedge (x \in B) \wedge [(x \in A)' \vee (x \in C)']\} \equiv$$

$$\equiv [(x \in A) \wedge (x \in B) \wedge (x \in C)'] \equiv$$

$$\equiv [(x \in A) \wedge (x \in B - C)] \equiv [x \in A \cap (B - C)]. \qquad (11)$$

5. Suppose A, B, C are three disjoint sets as shown in the Venn diagram:

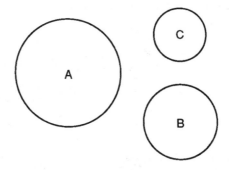

FIGURE 1

Then

$$A \cup (B - C) = A \cup B \qquad (12)$$

and

$$(A \cup B) - (A \cup C) = B \qquad (13)$$

Hence,

$$A \cup (B - C) \neq (A \cup B) - (A \cup C). \qquad (14)$$

43

In the theory of sets, we assume that all the sets under consideration are subsets of some fixed set, called the space (or universal set). For example, in geometry the space is the Euclidean space, and in analysis the space is the set of real or complex numbers.

Throughout this book, all the sets considered will belong to the space which we shall denote by (1). Hence

$$A \subset 1 \tag{1}$$

for each of the sets considered.

DEFINITION OF COMPLEMENT OF THE SET

The complement of the set A with respect to the given space 1 is denoted by A^C (or CA or $\sim A$) and defined by

$$A^C = 1 - A \tag{2}$$

See Figure 1.

■

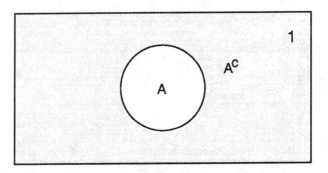

FIGURE 1.

We have

$$(x \in A^C) \equiv (x \in A)' \equiv (x \notin A) \tag{3}$$

Prove that:

$$1^C = \phi, \quad \phi^C = 1 \tag{4}$$

$$A^{CC} = A \tag{5}$$

$$A \cup A^C = 1, \quad A \cap A^C = \phi \tag{6}$$

SOLUTION:

From definition (2) we have

$$1^C = 1 - 1 = \phi \tag{7}$$

$$\phi^C = 1 - \phi = 1 \tag{8}$$

$$A^{CC} = (A^C)^C = (1 - A)^C =$$

$$= 1 - (1 - A) = A \tag{9}$$

$$A \cup A^C = A \cup (1 - A) = 1 \tag{10}$$

$$A \cap A^C = A \cap (1 - A) = \phi \tag{11}$$

● **PROBLEM 2–13**

Prove DeMorgan's Theorem:

1. $(A \cup B)^C = A^C \cap B^C$ (1)

2. $(A \cap B)^C = A^C \cup B^C$ (2)

SOLUTION:

1.
$$[x \in (A \cup B)^C] \equiv [x \in 1 - (A \cup B)] \equiv$$

$$\equiv [x \in A \cup B]' \equiv [(x \in A) \vee (x \in B)]' \equiv$$

$$\equiv [(x \in A)' \wedge (x \in B)'] \equiv [(x \in A^C) \wedge (x \in B^C)] \equiv$$

$$\equiv [x \in A^C \cap B^C]. \tag{3}$$

Hence,

$$(A \cup B)^C = A^C \cap B^C. \tag{4}$$

2. Similarly

$$[x \in (A \cap B)^C] \equiv [x \in 1 - (A \cap B)] \equiv$$

$$\equiv [x \in A \cap B]' \equiv [(x \in A) \wedge (x \in B)]' \equiv$$

$$\equiv [(x \in A)' \vee (x \in B)'] \equiv [(x \in A^C) \vee (x \in B^C)] \equiv$$

$$\equiv [x \in A^C \cup B^C]. \tag{5}$$

Hence,

$$(A \cap B)^C = A^C \cup B^C. \tag{6}$$

DeMorgan's theorem can be generalized to any finite number of sets.

$$(A_1 \cup A_2 \cup \ldots \cup A_n)^C = A_1^{\;C} \cap A_2^{\;C} \cap \ldots \cap A_n^{\;C} \tag{7}$$

$$(A_1 \cap A_2 \cap \ldots \cap A_n)^C = A_1^{\;C} \cup A_2^{\;C} \cup \ldots \cup A_n^{\;C}. \tag{8}$$

● **PROBLEM 2–14**

Prove that:

1. $A - B = A \cap B^C$ (1)

2. $A \subset B$ if and only if $B^C \subset A^C$ (2)

3. $A^C - B^C = B - A$ (3)

4. Prove Equation (4) of Problem 2–11 using the notion of complement.

SOLUTION:

1.
$$(x \in A \cap B^C) \equiv [(x \in A) \wedge (x \in B^C)] \equiv$$

$$\equiv [(x \in A) \wedge (x \in 1 - B)] \equiv [(x \in A) \wedge (x \in B)'] \equiv$$

$$\equiv (x \in A - B). \tag{4}$$

2. Here, we shall apply the law of contraposition:

$$\text{if } \alpha \Rightarrow \beta, \text{ then } \beta' \Rightarrow \alpha' \tag{5}$$

$$(A \subset B) \equiv [(x \in A) \Rightarrow (x \in B)] \equiv$$

$$\equiv [(x \in B)' \Rightarrow (x \in A)'] \equiv [(x \notin B) \Rightarrow (x \notin A)] \equiv$$

$$\equiv [(x \in 1 - B) \Rightarrow (x \in 1 - A)] \equiv [(x \in B^c) \Rightarrow (x \in A^c)] \equiv$$

$$\equiv (B^C \subset A^C). \tag{6}$$

Thus,

$$A \subset B \text{ iff } B^C \subset A^C. \tag{7}$$

3.

$$(x \in A^C - B^C) \equiv [(x \in A^c) \wedge (x \in B^c)'] \equiv$$

$$\equiv [(x \in 1 - A) \wedge (x \in 1 - B)'] \equiv [(x \in A)' \wedge (x \in B)''] \equiv$$

$$\equiv [(x \in B) \wedge (x \notin A)] \equiv (x \in B - A). \tag{8}$$

4. We have to prove that

$$(A \cap B) - (A \cap C) = A \cap (B - C) \tag{9}$$

By applying (1), we obtain

$$(A \cap B) - (A \cap C) = (A \cap B) \cap (A \cap C)^C =$$

$$= (A \cap B) \cap [1 - (A \cap C)] = (A \cap B) \cap (A^C \cup C^C) =$$

$$= (A \cap B \cap A^c) \cup (A \cap B \cap C^c) =$$

$$= (A \cap A^C \cap B) \cup (A \cap B \cap C^c) =$$

$$= A \cap B \cap C^C = A \cap (B - C). \tag{10}$$

In (10), we used DeMorgan's theorem.

● **PROBLEM 2-15**

Let $A_1, A_2, \ldots, A_n$ represent fixed subsets of the space 1. Let us denote

$$A_i^1 = 1 - A_i = A_i^C$$

$$A_i^0 = A_i. \tag{1}$$

A constituent of the space with respect to the sets $A_1, A_2, \ldots, A_n$ is any set of the form

$$A_1^{i_1} \cap A_2^{i_2} \cap \ldots \cap A_n^{i_n}$$

$$\text{where } i_1, \ldots, i_n = 1 \text{ or } 0. \tag{2}$$

1. Show that constituents are disjoint.

2. Prove that the union of all constituents is equal to 1.

SOLUTION:

1. Let $A_1^{i_1} \cap A_2^{i_2} \cap \ldots \cap A_n^{i_n}$ and $A_1^{k_1} \cap A_2^{k_2} \cap \ldots \cap A_n^{k_n}$ denote two constituents. Then at least one index i_l exists, such that

$$i_l \neq k_l. \tag{3}$$

Then the intersection of these two constituents is

$$A_1^{i_1} \cap A_2^{i_2} \ldots \cap A_n^{i_n} \cap A_1^{k_1} \cap A_2^{k_2} \cap \ldots \cap A_n^{k_n} \cap A_l^{i_l} \cap A_l^{k_l} =$$

$$= \ldots \cap A_l \cap A_l^C = \phi. \tag{4}$$

2. Now we shall show that the union of constituents is equal to 1.

Let $x \in 1$ represent an element of the space. We shall show that a constituent exists, such that x belongs to this constituent. Consider the sets $A_1, A_2, \ldots, A_n$. Two possibilities exist:

$$x \in A_1 \quad \text{or} \quad x \notin A_1.$$

If $x \in A_1 = A_1^0$, we set $i_1 = 0$. If $x \notin A_1$, then $x \in 1 - A_1 = A_1^1$ and we set $i_1 = 1$. We repeat the above procedure for each set $A_1, A_2, \ldots, A_n$ and obtain $i_1, i_2, \ldots, i_n$. The element x belongs to

$$x \in A_1^{i_1}$$

$$x \in A_2^{i_2}$$

$$\vdots$$

$$x \in A_n^{i_n} \tag{5}$$

Hence,

$$x \in A_1^{i_1} \cap A_2^{i_2} \cap \ldots \cap A_n^{i_n} \tag{6}$$

Since x is an arbitrary element of 1, we conclude that the union of constituents is equal to 1.

Let A, B, and C represent the subsets of the space.

1. Find the constituents of the space with respect to the sets A, B, and C.

2. Represent the set

$$A - (B \cup C) \qquad (1)$$

as the union of the constituents.

SOLUTION:

1. We shall consider all possible sets of the form

$$A^{i_1} \cap B^{i_2} \cap C^{i_3} \qquad (2)$$

where

$$i_1, i_2, i_3 = 0 \text{ or } 1 \text{ and}$$

$$A^0 = A, \ B^0 = B, \ C^0 = C$$

$$A^1 = A^C, \ B^1 = B^C, \ C^1 = C^C. \qquad (3)$$

The constituents are

$$A \cap B \cap C$$

$$A \cap B \cap C^C$$

$$A \cap B^C \cap C$$

$$A^C \cap B \cap C$$

$$A \cap B^C \cap C^C$$

$$A^C \cap B^C \cap C$$

$$A^C \cap B \cap C^C$$

$$A^C \cap B^C \cap C^C. \qquad (4)$$

2. Let us write $A - (B \cup C)$ in the form

$$A - (B \cup C) = (A - B) - C = (A \cap B^C) - C =$$

$$= A \cap B^C \cap C^C. \qquad (5)$$

Note that $A \cap B^C \cap C^C$ is itself a constituent.

The sets $A_1, A_2, \ldots, A_n$ are called independent if all the constituents are non-empty. Independent sets play an important role in the probability theory. Find the number of constituents of the independent sets $A_1, A_2, \ldots, A_n$.

SOLUTION:

The constituents are

$$A_1^{i_1} \cap A_2^{i_2} \cap \ldots \cap A_n^{i_n} \tag{1}$$

where $i_1, i_2, \ldots, i_n$ take one of two values, 0 or 1.

$$A_i^0 = A_i \text{ and } A_i^1 = 1 - A_i \tag{2}$$

The sets $A_1, A_2, \ldots, A_n$ are independent, therefore the constituents are non-empty sets. Setting

$$i_1 = i_2 = \ldots = i_n = 0 \tag{3}$$

in (1) we obtain

$$A_1 \cap A_2 \cap \ldots \cap A_n \neq \phi. \tag{4}$$

Thus, none of the sets $A_1, A_2, \ldots, A_n$ are empty sets. Similarly, setting $i_1 = i_2 = \ldots i_n = 1$, we find that

$$A_1^c \cap A_2^c \cap \ldots \cap A_n^c \neq \phi. \tag{5}$$

That is, none of the sets $A_1, A_2, \ldots, A_n$ are whole spaces.
Each of the superscripts $i_1, i_2, \ldots, i_n$ in

$$A_1^{i_1} \cap A_2^{i_2} \cap \ldots \cap A_n^{i_n}$$

takes up one of two values. Thus, the total number of combinations is

$$2 \times 2 \times \ldots \times 2 = 2^n. \tag{6}$$

The number of constituents is 2^n.

The symmetric difference of the sets A and B is defined by

$$A \div B = (A - B) \cup (B - A). \tag{1}$$

Prove the following formulas:

1. $A \doteq B = B \doteq A$ (2)

2. $A \doteq (B \doteq C) = (A \doteq B) \doteq C$ (3)
   The symmetric difference is associative.

3. $A \cup B = A \doteq B \doteq A \cap B$ (4)

SOLUTION:

We shall use

$$A - B = A \cap B^C \tag{5}$$

to write (1) in the form

$$A \doteq B = (A \cap B^C) \cup (A^C \cap B). \tag{6}$$

From (6) we obtain

$$(A \doteq B)^C = (A \cap B^C)^C \cap (A^C \cap B)^C =$$

$$= (A \cap B) \cup (A^C \cap B^C). \tag{7}$$

1. $A \doteq B = (A - B) \cup (B - A) = (B - A) \cup (A - B) =$

$$= B \doteq A. \tag{8}$$

2. $A \doteq (B \doteq C) = [A \cap (B \doteq C)^c] \cup [(B \doteq C) \cap A^c] =$ (9)

$$=\{A \cap [(C \cap B) \cup (C^C \cap B^C)]\} \cup \{[(B \cap C^c) \cup (B^C \cap C)] \cap A^c\} =$$

$$= (A \cap B \cap C) \cup (A \cap B^C \cap C^c) \cup (A^C \cap B \cap C^c) \cup (A^C \cap B^C \cap C).$$

Similarly,

$$(A \doteq B) \doteq C = [(A \cap B^C) \cup (A^C \cap B)] \doteq C = \tag{10}$$

$$= \{[(A \cap B^C) \cup (A^C \cap B)] \cap C^c\} \cup [(A \doteq B)^C \cap C] =$$

$$= (A \cap B^C \cap C^c) \cup (A^C \cap B \cap C^c) \cup \{[(A \cap B) \cup (A^C \cap B^C)] \cap C\} =$$

$$= (A \cap B^C \cap C^c) \cup (A^C \cap B \cap C^c) \cup (A \cap B \cap C) \cup (A^C \cap B^C \cap C).$$

Comparing (9) and (10) we find

$$(A \div B) \div C = A \div (B \div C) = A \div B \div C. \qquad (11)$$

3. $$A \div B \div A \cap B = (A \div B) \div (A \cap B) =$$

$$= [(A \cap B^c) \cup (A^c \cap B)] \div (A \cap B) =$$

$$= \{[(A \cap B^c) \cup (A^c \cap B)] \cap (A^c \cup B^c)\} \cup$$

$$\cup \{[(A^c \cup B) \cap (A \cup B^c)] \cap (A \cap B)\} =$$

$$= [(A \cap B^c) \cap (A^c \cup B^c)] \cup [(A^c \cap B) \cap (A^c \cup B^c)] \cup$$

$$\cup [(A^c \cap B^c) \cap (A \cap B)] \cup [(A \cap B) \cap (A \cap B)] =$$

$$= (A \cap B^c) \cup (A^c \cap B) \cup (A \cap B) =$$

$$= (A - B) \cup (B - A) \cup (A \cap B) = A \cup B. \qquad (12)$$

Similarly, we can show that

$$A - B = A \div A \cap B. \qquad (13)$$

● **PROBLEM 2-19**

Show that
$$(A_1 \cup \ldots \cup A_n) \div (B_1 \cup \ldots \cup B_n) \subset (A_1 \div B_1) \cup \ldots \cup (A_n \div B_n) \quad (1)$$
where
$$A \div B = (A - B) \cup (B - A). \qquad (2)$$

SOLUTION:

Let us denote
$$A = A_1 \cup \ldots \cup A_n$$

$$B = B_1 \cup \ldots \cup B_n. \qquad (3)$$

Suppose
$$x \in A \div B = (A - B) \cup (B - A). \qquad (4)$$

The sets $A - B$ and $B - A$ are disjoint sets (i.e., they have no common elements). We can assume that

$$x \in A - B, \tag{5}$$

then $x \in A$ and $(x \notin B) \equiv (x \in B)'$. We have

$$x \in A_1 \cup \ldots \cup A_n \tag{6}$$

and

$$x \notin B_1 \cup \ldots \cup B_n. \tag{7}$$

Therefore a set A_l exists, such that

$$x \in A_l \tag{8}$$

and

$$(x \in B)' \equiv (x \in B_1 \cup \ldots \cup B_n)' \equiv$$

$$\equiv [(x \in B_1) \vee (x \in B_2) \vee \ldots \vee (x \in B_n)]' \equiv$$

$$\equiv (x \in B_1)' \wedge (x \in B_2)' \wedge \ldots \wedge (x \in B_n)' \equiv$$

$$\equiv (x \notin B_1) \wedge (x \notin B_2) \wedge \ldots \wedge (x \notin B_n). \tag{9}$$

Thus,

$$x \in A_l \text{ and } x \notin B_l \tag{10}$$

and

$$x \in A_l \div B_l = (A_l - B_l) \cup (B_l - A_l). \tag{11}$$

Hence,

$$(A_1 \cap \ldots \cap A_n) \div (B_1 \cap \ldots \cap B_n) \subset (A_1 \div B_1) \cup \ldots \cup (A_n \div B_n). \tag{12}$$

● **PROBLEM 2-20**

DEFINITION OF A RING

The Operations + and · form a commutative ring, if they satisfy the conditions:

1. $x + y = y + x$, $x \cdot y = y \cdot x$ (1)

2. $x + (y + z) = (x + y) + z$, $x \cdot (y \cdot z) = (x \cdot y) \cdot z$ (2)

3. $x \cdot (y + z) = (x \cdot y) + (x \cdot z)$ (3)

4. An element 0 exists, such that

$$x + 0 = x \tag{4}$$

5. For every pair x, y, an element z exists, such that

$$x = y + z. \tag{5}$$

■

Show that the sets form a ring with respect to the operations $\div$ and $\cap$ but don't form a ring with respect to the operations $\cup$ and $\cap$.

SOLUTION:

Let us replace in (1) - (5) operation + by $\div$, and operation · by $\cap$. We obtain

$$A \div B = B \div A, \quad A \cap B = B \cap A \tag{6}$$

which are obviously true. Then

$$A \div (B \div C) = (A \div B) \div C, \quad A \cap (B \cap C) = (A \cap B) \cap C. \tag{7}$$

The relationship

$$A \cap (B \div C) = (A \cap B) \div (A \cap C) \tag{8}$$

is true.

For an empty set ϕ, we obtain

$$A \div \phi = A. \tag{9}$$

To prove condition 5, we have to show that

$$(A = B \div C) \Rightarrow (C = B \div A) \tag{10}$$

which is easy to prove because

$$A = B \div C = B \div B \div A = (B \div B) \div A = A. \tag{11}$$

Hence, operations $\div$ and $\cap$ form a ring.

Consider operations $\cup$ and $\cap$. We have

$$A \cup B = B \cup A, \quad A \cap B = B \cap A \tag{12}$$

$$A \cup (B \cup C) = (A \cup B) \cup C, \quad A \cap (B \cap C) = (A \cap = B) \cap C. \tag{13}$$

$$A \cap (B \cup C) = (A \cap B) \cup (A \cap C) \tag{14}$$

$$A \cup \{\phi\} = A \tag{15}$$

It is not true that for every pair A, B, a set C exists, such that

$$A = B \cup C. \qquad (16)$$

For example, let $A = \{\phi\}$ and $C \neq \{\phi\}$. Thus, $\cup$ and $\cap$ do not form a ring.

● PROBLEM 2-21

1. Show that the family of all subsets of a given set A is an ideal.

2. Show that the family of all sets B, such that $A \subset B \subset 1$, is a filter.

SOLUTION:

DEFINITION OF IDEAL

A family is a collection of sets. A non-empty family I of subsets of 1 is called ideal if

$$(A \in I) \wedge (B \subset A) \Rightarrow B \in I \qquad (1)$$

$$(A \in I) \wedge (B \in I) \Rightarrow (A \cup B \in I). \qquad (2)$$

1. Let P denote the family of all subsets of a given set A. If $D \in P$, then $D \subset A$. If $G \subset D$, then $G \subset A$ and $G \in P$. Hence, condition (1) is fulfilled.

If $D \in P$ and $G \in P$, then $D \subset A$ and $G \subset A$. Hence, $D \cup G \subset A$ and $D \cup G \in P$. Therefore, the family of all subsets of a given set is an ideal.

DEFINITION OF A FILTER

A non-empty family F is called a filter if

$$(A \in F) \wedge (A \subset B) \Rightarrow (B \in F) \qquad (3)$$

$$(A \in F) \wedge (B \in F) \Rightarrow (A \cap B \in F) \qquad (4)$$

■

2. Let A represent a given set and R the family of all sets B, such that $A \subset B \subset 1$. Suppose $D \in R$ and $D \subset G$, then $A \subset D \subset G \subset 1$ and $G \in R$. Condition (3) is fulfilled.

If $D \in R$ and $G \in R$, then $A \subset D$ and $A \subset G$. Therefore, $A \subset D \cap G$ and $D \cap G \in R$. Hence, R is a filter.

Show that a family of sets is a filter if and only if the family of the complements of these sets is an ideal.

SOLUTION:

Let $S = \{A, B, C, ...\}$ represent a family of sets. We shall prove that

$$\begin{pmatrix} S = \{A, B, C, ...\} \text{ is} \\ \text{a filter} \end{pmatrix} \Rightarrow \begin{pmatrix} P = \{A^c, B^c, C^c, ...\} \text{ is} \\ \text{an ideal} \end{pmatrix}$$

Let $A^c \in P$ and $B^c \subset A^c$, then $A \subset B$. But $A \in S$ (which is a filter); hence, $B \in S$. Therefore $B^c \in P$.

Let $A^c \in P$ and $B^c \in P$, then $A \in S$ and $B \in S$. Since S is a filter, $A \cap B \in S$.

But $A^c \cup B^c = (A \cap B)^c \in P$.

Similarly, we can show that

$$\begin{pmatrix} P = \{A^c, B^c, ...\} \text{ is} \\ \text{an ideal} \end{pmatrix} \Rightarrow \begin{pmatrix} S = \{A, B, C, ...\} \text{ is} \\ \text{a filter} \end{pmatrix}$$

Let $A \in S$ and $A \subset B$. Then $A^c \in P$ and $B^c \subset A^c$. Hence, $B^c \in P$ and $B \in S$.

Let $A \in S$ and $B \in S$. Then $A^c \in P$ and $B^c \in P$.

Since P is an ideal, $A^c \cup B^c \in P$ and $A \cap B$ $(A^c \cup B^c)^c \in S$.

In 1895, Georg Cantor created a theory of sets. At that time, it was accepted that a universal set (that is, "the set of all sets") exists. In 1902, Bertrand Russell showed in his famous paradox that the admission of a set of all sets leads to a contradiction. Explain Russell's paradox.

SOLUTION:

We shall show that the assumption that the universal set exists leads to contradictory statements.

THEOREM 1

Suppose that the set of all sets, denoted by W, exists. Let

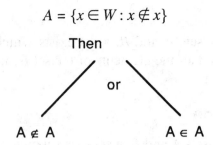

$$A = \{x \in W : x \notin x\}$$

Then

or

A ∉ A   A ∈ A

Indeed,

$(x \in A) \equiv (x \notin x)$; therefore $(A \in A) \equiv (A \notin A)$.

That leads to:

THEOREM 2

A set of all sets does not exist.

This fact was briefly described by Paul Halmos in this statement: "Nothing contains everything." New paradoxes of set theory appeared shortly after Russell's paradox. To overcome this difficulty, mathematicians created several variations of axiomatic set theory. None of these theories was completely satisfactory, nevertheless they eliminated most of the antinomies.

● PROBLEM 2–24

Consider the axiomatic formulation of the algebra of sets. It consists of the concept of element, of set and of the relation of an element belonging to a set $(x \in A)$ and four fundamental axioms. Using this system, show that

1. the null set exists.

2. for any two sets A and B, only one set $A \cup B$ exists.

SOLUTION:

We shall list the four fundamental axioms.

I. **UNIQUENESS AXIOM (Axiom of Extension).**

If the sets A and B have the same elements, then A and B are identical.

II. UNION AXIOM.

For two arbitrary sets A and B, a set exists which contains all the elements of the set A, and all the elements of the set B and which does not contain any other element.

III. DIFFERENCE AXIOM.

For two arbitrary sets A and B, a set exists which contains those and only those elements of the set A which are not elements of the set B.

IV. EXISTENCE AXIOM.

At least one set exists.

1. Let us define the null set by the formula

$$\phi = A - A.$$

The existence of at least one set is guaranteed by Axiom IV.

2. For given sets A and B, one and only one set satisfying Axiom II exists. Hence, the operation $\cup$ is unique.
   Since

$$A \cap B = A - (A - B),$$

it is not necessary to include an axiom on the existence of an intersection. The intersection can be defined in terms of the difference.

● **PROBLEM 2-25**

Explain the system of axioms called Boolean algebra.

SOLUTION:

Note that in most of the theorems of the algebra of sets the symbol $\in$ does not appear, although it occurs in their proofs. This suggests the possibility of establishing the system of axioms, which will enable us to prove the theorems without referring to the relation $\in$.

BOOLEAN ALGEBRA

The fundamental concepts are:

The set ϕ and the operations $\cup$, $\cap$, $-$.

We assume the following axioms:

1. $A \cup B = B \cup A$

2. $A \cap B = B \cap A$

3. $A \cup (B \cup C) = (A \cup B) \cup C$

4. $A \cap (B \cap C) = (A \cap B) \cap C$

5. $A \cup \phi = A$

6. $A \cap (A \cup B) = A$

7. $A \cup (A \cap B) = A$

8. $A \cap (B \cup C) = (A \cap B) \cup (A \cap C)$

9. $(A - B) \cup B = A \cup B$

10. $B \cap (A - B) = \phi$

11. $A \cap 1 = A$

12. $(A \subset B) \equiv (A \cup B = B)$

From the above axioms we can deduce all the theorems of the algebra of sets. Axiom 12 defines the Inclusion Theory of sets as only one of the applications of Boolean algebra. Replacing in axioms the notion of a set by sentences, we obtain the algebra of sentences. This explains the close relationship between the algebra of sentences and the algebra of sets.

Sentence		Sets
$\vee$	corresponds to	$\cup$
$\wedge$	corresponds to	$\cap$
$'$ (negation)	corresponds to	^{c} (complement)

By omitting axioms 8 - 12, we obtain the axiomatic formulation of the lattice theory, applied, for example, in electrical networks.

CHAPTER 3

EQUIVALENCE RELATIONS, AXIOMATIC SET THEORY, QUANTIFIERS

1. The set of all values of the variable x, for which $\varphi(x)$ is a true sentence, is denoted by the symbol

$$\{x : \varphi(x)\}. \tag{1}$$

The following equivalence holds:

$$\text{For every } a: [a \in \{x : \varphi(x)\}] \equiv \varphi(a). \tag{2}$$

Using (1) and (2), describe the set of the real numbers which are larger than zero and smaller than or equal to 5.

2. Prove the following formulas:

$$\{x : \varphi(x) \vee \zeta(x)\} = \{x : \varphi(x)\} \cup \{x : \zeta(x)\} \tag{3}$$

$$\{x : \zeta(x) \wedge \varphi(x)\} = \{x : \zeta(x)\} \cap \{x : \varphi(x)\} \tag{4}$$

$$\{x : \varphi(x) \wedge [\zeta(x)]'\} = \{x : \varphi(x)\} - \{x : \zeta(x)\}$$

$$\text{and } \{x : [\varphi(x)]'\} = \{x : \varphi(x)\}^C. \tag{5}$$

SOLUTION:

1. The set can be described as follows:

$$\{x : (x \in R) \wedge (x > 0) \wedge (x \leq 5)\}. \tag{6}$$

2.
$$a \in \{x : \varphi(x) \vee \zeta(x)\} \equiv \varphi(a) \vee \zeta(a) \equiv$$

$$\equiv [a \in \{x : \varphi(x)\}] \vee [a \in \{x : \zeta(x)\}] \equiv$$

$$\equiv a \in \{x : \varphi(x)\} \cup \{x : \zeta(x)\}. \tag{7}$$

We employ the strategy used in the solution of (3) to prove (4).

$$a \in \{x : \zeta(x) \wedge \varphi(x)\} \equiv \zeta(a) \wedge \varphi(a) \equiv$$

$$\equiv [a \in \{x : \zeta(x)\}] \wedge [a \in \{x : \varphi(x)\}] \equiv$$

$$\equiv a \in \{x : \zeta(x)\} \cap \{x : \varphi(x)\}. \tag{8}$$

Proving (5),

$$a \in \{x : \varphi(x) \wedge [\zeta(x)]'\} \equiv \varphi(a) \wedge [\zeta(a)]' \equiv$$

$$\equiv [a \in \{x : \varphi(x)\}] \wedge [a \notin \{x : \zeta(x)\}] \equiv$$

$$\equiv a \in \{x : \varphi(x)\} - \{x : \zeta(x)\}. \tag{9}$$

Proving (6),

$$a \in \{x : [\varphi(x)]'\} \equiv [\varphi(a)]' \equiv$$

$$\equiv (a \in 1) \wedge [a \notin \{x : \varphi(x)\}] \equiv$$

$$\equiv a \in \{x : \varphi(x)\}^c. \tag{10}$$

The symbol $\{x : \varphi(x)\}$ will be used very often, sometimes with some modifications, throughout this book.

● **PROBLEM 3–2**

1. Using qualifiers, write the sentence: For every natural number, a larger natural number exists.

2. Explain the quantifiers by using the calculus of sentences.

SOLUTION:

1. The quantifier "there exists" is denoted by $\exists$ and the quantifier "for each" is denoted by $\forall$.

Let A represent the domain of a general statement $p(x)$, i.e., $x \in A$. Then

$$\forall x \in A : p(x) \tag{1}$$

asserts that for all $x \in A$, the statement $p(x)$ is true. Similarly,

$$\exists x \in A : p(x) \tag{2}$$

means that at least one $x \in A$ exists, such that $p(x)$ is true.

By denoting the set of natural numbers by N, we write

$$\forall n \, \exists m : n < m \tag{3}$$

where $m, n \in N$.

2. Suppose the elements of the set A can be written in the form $a_1, a_2, a_3, \ldots$. The sentence $\forall x : p(x)$ means that $p(x)$ is true for each $a_1, a_2, a_3, \ldots$. Thus

$$\forall x : p(x) \text{ is equivalent to } p(a_1) \wedge p(a_2) \wedge \ldots \tag{4}$$

Similarly, the sentence $\exists x : p(x)$ means that $p(x)$ is true for at least one of the elements $a_1, a_2, a_3, \ldots$. Thus

$$\exists x : p(x) \text{ is equivalent to } p(a_1) \vee p(a_2) \vee p(a_3) \ldots \qquad (5)$$

Note that a universal quantifier $\forall$ is "stronger" than existential quantifier $\exists$.

$$[\forall x : p(x)] \Rightarrow [\exists x : p(x)]. \qquad (6)$$

● **PROBLEM 3-3**

1. Using quantifiers write the statement: For each x there exists a y such that for all z the statement $p(x, y, z)$ is true.

2. Write the negation of this statement.

SOLUTION:

We shall apply the following rule of quantifier negation:

$$[\forall x : p(x)]' \equiv \exists x : [p(x)]' \qquad (1)$$

$$[\exists x : p(x)]' \equiv \forall x : [p(x)]' \qquad (2)$$

Equations (1) and (2) are generalized DeMorgan formulas.

Note that by taking negation of (1), we can define a universal quantifier in terms of the existential quantifier and negation

$$\forall x : p(x) \equiv [\exists x : [p(x)]']'. \qquad (3)$$

Similarly, existential quantifiers can be defined in terms of the universal quantifier and negation

$$\exists x : p(x) \equiv [\forall x : [p(x)]']'. \qquad (4)$$

1. This statement can be written as

$$\forall x \, \exists y \, \forall z : p(x, y, z). \qquad (5)$$

2. The negation of (5) is

$$[\forall x \, \exists y \, \forall z : p(x, y, z)]' \equiv$$

$$\equiv [\forall x \, [\exists y \, \forall z : p(x, y, z)]]' \equiv$$

$$\equiv \exists x \, [\exists y \, \forall z : p(x, y, z)]'. \qquad (6)$$

63

Here we used (1).

$$\equiv \exists x \, [\exists y \, [\forall z : p(x, y, z)]]' \equiv$$

$$\equiv \exists x \, \forall y \, [\forall z : p(x, y, z)]' \equiv$$

$$\equiv \exists x \, \forall y \, \exists z : [p(x, y, z)]' \equiv \tag{7}$$

● **PROBLEM 3-4**

Using quantifiers, write down the definitions of:

1. Limit of a sequence of real numbers.

X

2. Continuous functions. *at a point*

3. Uniformly continuous functions.

SOLUTION:

1. Let (x_n) denote a sequence of real numbers. This sequence is said to have a limit x. We write

$$\lim_{n \to \infty} x_n = x \tag{1}$$

when

$$\forall \, \varepsilon > 0 \quad \exists \, m \in N \quad \forall \, n \geq m \quad |x_n - x| < \varepsilon. \tag{2}$$

X.
continuity
at a
point

2. Let $f : R^1 \to R^1$ be a function continuous at x_0. Then

$$\forall \, \varepsilon > 0 \quad \exists \, \delta > 0 \quad \forall \, x \in R^1 \quad |x - x_0| < \delta \Rightarrow |f(x) - f(x_0)| < \varepsilon. \tag{3}$$

Function $f(x)$ is continuous at x_0 if, for every $\varepsilon > 0$, $\delta > 0$ exists such that for every $x \in R^1$, if $|x - x_0| < \delta$, implies $|f(x) - f(x_0)| < \varepsilon$.

3. Function $f : R^1 \to R^1$ is said to be uniformly continuous on R^1 if

$$\forall \, \varepsilon > 0 \quad \exists \, \delta > 0 \quad \forall \, x, x' \in R^1 \quad |x - x'| < \delta \Rightarrow |f(x) - f(x')| < \varepsilon. \tag{4}$$

From (3) and (4) we conclude that if a function is uniformly continuous on R^1, then it is continuous on R^1. The opposite is not true. For example, function $f(x) = x^2$ is continuous on R^1 but not uniformly continuous.

1. Is the following definition of an ordered pair correct?

$$(a, b) = \{\{a\}, \{a, b\}\}. \qquad (1)$$

2. Find the Cartesian product $A \times B$, where $A = [1, 2]$ and $B = [2, 3]$.

3. Let Z denote the set of all positive and negative integers and zero. Find $A \times B$, where $A = B = Z$.

SOLUTION:

1. To each of two elements a, b there corresponds their ordered pair (a, b), which satisfies the condition

$$(a, b) = (c, d) \text{ if, and only if } a = c \text{ and } b = d. \qquad (2)$$

Hence, in general $(a, b) \neq (b, a)$. But

$$(a, b) = (b, a) \text{ iff } a = b. \qquad (3)$$

Definition (1) satisfies condition (2) because

$$\{a, \{a, b\}\} = \{c, \{c, d\}\} \text{ if, and only if } a = c \text{ and } b = d.$$

Thus, definition (1) is correct.

2. **DEFINITION**

Let A and B represent two sets. Their Cartesian product, $A \times B$, is the set of all ordered couples

$$A \times B = \{(a, b) : a \in A, \ b \in B\}. \qquad (4)$$

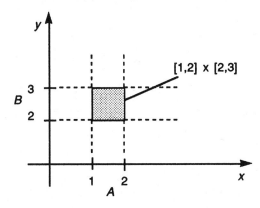

The Cartesian product is represented by the shaded area.

FIGURE 1.

65

3. The set $A \times B$, where $A = B = Z$ represents the set of lattice points in R^2.

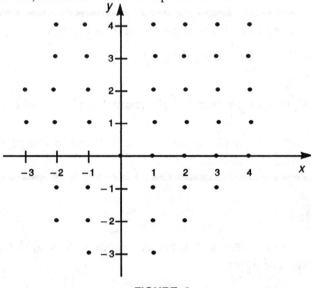

FIGURE 2.

Dots represent the elements of the set $Z \times Z$.

● PROBLEM 3-6

1. Let A represent the circle in the xy-plane, $x^2 + y^2 \leq 1$ and let B represent the set of points along the z coordinate, such that $0 \leq z \leq 1$. Find $A \times B$.

2. Show that,

$$(A \times B = \phi) \Leftrightarrow [(A = \phi) \vee (B = \phi)].$$ (1)

3. Show that,

if $A \times B \neq \phi$, then $A \times B \subset C \times D$ iff

$$(A \subset C) \wedge (B \subset D).$$ (2)

SOLUTION:

1. The set $A \times B$ is a cylinder of altitude 1, shown in Figure 1.

2. If $A \times B \neq \phi$, then an element exists $(a, b) \in A \times B$. Thus, $a \in A$ and $b \in B$ and $A \neq \phi$ and $B \neq \phi$.

Similarly, if $A \neq \phi$ and $B \neq \phi$, then elements a, b exist, such that $a \in A$ and $b \in B$. Therefore, $(a, b) \in A \times B$ and $A \times B \neq \phi$.

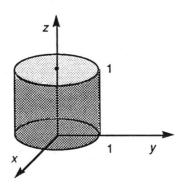

FIGURE 1.

3. We shall prove that

$$(A \times B \subset C \times D) \Leftrightarrow (A \subset C) \wedge (B \subset D). \tag{3}$$

$\Rightarrow$

Let $(a, b) \in A \times B$. Then $(a, b) \in C \times D$. Therefore, if $a \in A$, then $a \in C$. Similarly, if $b \in B$, then $b \in D$.

$\Leftarrow$

Let $(a, b) \in A \times B$. Then $a \in A$ and $a \in C$. Also $b \in B$ and $b \in D$. Therefore $(a, b) \in C \times D$ and

$$A \times B \subset C \times D. \tag{4}$$

● **PROBLEM 3-7**

Show that

$$A \times (B \cup C) = A \times B \cup A \times C \tag{1}$$

$$A \times (B \cap C) = A \times B \cap A \times C \tag{2}$$

$$A \times (B - C) = A \times B - A \times C \tag{3}$$

67

SOLUTION:

$$A \times (B \cup C) = \{(x, y) : (x \in A) \wedge (y \in B \cup C)\} =$$

$$= \{(x, y) : (x \in A) \wedge [(y \in B) \vee (y \in C)]\} =$$

$$= \{(x, y) : [(x \in A) \wedge (y \in B)] \vee [(x \in A) \wedge (y \in C)]\} =$$

$$= \{(x, y) : [(x, y) \in A \times B] \vee [(x, y) \in A \times C]\} =$$

$$= A \times B \cup A \times C. \tag{4}$$

Similarly, we prove (2).

$$A \times (B \cap C) = \{(x, y) : (x \in A) \wedge (y \in B \cap C)\} =$$

$$= \{(x, y) : (x \in A) \wedge [(y \in B) \wedge (y \in C)]\} =$$

$$= \{(x, y) : [(x, y) \in A \times B] \wedge [(x, y) \in A \times C]\} =$$

$$= A \times B \cap A \times C. \tag{5}$$

And (3).

$$A \times (B - C) = \{(x, y) : (x \in A) \wedge (y \in B - C)\} =$$

$$= \{(x, y) : (x \in A) \wedge [(y \in B) \wedge (y \notin C)]\} =$$

$$= \{(x, y) : [(x, y) \in A \times B] \wedge [(x, y) \notin A \times C]\} =$$

$$= A \times B - A \times C. \tag{6}$$

● **PROBLEM 3-8**

1. Prove that:

$$(A \times C) \cap (B \times D) = (A \cap B) \times (C \cap D) \tag{1}$$

$$(A \times C) \cup (B \times D) \subset (A \cup B) \times (C \cup D) \tag{2}$$

where $A, B \subset X$ and $C, D \subset Y$.

2.
$$A \times B = (A \times Y) \cap (X \times B) \tag{3}$$

$$(A \times B)^C = (A^C \times Y) \cup (X \times B^C) \tag{4}$$

68

where $A \subset X$ and $B \subset Y$. The symbols A^c and B^c denote the complements with respect to X and Y, and $(A \times B)^c$ denotes the complement with respect to $X \times Y$.

SOLUTION:

1.
$$[(x, y) \in (A \times C) \cap (B \times D)\,] \equiv$$

$$\equiv [(x, y) \in A \times C] \wedge [(x, y) \in B \times D] \equiv$$

$$\equiv (x \in A) \wedge (y \in C) \wedge (x \in B) \wedge (y \in D) \equiv$$

$$\equiv [(x \in A) \wedge (x \in B)] \wedge [(y \in C) \wedge (y \in D)] \equiv$$

$$\equiv (x \in A \cap B) \wedge (y \in C \cap D) \equiv$$

$$\equiv (x, y) \in (A \cap B) \times (C \cap D). \tag{5}$$

Applying Equation (1) of Problem 3-7 we find:
$$(A \cup B) \times (C \cup D) = [A \times (C \cup D)] \cup [B \times (C \cup D)] =$$

$$= (A \times C) \cup (A \times D) \cup (B \times C) \cup (B \times D). \tag{6}$$

Equation (2) follows from (6):
$$(A \times C) \cup (B \times D) \subset (A \cup B) \times (C \cup D). \tag{7}$$

2. Let us write (1) in the form:
$$(A \times Y) \cap (X \times B) = (A \cap X) \times (Y \cap B). \tag{8}$$

Since $A \subset X$, $A \cap X = A$. Similarly, since $B \subset Y$, $Y \cap B = B$.
Therefore,
$$A \times B = (A \times Y) \cap (X \times B). \tag{9}$$

To prove (4), we shall apply DeMorgan's law. Since
$$A \times B = (A \cap X) \times (Y \cap B) = (A \times Y) \cap (X \times B) \tag{10}$$

we obtain
$$(A \times B)^c = (A \times Y)^c \cup (X \times B)^c. \tag{11}$$

But
$$(A \times Y)^c = A^c \times Y \quad \text{and} \tag{12}$$

$$(X \times B)^c = X \times B^c. \tag{13}$$

69

Therefore

$$(A \times B)^C = (A^C \times Y) \cup (X \times B^C). \tag{14}$$

1. Give a definition of a family of sets and show that

$$\{A_n : n \in N\} \tag{1}$$

where $A_n =]-n, n[$ is a family of sets.

2. Is it true, that any set of sets can be converted to a family of sets?

SOLUTION:

DEFINITION OF A FAMILY OF SETS

The collection of sets

$$\{A_\lambda : \lambda \in \Lambda\} \tag{2}$$

is called a family of sets if, to each element λ of some set $\Lambda \neq \phi$, a set A_λ corresponds. Set Λ is called an indexing set for the family.

Unlike the definition of a set, where all elements have to be different, in a family of sets it is not required that sets with distinct indices be different. Equation (1) defines a family of sets, where N is the indexing set. Set A_n is an open interval

$$]-n, n[= \{x \in R' : -n < x < n\}. \tag{3}$$

2. Let G represent any set of sets. Then we can use G itself as an indexing set and assign the set it represents to each member of the set G.

Let

$$A_n =]-\frac{1}{n}, \frac{1}{n}[\tag{1}$$

where the indexing set is the set of all positive integers Z^+

$$Z^+ = \{1, 2, 3, \ldots \}. \qquad (2)$$

Find

$$\underset{n}{\cup} A_n$$

and

$$\underset{n}{\cap} A_n .$$

SOLUTION:

Let X represent a given set, and $\{A_\lambda : \lambda \in \Lambda\}$ a family of subsets of X. The union, denoted by $\underset{\lambda}{\cup} A_\lambda$, is the set

$$\{x \in X : \exists \lambda \in \Lambda; x \in A_\lambda\}. \qquad (3)$$

Hence, for the family of sets defined by (1), we obtain

$$\underset{n}{\cup} A_n = \underset{n}{\cup} \,] - \frac{1}{n}, \frac{1}{n} [\, = \,] - 1,1 \, [. \qquad (4)$$

The intersection, denoted by $\underset{\lambda}{\cap} A_\lambda$, is the set

$$\{x \in X : \forall \lambda \in \Lambda; x \in A_\lambda\}. \qquad (5)$$

Thus

$$\underset{n}{\cap} A_n = \underset{n}{\cap} \,] - \frac{1}{n}, \frac{1}{n} [\, = \{0\} \qquad (6)$$

since 0 is the only common point of all intervals $] - \frac{1}{n}, \frac{1}{n} [$.

Obviously, the union and intersection of a family of sets does not depend on how the family is indexed.

● **PROBLEM 3–11**

1. Show that $\underset{\lambda}{\cap}$ is distributive over $\cup$ and $\underset{\lambda}{\cup}$ is distributive over $\cap$. That is, show that:

$$[\cap \{A_\lambda : \lambda \in \Lambda\}] \cup [\cap \{B_\omega : \omega \in \Omega\}] =$$

$$= \cap \{A_\lambda \cup B_\omega : (\lambda, \omega) \in \Lambda \times \Omega\} \qquad (1)$$

$$[\cup \{A_\lambda : \lambda \in \Lambda\}] \cap [\cup \{B_\omega : \omega \in \Omega\}] =$$

$$= \cup \{A_\lambda \cap B_\omega : (\lambda, \omega) \in \Lambda \times \Omega\} \qquad (2)$$

2. Show that

$$\left(\bigcup_\lambda A_\lambda \right)^C = \bigcap_\lambda A_\lambda^C \tag{3}$$

and

$$\left(\bigcap_\lambda A_\lambda \right)^C = \bigcup_\lambda A_\lambda^C \tag{4}$$

where complements are taken with respect to X.

SOLUTION:

1.
$$x \in [\cap \{A_\lambda : \lambda \in \Lambda\}] \cup [\cap \{B_\omega : \omega \in \Omega\}] \equiv$$

$$\equiv [\forall \lambda \in \Lambda; x \in A_\lambda] \vee [\forall \omega \in \Omega; x \in B_\omega] \equiv$$

$$\equiv [\forall (\lambda, \omega) \in \Lambda \times \Omega : x \in A_\lambda \cup B_\omega] \equiv$$

$$\equiv x \in \cap \{A_\lambda \cup B_\omega : (\lambda, \omega) \in \Lambda \times \Omega\}. \tag{5}$$

In the same manner, we prove (2)

$$x \in [\cup \{A_\lambda : \lambda \in \Lambda\}] \cap [\cup \{B_\omega : \omega \in \Omega\}] \equiv$$

$$\equiv [x \in \cup A_\lambda] \wedge [x \in \cup B_\omega] \equiv$$

$$\equiv (\exists \lambda \in \Lambda : x \in A_\lambda) \wedge (\exists \omega \in \Omega : x \in B_\omega)$$

$$\equiv [\exists (\lambda, \omega) \in \Lambda \times \Omega : x \in A_\lambda \cap B_\omega] \equiv$$

$$\equiv x \in \cup \{A_\lambda \cap B_\omega : (\lambda, \omega) \in \Lambda \times \Omega\}. \tag{6}$$

2.
$$x \in \left(\bigcup_\lambda A_\lambda \right)^C \equiv x \in X - \left(\bigcup_\lambda A_\lambda \right) \equiv$$

$$\equiv x \notin \left(\bigcup_\lambda A_\lambda \right) \equiv [\exists \lambda \in \Lambda : x \in A_\lambda] \equiv$$

$$\equiv \forall \lambda \in \Lambda : x \notin A_\lambda \equiv \forall \lambda \in \Lambda : x \in A_\lambda^C \equiv$$

$$\equiv x \in \bigcap_\lambda A_\lambda^C. \tag{7}$$

Similarly,
$$x \in \left(\bigcap_\lambda A_\lambda \right)^C \equiv x \notin \left(\bigcap_\lambda A_\lambda \right) \equiv$$

$$\equiv [\forall \lambda \in \Lambda : x \in A_\lambda] \equiv [\exists \lambda \in \Lambda : x \notin A_\lambda] \equiv$$

$$\equiv [\exists \lambda \in \Lambda : x \in A_\lambda^C] \equiv x \in \bigcup_\lambda A_\lambda^C. \tag{8}$$

Let $\{A_n : n \in N\}$ be a family of sets and let

$$B_k = A_0 \cup A_1 \cup \dots \cup A_k, \quad k = 0, 1, 2, \dots \tag{1}$$

Show that the union

$$\overset{\infty}{\underset{0}{\cup}} A_n = A_0 \cup (A_1 - B_0) \cup \dots \cup (A_n - B_{n-1}) \cup \dots \tag{2}$$

is pairwise disjoint.

SOLUTION:

First, we shall prove (2). Suppose

$$x \in \overset{\infty}{\underset{0}{\cup}} A_n \tag{3}$$

and $x \in A_0$. Then $x \in A_0 \cup (A_1 - B_0) \cup \dots$. Now, suppose A_m is the first set in the sequence

$$A_0, A_1, \dots, A_m, \dots \tag{4}$$

such that

$$x \notin A_0, x \notin A_1, \dots, x \notin A_{m-1}, x \in A_m, \dots \tag{5}$$

Then

$$x \in \overset{\infty}{\underset{0}{\cup}} A_n \text{ and }$$
$$x \in A_m - B_{m-1} = A_m - (A_0 \cup \dots \cup A_{m-1}). \tag{6}$$

That proves (2).

To show that the union

$$A_0 \cup (A_1 - B_0) \cup \dots \cup (A_m - B_{m-1}) \cup \dots \tag{7}$$

consists of pairwise disjoint sets, suppose

$$x \notin A_0, x \notin A_1, \dots, x \notin A_{m-1}, x \in A_m, \tag{8}$$

Then

$$x \notin A_0, x \notin A_1 - B_0, x \notin A_2 - B_1, \dots \tag{9}$$

but

$$x \in A_m - B_{m-1}. \tag{10}$$

Again

$$x \notin A_{m+1} - B_m = A_{m+1} - (A_1 \cup \dots \cup A_m). \tag{11}$$

Hence the sets $A_0, A_1 - B_0, \dots, A_n - B_{n-1}, \dots$ in (7) are pairwise disjoint.

1. Let ρ denote the relation $<$ on $A \times B$, where $A = \{1, 3, 4, 5\}$ and $B = \{3, 4, 5\}$. Write all elements of ρ, its domain and range. Plot ρ on a coordinate diagram.

2. Let R be the set of real numbers and ρ the relation defined by

$$x \rho y \text{ iff both } x \in [n, n + 1] \text{ and } y \in [n, n + 1]$$

for some integer n.
   Plot the relation ρ.

SOLUTION:

We shall begin with a definition:

DEFINITION

A relation ρ is a subset $\rho \subset A \times B$. Often, $(a, b) \in \rho$ is written $a \rho b$.

1. A relation ρ consists of ordered pairs $(a, b) \in A \times B$, such that $a < b$. Therefore,

$$\rho = \{(1, 3), (1, 4), (1, 5), (3, 4), (3, 5), (4, 5)\}. \tag{1}$$

Note that the domain of ρ is the set of the first coordinates of the pairs in ρ. From (1) we have the

$$\text{domain of } \rho = \{1, 3, 4\}. \tag{2}$$

Similarly, the range of a relation is the set of the second coordinates of the pairs in this relation.
   Hence,

$$\text{range of } \rho = \{3, 4, 5\}. \tag{3}$$

Figure 1 depicts the sets A and B and the relation $<$.

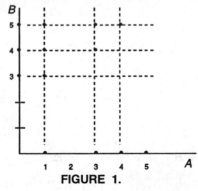

FIGURE 1.

74

2. Suppose $n = 0$, then

$$x \, \rho \, y \text{ if } x \in [0, 1] \quad \text{and} \quad y \in [0, 1],$$

which is a square shown in Figure 2.

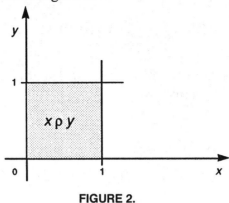

FIGURE 2.

For each integer n we obtain a square as shown in Figure 3.

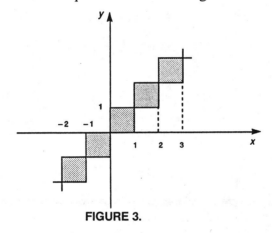

FIGURE 3.

● **PROBLEM 3-14**

1. Let Z denote the set of all integers. Find the diagonal of Z.

2. Let N denote the set of natural numbers and ρ, a relation defined by

$$n \, \rho \, m \text{ iff } n = 3m + 1. \tag{1}$$

Find the inverse of ρ, denoted by ρ^{-1}.

SOLUTION:

1. For any set A, its diagonal Δ is the relation of equality defined by

$$\Delta = \{(a, a) : a \in A\} \subset A \times A. \tag{2}$$

Thus, if Z is the set of all integers, its diagonal is the set

$$\Delta = \{(n, n) : n \in Z\} =$$

$$= \{\dots (-2, -2), (-1, -1), (0, 0), (1, 1), (2, 2), \dots\}. \tag{3}$$

2. Let ρ denote a relation defined as $A \times B$. The inverse of ρ, denoted by ρ^{-1}, is the subset of $B \times A$, defined by

$$\rho^{-1} = \{(b, a) : (a, b) \in \rho\}. \tag{4}$$

Relation ρ, defined by (1), consists of all the pairs

$$\rho = \{(1, 0), (4, 1), (7, 2), (10, 3), (-2, -1), (-5, -2) \dots\} \tag{5}$$

such that

$$(n, m) \in \rho \text{ if, and only if } n = 3m + 1. \tag{6}$$

The inverse ρ^{-1} consists of all the pairs $(m, n) \in \rho^{-1}$, such that

$$(m, n) \; \varepsilon \; \rho^{-1} \text{ iff } m = \frac{n - 1}{3}. \tag{7}$$

That leads to

$$\rho^{-1} = \{(0, 1), (1, 4), (2, 7), (3, 10), (-1, -2), (-2, -5) \dots\} \tag{8}$$

● **PROBLEM 3-15**

Consider three sets A, B, and C.

$$A = \{1, 2, 3, 4\}$$

$$B = \{5, 6, 7, 8\}$$

$$C = \{x, y, z\}. \tag{1}$$

Let ρ be a relation on $A \times B$

$$\rho = \{(1, 6), (1, 7), (3, 6), (3, 7), (3, 8), (4, 5)\} \tag{2}$$

and λ be a relation on $B \times C$

$$\lambda = \{(5, x), (6, y), (7, y), (7, z), (8, x)\}. \tag{3}$$

Find the composition of ρ and λ denoted by $\lambda \circ \rho$ (sometimes the symbol $\rho \circ \lambda$ is used instead).

SOLUTION:

We shall start with a definition of a composition of relations. Let A, B, C represent the sets, and $\rho \subset A \times B$ and $\lambda \subset B \times C$ represent two relations. The composition of ρ and λ, denoted by $\lambda \circ \rho$, is the relation on $A \times C$, such that

$$\lambda \circ \rho = \{(a, c) : a \in A, c \in C, \exists b \in B \text{ such that}$$

$$(a, b) \in \rho \wedge (b, c) \in \lambda\}. \tag{4}$$

Therefore, from (2), (3), and definition (4), we find

$(1, y) \in \lambda \circ \rho$ because $(1, 6) \in \rho$ and $(6, y) \in \lambda$

$(4, x) \in \lambda \circ \rho$ because $(4, 5) \in \rho$ and $(5, x) \in \lambda$

$(1, z) \in \lambda \circ \rho$ because $(1, 7) \in \rho$ and $(7, z) \in \lambda$

$(3, y) \in \lambda \circ \rho$ because $(3, 6) \in \rho$ and $(6, y) \in \lambda$

$(3, z) \in \lambda \circ \rho$ because $(3, 7) \in \rho$ and $(7, z) \in \lambda$

$(3, x) \in \lambda \circ \rho$ because $(3, 8) \in \rho$ and $(8, x) \in \lambda$

$(4, x) \in \lambda \circ \rho$ because $(4, 5) \in \rho$ and $(5, x) \in \lambda$

The composition $\lambda \circ \rho$ is the set consisting of

$$\lambda \circ \rho = \{(1, y), (4, x), (1, z), (3, y), (3, z), (3, x), (4, x)\}. \tag{5}$$

The relation $\lambda \circ \rho$ is illustrated in Figure 1.

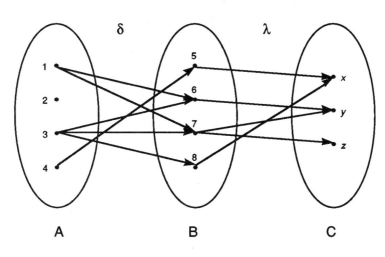

FIGURE 1.

For example $(3, x) \in \lambda \circ \rho$ because there is a "route" from 3 to x via 8.

77

Let A, B, C, and D represent the sets and $\rho \subset A \times B$, $\lambda \subset B \times C$, and $\delta \subset C \times D$ represent the relations. Prove that:

1. $(\delta \circ \lambda) \circ \rho = \delta \circ (\lambda \circ \rho)$ (1)

2. $(\lambda \circ \rho)^{-1} = \rho^{-1} \circ \lambda^{-1}$ (2)

SOLUTION:

1.

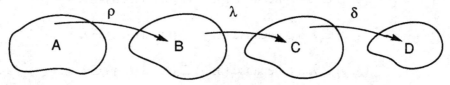

FIGURE 1.

$$(\delta \circ \lambda) \circ \rho = \{(a, d) : a \in A \wedge d \in D, \exists b \in B \text{ such that }$$

$$(a, b) \in \rho \wedge (b, d) \in \delta \circ \lambda\} =$$

$$= \{(a, d) : \exists b \in B, \exists c \in C \text{ such that }$$

$$(a, b) \in \rho, (b, c) \in \lambda, (c, d) \in \delta\} =$$

$$= \{(a, d) : \exists c \in C \text{ such that } (a, c) \in \lambda \circ \rho, (c, d) \in \delta\} =$$

$$= \delta \circ (\lambda \circ \rho). \tag{3}$$

Hence, we can write

$$(\delta \circ \lambda) \circ \rho = \delta \circ (\lambda \circ \rho) = \delta \circ \lambda \circ \delta. \tag{4}$$

2.
$$(\lambda \circ \rho)^{-1} = \{(c, a) : (a, c) \in \lambda \circ \rho\} =$$

$$= \{(c, a) : \exists b \in B, \text{ such that } (a, b) \in \rho, (b, c) \in \lambda\} =$$

$$= \{(c, a) : \exists b \in B, \text{ such that } (b, a) \in \rho^{-1}, (c, b) \in \lambda^{-1}\} =$$

$$= \{(c, a) : \exists b \in B, \text{ such that } (c, b) \in \lambda^{-1}, (b, a) \in \rho^{-1}\} =$$

$$= \rho^{-1} \circ \lambda^{-1}. \qquad (5)$$

Similarly,

$$(\delta \circ \lambda \circ \rho)^{-1} = [\delta \circ (\lambda \circ \rho)]^{-1} =$$

$$= (\lambda \circ \rho)^{-1} \circ \delta^{-1} = \rho^{-1} \circ \lambda^{-1} \circ \delta^{-1}. \qquad (6)$$

● **PROBLEM 3-17**

Let A represent the set of all people and R represent a relation defined as $A \times A$, such that for $a, b \in A$ and $a \, R \, b$ iff a knows b; that is, a is in relation to b if and only if a knows b. Is this relation an equivalence relation?

SOLUTION:

An equivalence relation is a very useful and frequently applied concept in mathematics. It enables us to divide the set into subsets (classes) according to some preassigned characteristics.

DEFINITION OF EQUIVALENCE RELATION

A relation R in A is called an equivalence relation if:

1. It is reflexive: $\forall a \in A, a \, R \, a$

2. It is symmetric: $(a \, R \, b) \Rightarrow (b \, R \, a)$

3. It is transitive: $(a \, R \, b) \wedge (b \, R \, c) \Rightarrow (a \, R \, c)$

If $a \, R \, b$, we say that a and b are equivalent.

The relation "knows" is reflexive, since obviously a knows a.

It is symmetric, because if a knows b, then b knows a.

It is not transitive, because if a knows b and b knows c, it does not guarantee that a knows c.

Therefore, the relation "knows" is not an equivalence relation.

Prove the following:

Let R denote a relation in A, that is, $R \subset A \times A$.

Then:

1. R is reflexive iff $\Delta \subset R$ (where Δ is the diagonal of A).

2. R is symmetric iff $R = R^{-1}$.

3. R is transitive iff $R \circ R \subset R$.

4. (R is symmetric) $\Rightarrow (R \circ R^{-1} = R^{-1} \circ R)$

5. (R is reflexive) $\Rightarrow \begin{pmatrix} 1. \ R \subset R \circ R \\ 2. \ R \circ R \ \text{is reflexive} \end{pmatrix}$

6. (R is transitive) $\Rightarrow (R \circ R$ is transitive)

SOLUTION:

1. The diagonal of A is

$$\Delta = \{(a, a) : a \in A\}. \qquad (1)$$

Thus R is reflexive iff,

$$\forall a \in A, \quad (a, a) \in R$$

iff $\Delta \subset R$.

2. $(R - \text{symmetric}) \Leftrightarrow (R = R^{-1})$
   $\Rightarrow$ If $(a, b) \in R$, then $(b, a) \in R$, therefore $R = R^{-1}$.
   $\Leftarrow$ Obvious.

3. Let $(a, c) \in R \circ R$. Then $\exists b \in A$, such that $(a, b) \in R$ and $(b, c) \in R$. Since R is transitive, and $(a, b), (b, c) \in R$ implies that $(a, c) \in R$, hence $R \circ R \subset R$.
   Conversely, if $R \circ R \subset R$, then $(a, b) \in R$ and $(b, c) \in R$ imply that $(a, c) \in R \circ R \subset R$. That is, R is transitive.

4. $\qquad R \circ R^{-1} = \{(a, c) : \exists b \in A, \text{ such that } (a, b) \in R^{-1} \wedge (b, c) \in R\} =$

$$= \{(a, c) : \exists b \in A, (a, b) \in R \land (b, c) \in R^{-1}\} =$$

$$= R^{-1} \circ R. \tag{2}$$

5. Let $(a, b) \in R$. Then

$$R \circ R = \{(a, c) : \exists b \in A, (a, b) \in R \land (b, c) \in R\}. \tag{3}$$

But, $(a, b) \in R$ and since R is reflexive, $(b, b) \in R$. Hence $(a, b) \in R \circ R$, that is, $R \subset R \circ R$.

Also,

$$\Delta \subset R \subset R \circ R \text{ implies that } R \circ R \text{ is reflexive.}$$

6. Let $(a, b) \in R \circ R$ and $(b, c) \in R \circ R$. Then from (3), $R \circ R \subset R$. Thus, $(a, b), (b, c) \in R$. So $(a, c) \in R \circ R$, and $R \circ R$ is transitive.

● **PROBLEM 3–19**

The set A consists of

$$A = \{1, 2, 3, 4, \alpha, \beta\}. \tag{1}$$

Find the smallest equivalence relation R, such that

$$(1, 2) \in R, \quad (2, \alpha) \in R, \quad (4, 2) \in R. \tag{2}$$

SOLUTION:

Relation R has to be reflexive, therefore,

$$(1, 1) \in R, \ (2, 2) \in R, \ (3, 3) \in R, \ (4, 4) \in R,$$

$$(\alpha, \alpha) \in R, \ (\beta, \beta) \in R. \tag{3}$$

It has to be symmetric, hence

$$(2, 1) \in R, \quad (\alpha, 2) \in R, \quad (2, 4) \in R. \tag{4}$$

Finally, R has to be transitive, thus,

$$(1, 2) \in R \land (2, \alpha) \in R \Rightarrow (1, \alpha) \in R$$

$$(1, 2) \in R \land (4, 2) \in R \Rightarrow (1, 4) \in R$$

$$(2, \alpha) \in R \wedge (4, 2) \in R \Rightarrow (4, \alpha) \in R. \tag{5}$$

The smallest equivalence relation is

$$R = \{(1, 1), (2, 2), (3, 3), (4, 4), (\alpha, \alpha), (\beta, \beta), (1, 2), (2, \alpha), (4, 2),$$

$$(2, 1), (\alpha, 2), (2, 4), (1, \alpha), (\alpha, 1), (1, 4), (4, 1), (4, \alpha), (\alpha, 4)\}. \tag{6}$$

The largest equivalence relation consists of all 36 elements of $A \times A$.

● **PROBLEM 3-20**

Let Z^+ denote the set of positive integers,

$$Z^+ = \{1, 2, \ldots\} \tag{1}$$

and let ρ represent the relation in $Z^+ \times Z^+$ defined by

$$(a, b) \, \rho \, (c, d) \text{ iff } a + d = b + c. \tag{2}$$

Show that ρ is an equivalence relation.

SOLUTION:

Element (a, b) is in relation ρ with itself. Indeed

$$(a, b) \, \rho \, (a, b) \text{ because } a + b = b + a. \tag{3}$$

Thus, ρ is reflexive.
Suppose:

$$(a, b) \, \rho \, (c, d), \text{ then } a + d = b + c \text{ or}$$

$$c + b = d + a. \tag{4}$$

Hence, $(c, d) \, \rho \, (a, b)$ and ρ are symmetric.
Suppose:

$$(a, b) \, \rho \, (c, d) \quad \text{and} \quad (c, d) \, \rho \, (e, f) \tag{5}$$

then

$$a + d = b + c \quad c + f = d + e. \tag{6}$$

Hence,

$$a + d + c + f = b + c + d + e \tag{7}$$

and

$$a + f = b + e. \tag{8}$$

From (8) we conclude that

$$(a, b) \, \rho \, (e, f). \tag{9}$$

Relation ρ is transitive. Thus, ρ is an equivalence relation.

● PROBLEM 3-21

Let $Z^+ \times Z^+$ denote the set of ordered pairs of positive integers. Let ρ denote the relation in $Z^+ \times Z^+$ defined by

$$(a, b) \, \rho \, (c, d) \text{ iff } ad = bc. \tag{1}$$

Show that ρ is an equivalence relation. Find the equivalence class of $(1,1)$.

SOLUTION:

Relation ρ is reflexive, because for every $(a, b) \in Z^+ \times Z^+$

$$(a, b) \, \rho \, (a, b), \text{ that is, } ab = ba. \tag{2}$$

Relation ρ is symmetric.
   Suppose

$$(a, b) \, \rho \, (c, d), \text{ that is, } ad = bc, \tag{3}$$

and $cb = da$. Therefore,

$$(c, d) \, \rho \, (a, b) \tag{4}$$

and ρ is symmetric.
   Suppose

$$(a, b) \, \rho \, (c, d) \quad \text{and} \quad (c, d) \, \rho \, (e, f). \tag{5}$$

That is,

$$ad = cb \text{ and } cf = de. \tag{6}$$

Hence,

$$adcf = cbde \tag{7}$$

and

$$af = be. \tag{8}$$

Therefore, $(a, b) \, \rho \, (e, f)$ and ρ is transitive.
   Relation ρ is reflexive, symmetric and transitive, thus it is an equivalence relation.

Suppose (a, b) is in relation ρ with $(1, 1)$

$$(a, b)\, \rho\, (1, 1) \quad \text{iff} \quad a = b. \tag{9}$$

The equivalence class of $(1, 1)$ consists of all pairs (a, a), $a \in Z^+$.

● **PROBLEM 3-22**

Consider a set X and its partition into disjoint classes

$$X = \bigcup_{i \in I} A_i. \tag{1}$$

This partition defines a relation $R \subset X \times X$, in such a way that $(x, y) \in R$ when $\exists i : x, y \in A_i$, that is, when the elements x, y belong to the same Class A_i. Show that R is an equivalence relation.

SOLUTION:

R is reflexive, because $(x, x) \in R$ means that x belongs to a certain class A_i, which is true, since

$$X = \bigcup_i A_i$$

That R is symmetric is obvious, because if $x, y \in A_i$, then $y, x \in A_i$. Hence

$$(x, y) \in R \Rightarrow (y, x) \in R. \tag{2}$$

Suppose $(x, y) \in R \wedge (y, z) \in R$. That means

$$\exists i\, \exists j\, (x, y \in A_i) \wedge (y, z \in A_j). \tag{3}$$

Thus

$$y \in A_i \cap A_j \neq 0 \tag{4}$$

which is a contradiction, because the classes are disjoint.

We proved that relation R is an equivalence.

In the next problem we shall prove the converse.

● **PROBLEM 3-23**

Prove the following:

THEOREM

Any equivalence relation in X defines a partition of X into classes of the form

$$A_x = \{y \in X : (x, y) \in R\} \tag{1}$$

■

SOLUTION:

Suppose R is an equivalence relation in X. Therefore R is reflexive

$$\forall x \in X \ (x, x) \in R \quad \text{and} \quad x \in A_x. \tag{2}$$

Hence

$$X = \bigcup_x A_x . \tag{3}$$

The classes cover the whole set X. We shall show that the classes are disjoint.
Suppose

$$A_x \cap A_y \neq \phi. \tag{4}$$

Then z exists, such that $z \in A_x \land z \in A_y$

$$[(z, x) \in R \land (z, y) \in R] \Rightarrow (x, y) \in R. \tag{5}$$

Therefore,

$$A_x = A_y. \tag{6}$$

That completes the proof.
The set of classes

$$\{A_x : x \in X\} \tag{7}$$

is called a quotient space and is denoted by $X/_R$. Mapping

$$\zeta : X \to \frac{X}{R} \tag{8}$$

where $\zeta(x) = A_x$, is called a canonical mapping or projection. The class A_x is denoted by $[x]$.

The principle of partition of a set into classes forms a basis of abstract thinking – one of the most interesting features of the human mind.

For example, abstract thinking leads us to the concept of temperature. Let X represent the set of all objects.

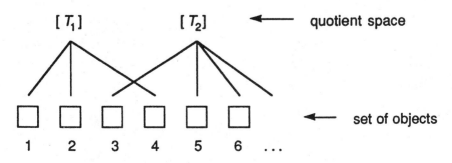

85

By using a thermometer, we establish an equivalence relation.

Two objects x and y are in relation T with each other, if they have the same temperature,

$$x\,T\,y. \tag{9}$$

It is easy to see that this is an equivalence relation. Neglecting all other properties like color, weight, age, etc., we classify objects into classes of equal temperature.

● **PROBLEM 3-24**

Let $\zeta(x, y)$ represent a given sentence function of two variables. Thus,

$$\forall y : \zeta(x, y) \quad \text{and} \quad \exists y : \zeta(x, y) \tag{1}$$

are sentence functions of one variable. Show that operation $\forall$ is commutative with respect to $\forall$, and $\exists$ is commutative with respect to $\exists$. That is, show that

$$\forall x\, \forall y\, \zeta(x, y) \equiv \forall y\, \forall x\, \zeta(x, y) \tag{2}$$

$$\exists x\, \exists y\, \zeta(x, y) \equiv \exists y\, \exists x\, \zeta(x, y) \tag{3}$$

Hence, the sequence of quantifiers is immaterial in (2) and (3).

On the other hand, the sequence of quantifiers $\forall$ and $\exists$ is significant. Prove the formula:

$$\exists x\, \forall y\, \zeta(x, y) \Rightarrow \forall y\, \exists x\, \zeta(x, y). \tag{4}$$

SOLUTION:

The sentence $\zeta(x, y)$ is true for all x and y, therefore, it is true for all y and x.

Similarly, if x and y exist, such that $\zeta(x, y)$ is true, then y and x exist, such that $\zeta(x, y)$ is true.

The left-hand side of (4) indicates that x_0 exists such that, for every y, $\zeta(x_0, y)$ is true. Therefore, for every y, an x (namely $x = x_0$) exists, such that $\zeta(x, y)$ is true. Hence, the left-hand side implies the right-hand side.

1. Show that the implication (4) of Problem 3–24 is not valid in the opposite direction.

2. Prove the formulas:

$$\exists x\ [\zeta(x) \wedge \varphi(x)] \Rightarrow \exists x\ \exists y\ [\zeta(x) \wedge \varphi(y)] \equiv$$

$$\equiv \exists x : \zeta(x) \wedge \exists y : \varphi(y) \equiv \exists x : \zeta(x) \wedge \exists x : \varphi(x) \tag{1}$$

and

$$\forall x : \zeta(x) \vee \forall x : \varphi(x) \equiv \forall x, y\ [\zeta(x) \vee \varphi(y)] \Rightarrow$$

$$\Rightarrow \forall x\ [\zeta(x) \vee \varphi(x)]. \tag{2}$$

SOLUTION:

1. Take, for instance, the set of real numbers. It is true that

$$\forall y\ \exists x : y < x. \tag{3}$$

On the other hand, the statement

$$\exists x\ \forall y : y < x \tag{4}$$

is not true. Hence, the implication in the opposite direction is not true.

2. Obviously the implication

$$\forall x : \zeta(x) \Rightarrow \exists x : \zeta(x) \tag{5}$$

is true. Setting $X = Y$, we can replace (5) by

$$\forall x\ \forall y : \zeta(x, y) \Rightarrow \forall x : \zeta(x, x) \Rightarrow \exists x : \zeta(x, x) \Rightarrow \exists x\ \exists y: \zeta(x, y). \tag{6}$$

Similarly, from

$$[\exists x : \zeta(x) \wedge \varphi(x)] \Rightarrow \exists x : \zeta(x) \wedge \exists x : \varphi(x) \tag{7}$$

we obtain

$$\exists x\ [\zeta(x) \wedge \varphi(x)] \Rightarrow \exists x\ \exists y\ [\zeta(x) \wedge \varphi(y)]. \tag{8}$$

The right-hand side is equivalent to

$$\exists x\ \exists y\ [\zeta(x) \wedge \varphi(y)] \equiv \exists x : \zeta(x) \wedge \exists y : \varphi(y) \equiv$$

$$\equiv \exists x : \zeta(x) \wedge \exists x : \varphi(x). \tag{9}$$

Similarly, from

$$[\forall x : \zeta(x) \vee \forall x : \varphi(x)] \Rightarrow \forall x \, [\zeta(x) \vee \varphi(x)] \qquad (10)$$

we obtain (2).

● **PROBLEM 3–26**

In Chapter 2, we discussed the axiomatic formulation of the algebra of sets. Then, four axioms:

1. Uniqueness Axiom

2. Union Axiom

3. Difference Axiom

4. Existence Axiom

were sufficient. What additional axioms are necessary for the purposes of Chapter 3?

SOLUTION:

We should add the following axioms:

5. For every sentence function $\zeta(x)$ and for every set A, a set exists consisting of those, and only those elements of A for which $\zeta(x)$ is true.
    This set is denoted by

$$\{x : \zeta(x), x \in A\}.$$

Without realizing it, we used this axiom frequently.
    For example, Axiom 5 guarantees the existence of the set $\{a, b\}$

$$\{a, b\} = \{x : x = a \vee x = b, x \in A\}.$$

6. For every set A, a set, whose elements are all the subsets of A, exists. This set is denoted by 2^A.

7. **AXIOM OF CHOICE.**
    For every family R of non-empty disjoint sets, a set exists which has one, and only one, element in common with each of the sets of the family R.

The last axiom is an existential axiom. In general, a set obtained by application of this axiom is not uniquely determined. In 1938, K. Gödel showed that if the axiomatic set theory without axiom of choice is consistent, then it is also consistent *with* the axiom of choice.

For some time, some mathematicians suspected that axiom of choice can be derived from the other axioms. In 1963, P.J. Cohen proved that this is not the case.

The axiom of choice is an independent axiom.

CHAPTER 4

MAPPINGS

Which of the diagrams depicts a function from $X = \{a, b, c\}$ to $Y = \{x, y, z\}$?

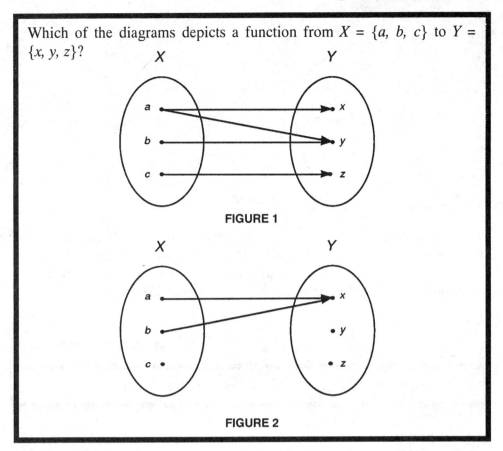

FIGURE 1

FIGURE 2

SOLUTION:

DEFINITION

f is a function from X to Y iff $f \not\subseteq X \times Y$, for all $x \in X$, there is a $y \in Y$, such that $(x, y) \in f$; $(x, y_1) \in f$ and $(x, y_2) \in f$ imply $y_1 = y_2$.

Function f is denoted

$$f : X \rightarrow Y. \tag{1}$$

The domain of f is X, the co-domain is Y.

The element $f(x) = y \in Y$ is called the image of x under f. Figure 1 illustrates the relation consisting of the pairs

$$(a, x), (a, y), (b, y), (c, z).$$

This relation assigns two elements, $x \in Y$ and $y \in Y$, to element $a \in X$. Hence, it is not a function.

Figure 2 also does not define a function. This relation is not defined for all elements of X. Element c is left "alone." It is easy to upgrade the relation shown in Figure 2 to obtain a function as shown in Figure 3.

91

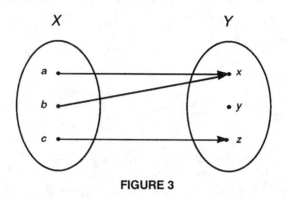

X   Y

FIGURE 3

Note that a function $f : X \rightarrow Y$ has to be defined for all elements of X, but it does not have to "cover" all elements of Y, as shown in Figure 3. In the extreme case, a function f can assign all elements of X to one element of Y.

Sometimes a word mapping or map is used instead of a function.

● PROBLEM 4-2

Prove that not every injection of a set into itself is a bijection.

SOLUTION:

We shall start with some definitions.

DEFINITION (SURJECTION (ONTO) FUNCTION)

A function f from X to Y, $f : X \rightarrow Y$ is an onto function (also called surjection) if, and only if, for all $y \in Y$, an $x \in X$ exists such that $y = f(x)$.

For example, function $f : R \rightarrow R$, where R is the set of real numbers, is a surjection when $f(x) = 3x + 1$ and is not a surjection when $f(x) = x^2$.

DEFINITION (INJECTION – ONE-TO-ONE FUNCTION)

A function $f : X \rightarrow Y$ is a one-to-one function (also called injection) iff $f(x_1) = f(x_2)$ implies $x_1 = x_2$.

DEFINITION (BIJECTION)

A function $f : X \rightarrow Y$ is a bijection iff f is both a surjection and injection.

For example, the function shown in Figure 1

92

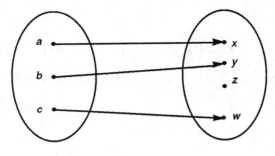

FIGURE 1

is one-to-one, but it is not onto, hence it is not a bijection.

We shall find function $f : X \rightarrow X$ which is an injection, but not a surjection. Let Z^+ denote the set of positive integers, then

$$f : Z^+ \rightarrow Z^+$$

such that

$$f(n) = n + 1$$

is one-to-one but not onto. Indeed no positive integer n exists such that

$$f(n) = 1.$$

Hence, $f(n) = n + 1$ is not a bijection.

● PROBLEM 4-3

1. Find the graph of the function shown in Figure 1.

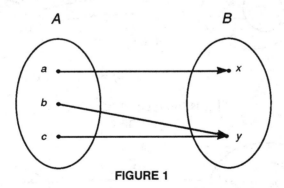

FIGURE 1

2. Let $f : R \rightarrow R$ and $g : R \rightarrow R$, where $f(x) = (x + 1)^2$, and $g(x) = x^2 + 2x + 1$ and R is the set of real numbers. Are these functions equal?

3. Define the restriction and extension of a function.

SOLUTION:

1. The set G

$$G = \{(x, y) : f(x) = y\} \subset X \times Y \tag{1}$$

is called the graph of f. From Figure 1, we find the graph of the function to be

$$\{(a, x), (b, y), (c, y)\}. \tag{2}$$

2. **DEFINITION**

    Two functions $f : A \rightarrow B$ and $g : A \rightarrow B$ are equal, $f = g$ iff

    $$f(a) = g(a) \text{ for every } a \in A.$$

    Obviously, for all real numbers

    $$(x + 1)^2 = x^2 + 2x + 1. \tag{3}$$

Hence $f(x) = g(x)$.

3. **DEFINITION**

    Let $f : X \rightarrow Y$ and $A \subset X$; the function f considered only on A is called the restriction of f to A, written $f \,|\, A$, if

    $$f \,|\, A = f \cap (A \times Y). \tag{4}$$

Similarly, if $A \subset X$ and $g : A \rightarrow Y$ is a given function, then

$$G : X \rightarrow Y$$

such that $G \,|\, A = g$ is called an extension of g over X.

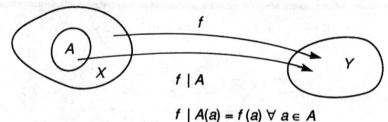

$$f \,|\, A(a) = f(a) \; \forall \; a \in A$$

$f \,|\, A$ is the restriction of f to A

FIGURE 2

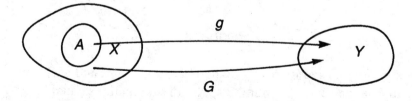

G is an extension of g.

FIGURE 3

Let f^{-1} be the inverse of the function f. In general, f^{-1} is only a relation. Find the conditions for f^{-1} to be a function.

SOLUTION:

We defined a function $f: X \rightarrow Y$ as a subset of $X \times Y$, $f \subset X \times Y$, such that

$$[(x, y_1) \in f \wedge (x, y_2) \in f] \Rightarrow [y_1 = y_2]. \qquad (1)$$

Since f is a relation, we can always find f^{-1}.

Function f must be a surjection, that is, $f(X) = Y$, then f^{-1} is defined for all $y \in Y$. In order for f^{-1} to be a function, function f must be an injection. Indeed, if

$$[(x_1, y) \in f \wedge (x_2, y) \in f] \Rightarrow [x_1 = x_2]$$

then we have

$$[(y, x_1) \in f' \wedge (y, x_2) \in f'] \Rightarrow [x_1 = x_2].$$

We conclude that only a function, which is a bijection, has the inverse function. It is easy to show that

$$(f^{-1})^{-1} = f.$$

Let the functions

$$f: R \rightarrow R \text{ and } g: R \rightarrow R$$

be defined by

$$f(x) = 3x - 1 \quad g(x) = x^3 + 1.$$

Find their compositions

$$f \circ g \quad \text{and} \quad g \circ f.$$

SOLUTION:

DEFINITION OF A COMPOSITION

Let $f: A \rightarrow B$ and $g: B \rightarrow C$. The composition of f and g is the function $g \circ f$ defined by:

$$(g \circ f)(a) = g(f(a)). \qquad (1)$$

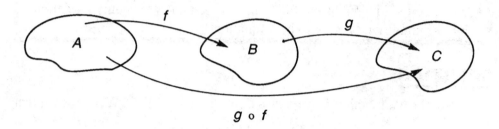

$g \circ f$

FIGURE 1

Let us find the composition $g \circ f$

$$(g \circ f)(x) = g(f(x)) = g(3x - 1) = (3x - 1)^3 + 1. \tag{2}$$

Similarly,

$$(f \circ g)(x) = f(g(x)) = f(x^3 + 1) = 3(x^3 + 1) - 1. \tag{3}$$

From (2) and (3) we conclude that

$$f \circ g \neq g \circ f. \tag{4}$$

● PROBLEM 4-6

Prove that the composition of functions is associative. That is, prove that if

$$f : A \to B, \quad g : B \to C, \quad h : C \to D$$

$g \circ f$

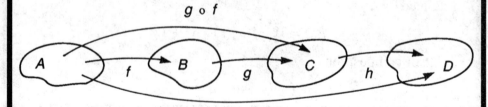

$h \circ g \circ f$

FIGURE 1

then

$$(h \circ g) \circ f = h \circ (g \circ f). \tag{1}$$

SOLUTION:

compositions of [defined p77 #4]   X

In Chapter 3 we proved that relations are associative and since functions are relations, this result holds for functions as well. Here is another way of showing it:

$$[(h \circ g) \circ f] (a) = (h \circ g) [f(a)] =$$

$$= h[g(f(a))] \quad \text{for every } a \in A. \tag{2}$$

Similarly,

$$[h \circ (g \circ f)] (a) = h [(g \circ f) (a)] =$$

$$= h[g(f(a))] \quad \text{for every } a \in A. \tag{3}$$

Thus

$$(h \circ g) \circ f = h \circ (g \circ f) = h \circ g \circ f. \tag{4}$$

● **PROBLEM 4-7**

1. Let $f : A \rightarrow B$, $g : B \rightarrow C$. Prove that

   a) If f and g are onto, then

   $$g \circ f : A \rightarrow C \text{ is onto;}$$

   b) If f and g are one-to-one, then

   $$g \circ f \text{ is one-to-one.}$$

2. Determine which of the following functions are onto, one-to-one or bijection.

$$f : A \rightarrow A, \quad A = [-1, 1]$$

$$f(x) = x^2, \quad f(x) = \sin x$$

$$f(x) = x^3, \quad f(x) = \sin \pi x \tag{1}$$

SOLUTION:

1. We shall show that $g \circ f$ is onto when both g and f are onto.
   Let $c \in C$, since g is onto

$$\exists b \in B, \text{ such that } g(b) = c.$$

Also, f is onto, therefore,

$$\exists a \in A, \text{ such that } f(a) = b.$$

We conclude

$$\forall c \in C \quad \exists a \in A, \text{ such that } (g \circ f)(a) = c \tag{2}$$

hence $g \circ f$ is onto.

Now we show that if f and g are one-to-one, then $g \circ f$ is one-to-one. Suppose it is not the case

$$(g \circ f)(a) = (g \circ f)(a') \tag{3}$$

or

$$g(f(a)) = g(f(a')). \tag{4}$$

But g is one-to-one. Hence

$$f(a) = f(a'). \tag{5}$$

Also f is one-to-one, hence

$$a = a'.$$

We conclude that $g \circ f$ is one-to-one.

2. Function $f(x) = x^2$ (Figure 1) is

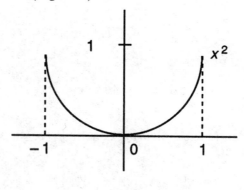

FIGURE 1

not onto because $f([-1, 1]) = [0,1]$. It is not one-to-one because, for example,

$$(-1)^2 = (1)^2.$$

Function $f(x) = \sin x$ (Figure 2) is one-to-one, but is not onto.

Function $f(x) = x^3$ (Figure 3) is one-to-one and onto, hence it is a bijection.

Function $f(x) = \sin \pi x$ (Figure 4) is onto but not one-to-one.

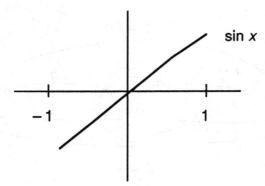

FIGURE 2

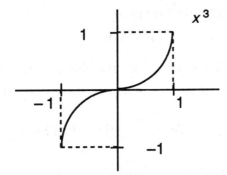

FIGURE 3

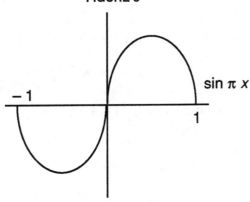

FIGURE 4

● **PROBLEM 4-8**

Prove that:

If $f : A \rightarrow B$ and $g : B \rightarrow C$ are one-to-one and onto, then

$$(g \circ f)^{-1} = f^{-1} \circ g^{-1}. \qquad (1)$$

SOLUTION:

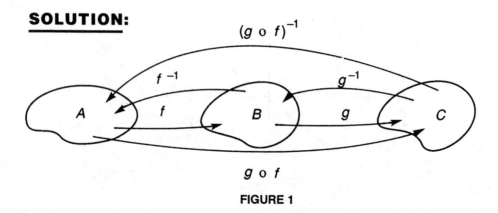

FIGURE 1

Let 1_A be the identity function on A

$$1_A(a) = a, \forall a \in A. \tag{2}$$

By applying the associative law of composition of functions, we find

$$(f^{-1} \circ g^{-1}) \circ (g \circ f) = f^{-1} \circ [g^{-1} \circ (g \circ f)] =$$

$$= f^{-1} \circ [(g^{-1} \circ g) \circ f] = f^{-1} \circ (1 \circ f) =$$

$$= f^{-1} \circ f = 1_A \tag{3}$$

since $g^{-1} \circ g = 1_A$ and $1_A \circ f = f \circ 1_A = f$.
   Similarly,

$$(g \circ f) \circ (f^{-1} \circ g^{-1}) = g \circ [f \circ (f^{-1} \circ g^{-1})] =$$

$$= g \circ [(f \circ f^{-1}) \circ g^{-1}] =$$

$$= g \circ (1 \circ g^{-1}) = g \circ g^{-1} = 1_A. \tag{4}$$

Therefore,

$$(g \circ f)^{-1} = f^{-1} \circ g^{-1}. \tag{5}$$

● **PROBLEM 4–9**

1. Let $X = [-1, 1]$ and $Y = R$, and

$$f : X \to R \text{ be } f(x) = x^2. \tag{1}$$

Find the image of X in R under f.

2. Find the inverse image of $[1, 4]$ in X under f, $f(x) = x^2$.

3. Prove that:

If $f : X \rightarrow Y$ and $A \subset X$, $B \subset X$, then

$$f(A \cup B) = f(A) \cup f(B) \tag{2}$$

$$f(A) - f(B) \subset f(A - B). \tag{3}$$

SOLUTION:

1. Let $f : X \rightarrow Y$. The image of $A \subset X$, denoted by $f(A)$, is the set of images of points of A

$$f(A) = \{f(x) : x \in A\}. \tag{4}$$

Hence, the image of $[-1, 1]$ is

$$f([-1, 1]) = [0, 1]. \tag{5}$$

2. The inverse image $f^{-1}(B)$ of any subset $B \subset Y$, is the set

$$f^{-1}(B) = \{x : x \in X, f(x) \in B\}. \tag{6}$$

The inverse image of $[1, 4]$ is

$$f^{-1}([1, 4]) = [1, 2] \tag{7}$$

or

UNION ? X

$$f^{-1}([1, 4]) = [-1, -2]. \tag{8}$$

3. First, we shall show that

$$f(A \cup B) \subset f(A) \cup f(B). \tag{9}$$

Let $y \in f(A \cup B)$, that is, $x \in A \cup B$ exists such that $f(x) = y$. We have

$$x \in A \quad \text{or} \quad x \in B \quad \text{then}$$

$$x \in A \Rightarrow f(x) = y \in f(A) \text{ or } x \in B \Rightarrow f(x) = y \in f(B). \tag{10}$$

Hence

$$y = f(x) \quad \text{and} \quad y \in f(A) \cup f(B). \tag{11}$$

Now we prove

$$f(A) \cup f(B) \subset f(A \cup B). \tag{12}$$

Let $y \in f(A) \cup f(B)$ then

$$y \in f(A) \quad \text{or} \quad y \in f(B)$$

101

$$y \in f(A) \Rightarrow \exists x \in A, \text{ such that } f(x) = y$$

$$y \in f(B) \Rightarrow \exists x \in B, \text{ such that } f(x) = y. \tag{13}$$

Hence

$$y = f(x) \quad \text{and} \quad x \in A \cup B, \text{ that is,}$$

$$y \in f(A \cup B).$$

Let $y \in f(A) - f(B)$. Then $\exists x \in A$, such that $f(x) = y$ but $y \notin \{f(x) : x \in B\}$. Hence $x \notin B$ and $x \in A - B$. Then, $y \in f(A - B)$.

● **PROBLEM 4–10**

Prove the following:
Let $f : X \rightarrow Y$, then for any subsets A and B of X,

1. $f(A \cap B) \subset f(A) \cap f(B)$ \hfill (1)

2. $A \subset B \Rightarrow f(A) \subset f(B)$ \hfill (2)

3. For any family A_α of subsets of X

$$f(\bigcup_\alpha A_\alpha) = \bigcup_\alpha f(A_\alpha) \tag{3}$$

$$f(\bigcap_\alpha A_\alpha) \subset \bigcap_\alpha f(A_\alpha) \tag{4}$$

4. Show that inclusion in (1) cannot be replaced, in general, by equality.

SOLUTION:

1. Suppose $y \in f(A \cap B)$. Hence $x \in A \cap B$ exists, such that $f(x) = y$. But $x \in A$ and $x \in B$, therefore

$$y = f(x) \in f(A) \quad \text{and} \quad y = f(x) \in f(B) . \tag{5}$$

Thus

$$y \in f(A) \cap f(B) \tag{6}$$

and

$$f(A \cap B) \subset f(A) \cap f(B). \tag{7}$$

2. Suppose $A \subset B$ and $y \in f(A)$. Then an x exists such that $f(x) = y$ and $x \in$

A. Hence $x \in B$ and $f(x) = y \in f(B)$. Therefore,

$$A \quad B \Rightarrow f(A) \quad f(B).$$ (8)

3. In Problem 4-9 we proved that

$$f(A \cup B) = f(A) \cup f(B).$$ (9)

Similarly,

$$y \in \bigcup_{\alpha} f(A_{\alpha}) \equiv \exists \, \alpha : y \in f(A_{\alpha}) \equiv$$

$$\equiv \exists \, x \in A_{\alpha} : f(x) = y \equiv \exists \, x \in \bigcup_{\alpha} A_{\alpha} : f(x) = y \equiv$$

$$\equiv y = f(x) \in f(\bigcup_{\alpha} A_{\alpha}).$$ (10)

That proves (3).

Suppose $y \in f(\bigcap_{\alpha} A_{\alpha})$, then

$$y \in f(\bigcap_{\alpha} A_{\alpha}) \equiv \exists x \in \bigcap_{\alpha} A_{\alpha} : f(x) = y \equiv$$

$$\equiv \exists \, x : \forall \, \alpha : x \in A_{\alpha} : f(x) = y \Rightarrow$$

$$\exists \, x : \forall \, \alpha : y \in f(A_{\alpha}) \equiv y \in \bigcap_{\alpha} f(A_{\alpha}).$$ (11)

4. Consider the function $f(x) = x^2$. Let $A = [-1, 0]$ and $B = [0, 1]$ then

$$A \cap B = [0] \quad \text{and} \quad f(A \cap B) = f(0) = [0]$$

$$f(A) = f([-1, 0]) = [0, 1].$$

$$f(B) = [0, 1].$$

$$f(A) \cap f(B) = [0, 1] \neq f(A \cap B) = [0].$$ (12)

● **PROBLEM 4–11**

Prove that:

$$\begin{pmatrix} f : X \to Y \\ \text{is one – to – one} \end{pmatrix} \Leftrightarrow \begin{pmatrix} \forall A, \forall B : A, B \subset X \\ f(A \cap B) = f(A) \cap f(B) \end{pmatrix}$$ (1)

SOLUTION:

First we shall prove $\Rightarrow$

Suppose $f : X \to Y$ is one-to-one. Let A and B be any subsets of X.

Then, for any $y \in f(A \cap B)$

$$y \in f(A \cap B) \equiv \exists x : x \in A \cap B, f(x) = y \equiv \qquad (2)$$

(Note that there is only one $x \in A \cap B$, such that $f(x) = y$.)

$$\equiv x \in A \wedge x \in B \wedge f(x) = y \equiv$$

$$\equiv y \in f(A) \wedge y \in f(B) \equiv y \in f(A \cap B). \qquad (3)$$

$\Leftarrow$ Now, suppose for any sets $A \subset X$ and $B \subset X$

$$f(A \cap B) = f(A) \cap f(B) \qquad (4)$$

and suppose $f : X \rightarrow Y$ is not one-to-one. Then x_1 and x_2, $x_1 \neq x_2$, exist such that

$$y = f(x_1) = f(x_2). \qquad (5)$$

Let $A = x_1$ and $B = x_2$; we have

$$f(A) \cap f(B) = f(x_1) \cap f(x_2) = y \neq f(A \cap B) =$$

$$= f(\phi). \qquad (6)$$

● **PROBLEM 4-12**

Let $f : X \rightarrow Y$ and let $P(X)$ and $P(Y)$ denote power sets of X and Y respectively. Function f induces functions

$$f : P(X) \rightarrow P(Y) \qquad (1)$$

by $X \supset A \rightarrow f(A)$ and

$$f^{-1} : P(Y) \rightarrow P(X) \qquad (2)$$

by

$$Y \supset B \rightarrow f^{-1}(B).$$

Prove the following:

THEOREM

Let $f : X \rightarrow Y$ denote one-to-one. Then the induced function

$$f : P(X) \rightarrow P(Y)$$

is also one-to-one.

SOLUTION:

If $X = \phi$, then $P(X) = \{\phi\}$, and

$$f : P(X) \rightarrow P(Y)$$

is one-to-one, because no two different elements of $P(X)$ can have the same image, as the set $P(X)$ consists of only one element.

Suppose $X \neq \phi$, then $P(X)$ has at least two elements. Let $A \in P(X)$ and $B \in P(X)$, but $A \neq B$. Then $x \in X$ exists, such that $x \in A$ and $x \notin B$. Hence, or $x \in B$ & $x \notin A$

$$f(x) \in f(A) \quad \text{and} \quad f(x) \notin f(B)$$

because f is one-to-one. We find

$$f(A) \neq f(B)$$

therefore, the induced function is also one-to-one.

● PROBLEM 4-13

Prove the following theorem:

THEOREM

Let $f : X \rightarrow Y$, then the induced function $f^{-1} : P(Y) \rightarrow P(X)$ preserves the elementary set operations

1. $f^{-1}(\bigcup_\alpha B_\alpha) = \bigcup_\alpha f^{-1}(B_\alpha)$ (1)

2. $f^{-1}(\bigcap_\alpha B_\alpha) = \bigcap_\alpha f^{-1}(B_\alpha)$ (2)

3. $f^{-1}(B_1 - B_2) = f^{-1}(B_1) - f^{-1}(B_2)$ (3)

SOLUTION:

1.
$$x \in f^{-1}(\bigcup_\alpha B_\alpha) \equiv f(x) \in \bigcup_\alpha B_\alpha \equiv$$

$$\equiv \exists \, \alpha : f(x) \in B_\alpha \equiv \exists \, \alpha : x \in f^{-1}(B_\alpha) \equiv$$

$$\equiv x \in \bigcup_\alpha f^{-1}(B_\alpha). \tag{4}$$

2.
$$x \in f^{-1}(\bigcap_\alpha B_\alpha) \equiv f(x) \in \bigcap_\alpha B_\alpha \equiv$$

$$\equiv \forall \, \alpha : f(x) \in B_\alpha \equiv \forall \, \alpha : x \in f^{-1}(B_\alpha) \equiv$$

$$\equiv x \in \bigcap_\alpha f^{-1}(B_\alpha). \tag{5}$$

3.
$$x \in f^{-1}(B_1 - B_2) \equiv f(x) \in B_1 - B_2 \equiv$$

$$\equiv f(x) \in B_1 \wedge f(x) \notin B_2 \equiv$$

$$\equiv x \in f^{-1}(B_1) \wedge x \notin f^{-1}(B_2) \equiv$$

$$\equiv x \in f^{-1}(B_1) - f^{-1}(B_2). \tag{6}$$

We conclude that the induced function f^{-1} preserves the elementary set operations. The induced function $f: P(X) \rightarrow P(Y)$ preserves unions, but it does not preserve intersections in general.

● **PROBLEM 4-14**

Show that:
If $f: X \rightarrow Y$, then:

1. For each $A \subset X$,

$$A \subset f^{-1}[f(A)]. \tag{1}$$

2. For each $A \subset X$ and $B \subset Y$,

$$B \cap f(A) = f[f^{-1}(B) \cap A] \tag{2}$$

in particular,

$$B \cap f(X) = f[f^{-1}(B)]. \tag{3}$$

SOLUTION:

1. Let A represent any subset of X and $x \in A$. Function f maps x into $f(x) \in Y$. Since function f is not necessarily one-to-one, it is possible that elements $x, x_1, x_2, \ldots$ exist, such that

$$f(x) = f(x_1) = f(x_2) \ldots \text{ (See Figure 1).} \tag{4}$$

Therefore,

$$x \in f^{-1}[f(x)] \tag{5}$$

Hence

$$A \subset f^{-1}[f(A)]. \tag{6}$$

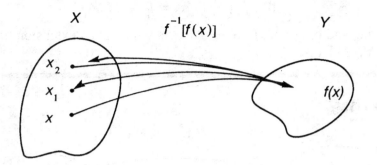

FIGURE 1

2. Suppose $y \in Y$ and

$$y \in f[f^{-1}(B) \cap A] \equiv \exists x : [x \in f^{-1}(B) \cap A] \wedge [f(x) = y] \equiv$$

$$\equiv \exists x : x \in A \wedge x \in f^{-1}(B) \wedge f(x) = y \equiv$$

$$\equiv \exists x : f(x) \in f(A) \wedge f(x) \in B \wedge f(x) = y \equiv$$

$$\equiv y \in f(A) \wedge y \in B \equiv y \in f(A) \cap B. \tag{7}$$

Therefore,

$$f[f^{-1}(B) \cap A] = f(A) \cap B. \tag{8}$$

In particular, setting $A = X$ we obtain

$$f[f^{-1}(B)] = f(X) \cap B \tag{9}$$

because

$$f^{-1}(B) \cap X = f^{-1}(B). \tag{10}$$

● **PROBLEM 4–15**

Prove this theorem:
THEOREM

Let X represent any set and $\{A_\alpha : \alpha \in \Omega\}$ its covering, i.e.

$$X = \bigcup_\alpha A_\alpha \tag{1}$$

Furthermore, let

$$\forall \alpha \quad \exists f_\alpha : A_\alpha \to Y \tag{2}$$

such that

107

$$\forall \alpha, \beta \quad f_\alpha | A_\alpha \cap A_\beta = f_\beta | A_\alpha \cap A_\beta. \tag{3}$$

Then one, and only one, $f : X \rightarrow Y$ exists which is an extension of each f_α, i.e. $\forall \alpha : f | A_\alpha = f_\alpha$.

SOLUTION:

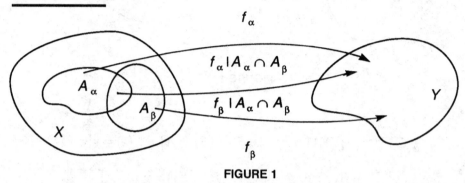

FIGURE 1

Let us define for each $x \in X$, $f(x) = f_\alpha(x)$, where $x \in A_\alpha$. If also $x \in A_\beta$, then $f(x) = f_\beta(x)$ but

$$f_\alpha | A_\alpha \cap A_\beta(x) = f_\beta | A_\alpha \cap A_\beta(x). \tag{4}$$

Hence, this is a correct definition of a function: it assigns to an element of X only one element of Y.

$$f(x) = f_\alpha(x) = f_\beta(x). \tag{5}$$

The function

$$f : X \rightarrow Y \tag{6}$$

is an extension of each f_α over X. Function f is unique, each x belongs to some A_α (A_α is a covering of X) and the value $f_\alpha(x)$ is uniquely defined.

● **PROBLEM 4–16**

Let

$$\{A_\alpha : \alpha \in \Omega\}$$

be a covering of X, i.e.

$$X = \bigcup_\alpha A_\alpha$$

such that if $\alpha \neq \beta$, then $A_\alpha \cap A_\beta = \phi$; then the family $\{A_\alpha : \alpha \in \Omega\}$ is called a partition of X.

Show that:

If $\{A_\alpha : \alpha \in \Omega\}$ is a partition of X and for each A_α, f_α is given, such that $f_\alpha : A_\alpha \to Y$, then $f : X \to Y$, which is an extension of each f_α, is uniquely defined by

$$f(x) = f_\alpha(x) \quad \text{for} \quad x \in A_\alpha. \tag{1}$$

SOLUTION:

We shall use the theorem proved in Problem 4-15. Since $\{A_\alpha : \alpha \in \Omega\}$ is a partition of X, the condition

$$f_\alpha | A_\alpha \cap A_\beta = f_\beta | A_\alpha \cap A_\beta \tag{2}$$

is satisfied. Therefore, (1) defines a function on X which is unique.

● **PROBLEM 4-17**

Let

$$\{A_\alpha : \alpha \in \Omega\} \text{ and } \{B_\beta : \beta \in \Lambda\} \tag{1}$$

denote two coverings (partitions) of X. Show that

$$\{A_\alpha \cap B_\beta : (\alpha, \beta) \in \Omega \times \Lambda\} \tag{2}$$

is also a covering (partition) of X.

SOLUTION:

Both $\{A_\alpha\}$ and $\{B_\beta\}$ are coverings of X. Let $x \in X$ represent any element of X. Since

$$X = \bigcup_\alpha A_\alpha$$

α' exists, such that $x \in A_{\alpha'}$. Also, β' exists, such that $x \in A_{\beta'}$. Therefore, $x \in A_{\alpha'} \cap B_{\beta'}$.

Hence,

$$\bigcup_{(\alpha, \beta)} A_\alpha \cap A_\beta = X \tag{3}$$

and (2) is a covering of X. Now we shall show that if both $\{A_\alpha\}$ and $\{B_\beta\}$ are partitions, then $\{A_\alpha \cap B_\beta\}$ is also a partition.

Consider,

$$(A_\alpha \cap B_\beta) \cap (A_{\alpha'} \cap B_{\beta'}) =$$

$$= (A_\alpha \cap A_{\alpha'}) \cap (B_\beta \cap B_{\beta'}). \tag{4}$$

109

Then

$$A_\alpha \cap A_{\alpha'} = \phi \quad \text{for } \alpha \neq \alpha'$$

and

$$B_\beta \cap B_{\beta'} = \phi \quad \text{for } \beta \neq \beta'.$$

Therefore, $\{A_\alpha \cap B_\beta\}$ is a partition.

● **PROBLEM 4-18**

Let $\{A_\alpha : \alpha \in \Omega\}$ and $\{B_\beta : \beta \in \Lambda\}$ denote coverings (partitions) of sets X and Y, respectively. Show that

$$\{A_\alpha \times B_\beta : (\alpha, \beta) \in \Omega \times \Lambda\} \tag{1}$$

is a covering (partition) of $X \times Y$.

SOLUTION:

Let $(x, y) \in X \times Y$. Then $x \in X$ and $y \in Y$. $\{A_\alpha\}$ is a covering of X. Hence

$$X = \bigcup_\alpha A_\alpha \tag{2}$$

and

$$\exists \alpha' \in \Omega : x \in A_{\alpha'}. \tag{3}$$

$\{B_\beta\}$ is a covering of Y. Hence

$$Y = \bigcup_\beta B_\beta \tag{4}$$

and

$$\exists \beta' \in \Lambda : y \in B_{\beta'}. \tag{5}$$

Thus,

$$(x, y) \in A_{\alpha'} \times B_{\beta'} \tag{6}$$

and (1) is a covering of $X \times Y$

$$X \times Y = \bigcup_{(\alpha, \beta)} A_\alpha \times B_\beta. \tag{7}$$

Suppose $\{A_\alpha\}$ and $\{B_\beta\}$ are partitions of X and Y, respectively; then

$$(A_\alpha \times B_\beta) \cap (A_{\alpha'} \times B_{\beta'}) = (A_\alpha \cap A_{\alpha'}) \times (B_\beta \cap B_{\beta'}) =$$

$$= \phi \times \phi = \phi. \tag{8}$$

Hence, $\{A_\alpha \times B_\beta\}$ is also a partition.

Let T be a family of subsets of X. Show that the smallest additive family U exists, such that $T \subset U$.

SOLUTION:

DEFINITION (ADDITIVE FAMILY)

The family of sets U is additive if

$$(A \in U, B \in U) \Rightarrow (A \cup B \in U) \tag{1}$$

Let N represent the class of all additive families S, such that

$$T \subset S. \tag{2}$$

Since the family of all subsets of X is an element of N, $N \neq \phi$. Let

$$U = \cap N. \tag{3}$$

We shall show that U defined by (3) is additive.

Let $A \in U$ and $B \in U$, then for every $S \in N$, $A \in S$ and $B \in S$. The set N consists of additive families, therefore,

$$A \cup B \in S. \tag{4}$$

Equation (4) holds for every $S \in N$, hence

$$A \cup B \in U. \tag{5}$$

U is additive.

Now, we shall show $T \subset U$. By assumption:

$$\text{for every } S \in N, T \subset S.$$

Hence, if $A \in T$, then $A \in S$ and

$$A \in U. \tag{6}$$

Therefore,

$$A \in T \Rightarrow A \in U \tag{7}$$

or

$$T \subset U.$$

The family $U = \cap N$ is the smallest additive family containing the family T, because U is the intersection of all families with this property.

We define:

The family S of sets is multiplicative, if

$$(A \in S \wedge B \in S) \Rightarrow (A \cap B \in S). \tag{8}$$

For multiplicative families, a similar theorem exists.

THEOREM

For every family T of subsets of X, the smallest multiplicative family U exists, such that $T \subset U$.

● **PROBLEM 4–20**

Consider the set X and the family T of all its one-element subsets. Find the smallest additive family U, such that $T \subset U$.

SOLUTION:

Since T is the family of all one-element subsets of X, we have

$$\forall \{a\} \ \forall \{b\}, \ \{a\} \in T, \ \{b\} \in T$$

$$\{a\} \cup \{b\} = \{a, b\} \in U \tag{1}$$

Because U is additive and $T \subset U$, U contains all one- and two-element subsets. By the same token, we can show that U consists of all finite subsets of X.

From the above considerations, it follows that a necessary and sufficient condition for a set X to be finite is that the family of all its non-empty subsets must be identical with U.

Note that by using this property, we can define a finite set without referring to the concept of natural numbers.

● **PROBLEM 4–21**

Let S denote a family of sets and S' denote the family of all sets of the form $D' = A - B$, where $A, B \in S$. Prove that

$$S' \subset S''. \tag{1}$$

SOLUTION:

The set S' consists of

$$S' = \{D' : D' = A - B, A \in S \wedge B \in S\} \tag{2}$$

112

while S'' consists of

$$S'' = \{D'' : D'' = E' - F', E' \in S' \wedge F' \in S'\}. \tag{3}$$

We understand that

$$S'' = (S')' \tag{4}$$

From (2) and (3), we obtain

$$S'' = \{D'' : D'' = (A - B) - (G - H), A, B, G, H \in S\} \tag{5}$$

where $E' = A - B$ and $F' = G - H$.

Suppose $A' \in S'$, then

$$A' = A - B \tag{6}$$

where $A, B \in S$.

Since S' consists of all sets of the form $A - B$, we conclude that

$$\phi \in S' \tag{7}$$

Therefore,

$$(A - B) - (A - A) = (A - B) - \phi = A - B \in S'' \tag{8}$$

and

$$S' \subset S''. \tag{9}$$

The inverse inclusion, in general, is false.

● **PROBLEM 4-22**

Prove that:

$$(A^B = \phi) \Leftrightarrow (A = \phi \wedge B \neq \phi). \tag{1}$$

Remember that $A^B = \{f : f : B \rightarrow A\}$.

SOLUTION:

$\Leftarrow$ Suppose $A = \phi$ and $B \neq \phi$. Then $f \in A^B$ implies $f \subset B \times A = B \times \phi = \phi$. But ϕ is not a function $f : B \rightarrow A$, $B \neq \phi$ and $A = \phi$. Therefore,

$$A^B = \phi. \tag{2}$$

$\Rightarrow$ Suppose $A \neq \phi$. Then $\exists a : a \in A$ and

$$\{(x, a) : x \in B\}$$

is a function from B to A. Hence

$$A^B \neq \phi. \tag{3}$$

113

Now, suppose $B = \phi$, then ϕ is a function from B to A and

$$A^B \neq \phi.$$

● **PROBLEM 4-23**

Prove that:

$$(A^B = B^A) \Rightarrow (A = B). \tag{1}$$

SOLUTION:

Suppose $A^B \neq \phi$, then a function f exists, such that

$$f : B \to A$$

$F \in A^B = B^A$. Since $f \in A^B$, the domain of f is B.
On the other hand, since $F \in B^A$, the domain of f is A. Thus,

$$A = B. \tag{2}$$

Now, suppose $A^B = \phi$ then

$$(A^B = \phi) \Rightarrow (A = \phi \wedge B \neq \phi). \tag{3}$$

Hence $A = \phi$.
But $A^B = B^A = \phi$ and $B = \phi$. Therefore,

$$A = B. \tag{4}$$

● **PROBLEM 4-24**

Let $F(X, R)$ denote the set of all real-valued functions $f \in F(X, R)$

$$f : X \to R. \tag{1}$$

Show that $F(X, R)$, with the usual algebraic operations, forms a real linear vector space.

SOLUTION:

We shall define algebraic operations on the elements $F(X, R)$. Let $f : X \to R$ and $g : X \to R$ and $a \in R$, then we define

$$(f + g)(x) = f(x) + g(x) \tag{2}$$

$$(a \cdot f)(x) = a \cdot f(x) \tag{3}$$

$$(f \cdot g)(x) = f(x) \cdot g(x) \tag{4}$$

$$(f + a)(x) = f(x) + a \tag{5}$$

We shall verify that $F(X, R)$ satisfies the axioms of the real linear vector space:

1. $(f + g) + (h) = f + (g + h)$

2. $f + g = g + f$

3. $\exists 0 \in F(X, R)$, such that $\forall f \in F(X, R), f + 0 = f$

4. $\forall f \in F(X, R)$ $\exists -f \in F(X, R)$, such that $f + (-f) = 0$

5. $a \cdot (b \cdot f) = (ab) \cdot f$ for any real numbers a and b

6. $1 \cdot f = f$

7. $a \cdot (f + g) = af + ag$

8. $(a + b)f = af + bf$

● **PROBLEM 4-25**

Let X represent the space (universal set), and A any subset of X. The characteristic function of A is defined by

$$X_A(x) = \begin{cases} 1 & \text{if } x \in A \\ 0 & \text{if } x \notin A \end{cases} \tag{1}$$

Show that

1. $X_{A \cap B} = X_A X_B$ (2)

2. $X_{A \cup B} = X_A + X_B - X_{A \cap B}$ (3)

3. $X_{A - B} = X_A - X_{A \cap B}$ (4)

SOLUTION:

1. We shall prove that

$$X_{A \cap B} = X_A \cdot X_B \tag{5}$$

115

Suppose $x \in A \cap B$, then

$$X_{A \cap B} = 1 \text{ and } X_A = X_B = 1.$$

If, on the other hand, $x \notin A \cap B$, then $X_{A \cap B} = 0$ and at least one of the functions X_A or X_B is equal to zero, $X_A \cdot X_B = 0$.

2. $X_{A \cup B} = 1$ when

(a) $x \in A \wedge x \notin B$

(b) $x \notin A \wedge x \in B$

(c) $x \in A \wedge x \in B$

For (a) we have $X_A + X_B - X_{A \cap B} = 1 + 0 - 0 = 1$; for (b) we have $X_A + X_B - X_{A \cap B} = 0 + 1 - 0 = 1$; for (c) we have $X_A + X_B - X_{A \cap B} = 1 + 1 - 1 = 1$.
If $X_{A \cup B} = 0$, then $x \notin A$ and $x \notin B$ hence $X_A + X_B - X_{A \cap B} = 0$.

3. Suppose $X_{A-B} = 1$, then $x \in A$ and $x \notin B$,

$$X_A - X_{A \cap B} = 1 - 0 = 1$$

If $X_{A-B} = 0$, then

(a) $x \notin A$, or

(b) $x \in A \wedge x \in B$

For (a) $X_A - X_{A \cap B} = 0 - 0 = 0$; for (b) $X_A - X_{A \cap B} = 1 - 1 = 0$.

● PROBLEM 4-26

Prove the following:
THEOREM

Let $f: A \rightarrow B$ represent relation-preserving, and let R represent relation in A, and S relation in B. Then, there is one, and only one function

(1)

such that
$$F: \frac{A}{R} \rightarrow \frac{B}{S},$$

$$p_B \circ f = F \circ p_A \qquad (2)$$

F is called the function induced by f and $p_A : A \rightarrow {}^A/_R$ and $p_B : B \rightarrow {}^B/_R$.

SOLUTION:

Let us start with the definition.

DEFINITION (RELATION-PRESERVING FUNCTION)

Let A, B denote two sets with equivalence relations R and S respectively. A function $f : A \to B$ is called relation-preserving if

$$(a, b) \in R \Rightarrow [f(a), f(b)] \in S. \qquad (3)$$

Now we shall show that the diagram

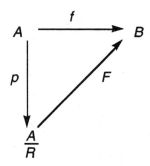

FIGURE 1

commutes. An equivalence class (see Problem 3–23) is denoted by $[a]$. Let us define function F by

$$F : [a] \to [f(a)] \qquad (4)$$

where $[a] \in {}^{A}/_{R}$ and $[f(a)] \in {}^{B}/_{S}$. Note that the equivalence class, $[f(a)] \in {}^{B}/_{S}$ is independent of the representative $a \in [a]$.

Take for example, $a' \in [a]$, then $(a, a') \in R$; but since f is relation preserving,

$$(a, a') \in R \Rightarrow [f(a), f(a')] \in S. \qquad (5)$$

Therefore,

$$f(a) \in [f(a)] \quad \text{and} \quad f(a') \in [f(a)]$$

and F defined by (4) is uniquely defined.

Function F is unique because p_A is a surjective. Suppose F^* is another function for which the diagram commutes, then for at least one $[a] \in {}^{A}/_{R}$,

$$F^*([a]) \neq f([a]). \qquad (6)$$

But since p_A is a surjective, (6) leads to

$$F^* \circ p_A(a) \neq F \circ p_A(a) \qquad (7)$$

117

which is impossible, because of commutativity.
That the diagram commutes follows from

$$p_B \circ f(a) = [f(a)] = F([a]) = F \circ p_A(a). \tag{8}$$

Let $f : A \to B$. Show that

$$(a, b) \in R \Leftrightarrow f(a) = f(b) \tag{1}$$

is an equivalence relation and show that a function $F : {}^A/_R \to B$ exists, such that

$$f = F \circ p \tag{2}$$

i.e. show that the diagram (Figure 1) commutes.

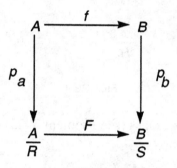

FIGURE 1

SOLUTION:

We shall show that (1) defines an equivalence relation.

1. $\forall a \in A : f(a) = f(a) \Rightarrow (a, a) \in R.$

2. $\forall a, b \in A : (a, b) \in R \Leftrightarrow f(a) = f(b) \Leftrightarrow (b, a) \in R.$

3. $\forall a, b, c \in A : (a, b) \in R \wedge (b, c) \in R \Leftrightarrow f(a) = f(b) \wedge f(b) = f(c) \Rightarrow$
    $\Rightarrow f(a) = f(c) \Leftrightarrow (a, c) \in R$

Hence, R is an equivalence relation.
Let us define $F : {}^A/_R \to B$ by

$$F([a]) = f(a). \tag{3}$$

Function F is uniquely defined. Indeed, suppose $b \in [a]$, then

$$(a, b) \in R \Rightarrow f(a) = f(b). \tag{4}$$

The diagram commutes because

$$(F \circ p)(a) = F(p(a)) = F([a]) = f(a). \tag{5}$$

● PROBLEM 4–28

Let $\{A_\alpha : \alpha \in \Omega\}$ be a family of sets. Define the Cartesian product of $\{A_\alpha\}$.

SOLUTION:

The Cartesian product of two sets A_1 and A_2, denoted $A_1 \times A_2$, was defined as a set of all ordered pairs

$$A_1 \times A_2 = \{(a, b) : a \in A_1 \wedge b \in A_2\} \tag{1}$$

On the other hand, $A_1 \times A_2$ can be considered to be the set of functions f of the index set $\{1, 2\}$ into $A_1 \cup A_2$, such that

$$f : \{1, 2\} \to A_1 \cup A_2$$

$$f(1) \in A_1 \quad \text{and} \quad f(2) \in A_2. \tag{2}$$

DEFINITION

Let $\{A_\alpha : \alpha \in \Omega\}$ represent the family of sets. The Cartesian product, denoted $X A_\alpha$, is the set of all maps

$$F : \Omega \to \bigcup_\alpha A_\alpha \tag{3}$$

such that

$$\forall \alpha \in \Omega : F(\alpha) \in A_\alpha.$$

An element $f \in X_\alpha A_\alpha$ is usually denoted as $\{a_\alpha\}$, where $\forall \alpha \in \Omega : f(\alpha) \in a_\alpha$.

The element $a_\alpha \in A_\alpha$ is called the αth coordinate of $\{a_\alpha\}$.

The set A_α is called the αth factor of $X_\alpha A_\alpha$.

When the index set of Ω is the set of natural numbers and all A_n are $A_n = R$, where R is the set of real numbers, we obtain

$$\mathop{X}_{m=1}^{n} R = R^n$$

the n-dimensional Euclidean space; and the product,

119

$$\overset{\infty}{\underset{n=1}{X}} R_n$$

is the extension of the n-dimensional space R^n to an infinite number of dimensions.

Prove that the following properties are equivalent:

1. If each $A_\alpha \neq \phi$ and $\{A_\alpha : \alpha \in \Omega\}$ is a non-empty family, then $\underset{\alpha}{X} A_\alpha \neq \phi$

2. The axiom of choice.

3. If each $A_\alpha \neq \phi$ and $\{A_\alpha : \alpha \in \Omega\}$ is a non-empty family of sets, then a function $F : \Omega \to \underset{\alpha}{\bigcup} A_\alpha$ exists such that $\forall \alpha \in \Omega : F(\alpha) \in A_\alpha$. F is called the choice function.

SOLUTION:

$(1) \Rightarrow (2)$.

Suppose $\{A_\alpha : \alpha \in \Omega\}$ is a non-empty family of sets. Since $\underset{\alpha}{X} A_\alpha \neq \phi$, an element $F = \{a_\alpha\}$ exists and

$$F(\Omega)$$

is a set which satisfies the axiom of choice.

$(2) \Rightarrow (3)$.

Let us define for each $\alpha \in \Omega$

$$A'_\alpha = \{\alpha\} \times A_\alpha$$

Each A'_α is a non-empty set.

According to the axiom of choice, a set T exists consisting of exactly one element from each A'_α. Hence, for each α, there is a unique $(\alpha, a_\alpha) \in T$, $a_\alpha \in A_\alpha$

$$T \subset \underset{\alpha}{\bigcup}(\{\alpha\} \times A_\alpha) \subset \underset{\alpha}{\bigcup}(\Omega \times A_\alpha) = \Omega \times \underset{\alpha}{\bigcup} A_\alpha$$

and T is a function

$$T : \Omega \to \underset{\alpha}{\bigcup} A_\alpha .$$

$(3) \Rightarrow (1)$.

If $F : \Omega \to \underset{\alpha}{\bigcup} A_\alpha$, then F is an element of $\underset{\alpha}{X} A_\alpha$.

120

Let $\{A_\alpha : \alpha \in \Omega\}$ represent a family of non-empty sets, and let

$$\Lambda \subset \Omega.$$

Show that the function

$$P : X_\alpha \{A_\alpha : \alpha \in \Omega\} \to X_\alpha \{A_\alpha : \alpha \in \Lambda\} \qquad (1)$$

defined by

$$P(F) = F \mid \Lambda \qquad (2)$$

is onto and that each projection

$$P_\beta : X_\alpha A_\alpha \to A_\beta \qquad (3)$$

is onto.

SOLUTION:

Let f be any element

$$f \in X \{A_\alpha : \alpha \in \Lambda\}. \qquad (4)$$

We shall show that

$$F \in X \{A_\alpha : \alpha \in \Omega\} \qquad (5)$$

exists, such that

$$P(F) = f. \qquad (6)$$

By Problem 4-29 part 3, a choice function exists such that

$$G : \Omega - \Lambda \to \cup \{A_\alpha : \alpha \in \Omega - \Lambda\} \qquad (7)$$

Then

$$F : \Omega \to \cup \{A_\alpha : \alpha \in \Omega\} \qquad (8)$$

defined by

$$F \mid \Lambda = f$$

$$F \mid \Omega - \Lambda = G \qquad (9)$$

is an element

$$F \in X \{A_\alpha : \alpha \in \Omega\} \qquad (10)$$

and

$$P(F) = F \mid \Lambda = f.$$

Setting $\Lambda = \{\beta\}$, $\beta \in \Omega$ we obtain $p_\beta = P$, which completes the proof.

Two families are given $\{A_\alpha : \alpha \in \Omega\}$ and $\{B_\beta : \beta \in \Lambda\}$. Prove that

$$\left(\bigcup_{\alpha \in \Omega} A_\alpha\right) \times \left(\bigcup_{\beta \in \Lambda} B_\beta\right) = \bigcup_{(\alpha, \beta) \in \Omega \times \Lambda} (A_\alpha \times B_\beta). \tag{1}$$

SOLUTION:

Suppose

$$(a, b) \in \left(\bigcup_\alpha A_\alpha\right) \times \left(\bigcup_\beta B_\beta\right) \tag{2}$$

then

$$a \in \bigcup_{\alpha \in \Omega} A_\alpha \quad \text{and} \quad b \in \bigcup_{\beta \in \Lambda} B_\beta. \tag{3}$$

Hence, for some $\alpha \in \Omega$, $a \in A_\alpha$ and for some $\beta \in \Lambda$, $b \in B_\beta$. Therefore,

$$a \in A_\alpha \text{ and } b \in B_\beta \text{ for some } (\alpha, \beta) \in \Omega \times \Lambda.$$

Therefore,

$$(a, b) \in A_\alpha \times B_\beta \text{ for some } (\alpha, \beta) \in \Omega \times \Lambda \tag{4}$$

and finally

$$(a, b) \in \bigcup_{(\alpha, \beta) \in \Omega \times \Lambda} A_\alpha \times B_\beta. \tag{5}$$

CHAPTER 5

POWER OF A SET, EQUIVALENT SETS

Let $N = \{1, 2, 3, ...\}$ and $P = \{2, 4, 6, ...\}$. Show that N and P are equivalent sets. Are they infinite?

SOLUTION:

We shall start with a very important definition.

DEFINITION OF EQUIVALENT SETS

Two sets A and B are called equivalent, written $A \sim B$, if a function $f : A \rightarrow B$ exists, which is one-to-one and onto. Function f defines a one-to-one correspondence between the sets A and B.

Let us define

$$f : n \rightarrow 2n, \quad f : N \rightarrow P. \tag{1}$$

Function (1) is one-to-one and onto. Therefore, sets N and P are equivalent.
Here is another definition.

DEFINITION OF FINITE SETS

A set is finite iff it is empty or equivalent to $\{1, 2, ..., n\}$ for some $n \in N$. Otherwise, the set is called infinite.

The set N is infinite because it is not empty and not equivalent to any set $\{1, 2, ..., n\}$. Since N and P are equivalent, P is also infinite.

● **PROBLEM 5-2**

Determine which of the following pairs are equivalent sets.

1. $\{1, 2, 3\}$ and $\{a, b, 4\}$.

2. Points of two circles.

SOLUTION:

1. There is a one-to-one and onto correspondence between the sets $\{1, 2, 3\}$ and $\{a, b, 4\}$.

$$f : 1 \rightarrow a$$
$$2 \rightarrow b$$

$3 \rightarrow 4$

Two finite sets are equivalent if, and only if, they contain the same number of elements.

2. We can assume that circles are concentric.

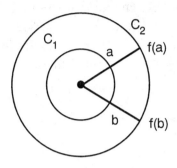

FIGURE 1

If they have equal radii, then equivalence is obvious. If $r_1 \neq r_2$, we can establish a one-to-one and onto correspondence in the following way:

Let a represent any point on the circle C_1. Then $f(a)$ is the point of intersection of the radius from the center, through a, to C_2, as shown in Figure 1. Function f defined above is one-to-one and onto. Hence the sets consisting of the points of C_1 and C_2 are equivalent.

● PROBLEM 5-3

Show that the open interval $]-1, 1[$ and the real axis R are equivalent.

SOLUTION:

Consider the "structure" shown in Figure 1.

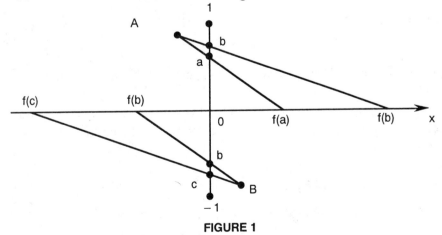

FIGURE 1

125

The "upper" half of the interval is mapped on the positive part of the real axis in the following manner:

Let a represent any point, such that $a \in \,]1, 0]$. We will draw a line from point A (which is at the same level as point 1 of the interval, and to the left of it) through point a to the x-axis. The point of intersection is $f(a)$. In the same manner, we assign a unique point of the negative axis $(x < 0)$ to each point of $]0, -1[$. The determined function is one-to-one and onto. Hence, the open interval $]-1, 1[$ is equivalent to R.

Another way to show it is to define function f, such that

$$f:]-1,1[\;\rightarrow R \text{ and } f(x) = \frac{x}{1-|x|}.$$

Function f is onto and one-to-one. The desired result follows.

● **PROBLEM 5-4**

1. Show that the set of elements of any infinite sequence of distinct terms is denumerable.

2. Prove that the relation $A \sim B$ *(A is equivalent to B)* is an equivalence relation.

SOLUTION:

1. We shall start with a definition.

DEFINITION OF DENUMERABLE SETS

Set A is called denumerable if it is equivalent to the set of positive integers $N = \{1, 2, 3, \ldots\}$. Such a set is said to have cardinality aleph-null, denoted $\aleph_0$ or card N.

Let us define the function $f: N \rightarrow \{a_n\}$ by

$$f: n \rightarrow a_n.$$

Its domain is the set N and its range is the set of all elements of the sequence (a_n).

Since the elements of (a_n) are distinct, the function is one-to-one and onto. Hence, the set of elements of an infinite sequence of distinct terms is denumerable.

DEFINITION OF COUNTABLE SETS

A set is called countable if it is finite or denumerable.

Any sequence is a countable set.

2. We shall show that $A \sim B$ is an equivalence relation. The identity function , i.e.

$$1 : x \rightarrow x$$

is a one-to-one mapping of the set A onto itself. Since the inverse of a one-to-one mapping is one-to-one, it follows that

$$(A \sim B) \Rightarrow (B \sim A).$$

The composition of two one-to-one and onto mappings is one-to-one and onto. Therefore,

$$(A \sim B) \wedge (B \sim C) \Rightarrow A \sim C.$$

Remark:

Relation $\sim$ is an equivalence relation. Hence we can classify the sets with respect to their power.

To each set A we assign a cardinal number (denoted sometimes by $\bar{\bar{A}}$ or card A) in such a way that the same cardinal number is assigned to two distinct sets if these sets are equivalent (have the same power).

● **PROBLEM 5–5**

1. Show that the lattice points (n, m) in a plane form a denumerable set, where n and m are integers.

2. Show that the set of positive rational numbers is denumerable.

SOLUTION:

1. Let us order the lattice points in the manner shown in Figure 1.

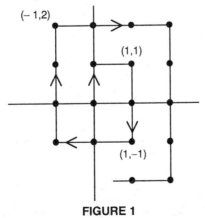

FIGURE 1

We order the set of lattice points into a sequence. That is, we establish a one-to-one and onto correspondence between the set of natural numbers, and the set of lattice points.

The lattice points are ordered as follows:

$(0,0)$ — 1
$(0,1)$ — 2
$(1,1)$ — 3
$(1,0)$ — 4
$(1,-1)$ — 5 etc.

The ordering starts at the origin.

2. The set of rational numbers is arranged as shown in Figure 2.

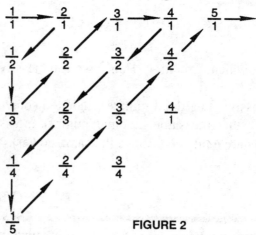

FIGURE 2

The diagram is traversed in the diagonally arrowed path. Every positive rational number will be included. This procedure establishes a one-to-one and onto mapping between the set of positive integers N and the set of rational positive numbers

$1 - \frac{1}{1}$
$2 - \frac{2}{1}$
$3 - \frac{1}{2}$
$4 - \frac{1}{3}$
$5 - \frac{2}{3}$
$\vdots$

Hence, the set of rational numbers is denumerable.

● **PROBLEM 5–6**

Prove the following:
THEOREM

The set of all real numbers is non-countable.

SOLUTION:

To prove this theorem, it suffices to prove that for every sequence of real numbers $a_1, a_2, \ldots, a_n, \ldots$ a real number, a, which does not belong to the sequence, exists.

We define a sequence of closed intervals

$$\alpha_1 \beta_1, \alpha_2 \beta_2, \ldots, \alpha_n \beta_n, \ldots \tag{1}$$

such that

$$\beta_n - \alpha_n = \frac{1}{3^n} \quad \text{and} \quad \alpha_n \beta_n \subset \alpha_{n-1} \beta_{n-1}, \ a_n \notin \alpha_n \beta_n \tag{2}$$

We divide the closed interval $[0, 1]$ into three $[0, \frac{1}{3}]$, $[\frac{1}{3}, \frac{2}{3}]$, $[\frac{2}{3}, 1]$ and choose one which does not contain the point a_1. That interval is $\alpha_1 \beta_1$. Similarly, we divide $\alpha_1 \beta_1$ into three intervals and choose the one which does not contain a_2. That interval is $\alpha_2 \beta_2$. So, in the closed interval $\alpha_{n-1} \beta_{n-1}$ we determine a closed interval $\alpha_n \beta_n$ of length $\frac{1}{3^n}$, which does not contain the point a_n. Let a denote the common point of all the closed intervals $\alpha_n \beta_n$

$$a = \bigcap_{n=1}^{\infty} \alpha_n \beta_n \tag{3}$$

Then

$$a = \lim_{n \to \infty} \alpha_n = \lim_{n \to \infty} \beta_n. \tag{4}$$

For every n, $a \in \alpha_n \beta_n$ while $a \neq a_n$, because $a_n \notin \alpha_n \beta_n$.

We have a sequence of real numbers $(a_1, a_2, a_3, \ldots)$. A real number, a, exists, which does not belong to the sequence.

Hence, the set of real numbers is not countable.

● **PROBLEM 5-7**

Show that the union $A \cup B$ of two countable sets A and B is countable.

SOLUTION:

The set A is countable, therefore its elements can be written in the form of an infinite sequence

$$a_1, a_2, a_3, \ldots, a_n, \ldots \tag{1}$$

Similarly, the elements of B can be written in the form of a sequence

$$b_1, b_2, \ldots, b_n, \ldots \tag{2}$$

We form the sequence

$$a_1, b_1, a_2, b_2, a_3, b_3, \ldots, a_n, b_n, \ldots \tag{3}$$

The terms of sequence (3) form the set $A \cup B$. Hence $A \cup B$ is a countable set.

From this, we conclude that the set of all integers is countable.

The set of all positive integers is countable and the set of all negative integers is countable.

● PROBLEM 5-8

Prove that the set of all rational numbers is denumerable.

SOLUTION:

Let Q^+ denote the set of positive rational numbers and let Q^- be the set of negative rational numbers. Then the set of all rational numbers Q is

$$Q = Q^+ \cup Q^- \cup \{0\}. \tag{1}$$

In Problem 5-5 we proved that the set of positive rational numbers Q^+ is denumerable. By the same token, we can show that the set of negative rational numbers Q^- is denumerable. The sets Q^+, Q^- and $\{0\}$ are countable.

From Problem 5-7, we conclude that the set Q is denumerable as the union of Q^+, Q^- and $\{0\}$.

● PROBLEM 5-9

Prove the following:

THEOREM

The Cartesian product of two (or, more generally, of a finite number of) countable sets is a countable set.

SOLUTION:

First we shall prove that the set of pairs (m, n), where m and n are natural numbers, is countable.

We will arrange the pairs in a sequence, in such a way that if

$$m + n < m' + n' \tag{1}$$

then (m, n) comes before (m', n'). If $m + n = m' + n'$, then the pair with the smaller antecedent comes first. We obtain the sequence of pairs

$$(1, 1), (1, 2), (2, 1), (1, 3), (2, 2), (3, 1), (1, 4), (2, 3), \ldots \tag{2}$$

Similarly, using the indexes of a_n and b_m, we write

$$(a_1, b_1), (a_1, b_2), (a_2, b_1), (a_1, b_3), (a_2, b_2), (a_3, b_1), (a_1, b_4), \ldots \qquad (3)$$

Suppose A and B are countable sets, then

$$A = \{a_1, a_2, a_3, a_4, \ldots\} \text{ and } B = \{b_1, b_2, b_3, \ldots\} \qquad (4)$$

The Cartesian product of A and B is

$$A \times B = \{(a_n, b_m) : a_n \in A \wedge b_m \in B\}. \qquad (5)$$

The elements of $A \times B$ can be arranged in a sequence as shown in (3).
   If the sets A, B, and C are countable, then $A \times B$ is countable, and

$$(A \times B) \times C = A \times B \times C$$

is a countable set. Hence, the Cartesian product of a finite number of countable sets is a countable set.
   It is easy to conclude from the above theorem that the set of all rational numbers is countable.

● **PROBLEM 5-10**

Prove:
The collection P of all polynomials with integer coefficients

$$p(x) = a_0 + a_1 x + a_2 x^2 + \ldots + a_n x^n \qquad (1)$$

is a denumerable set.

SOLUTION:

Let us define for each pair

$$(m, n) \in N \times N \qquad (2)$$

the set P_{mn} of polynomials $p(x)$ of degree n, such that

$$|a_0| + |a_1| + \ldots + |a_n| = m. \qquad (3)$$

For each pair (m, n), the set P_{mn} is finite.
   The set of all polynomials P can be represented as

$$P = \bigcup_{m, n} \{P_{mn} : (m, n) \in N \times N\}. \qquad (4)$$

The set P is a countable union of countable sets, therefore, P is a countable set. Since P is not finite, P is denumerable.

Prove that the set Ω of algebraic numbers is denumerable.

SOLUTION:

A real number q is called an algebraic number if q is a solution to a polynomial equation

$$a_m x^m + \dots + a_1 x + a_0 = 0 \tag{1}$$

with integral coefficients.

The set P of all polynomials with integer coefficients is a denumerable set (see Problem 5-10). Hence

$$P = \{f_1(x) = 0, f_2(x) = 0, f_3(x) = 0, \dots\} \tag{2}$$

where $f_1, f_2, f_3, \dots$ are polynomials.

Let us denote by Q_i

$$Q_i = \{x : f_i(x) = 0\} \tag{3}$$

the set of the solutions of $f_i(x) = 0$. Since a polynomial of degree m has, at most, m solutions, each Q_i is finite. Therefore

$$\Omega = \bigcup_{i \in N} Q_i \tag{4}$$

is denumerable.

The set of all real numbers is non-countable. Each real number is either algebraic or transcendental (i.e. non-algebraic),

$$R = \Omega \cup T$$

where T is the set of transcendental numbers. We conclude that transcendental numbers exist and that the set T of transcendental numbers is non-countable. The members e and π are transcendental numbers.

● **PROBLEM 5–12**

Use the axiom of choice to prove:

THEOREM

Every infinite set A contains a subset B which is denumerable.

SOLUTION:

Let $P(A)$ represent the power set of A, that is, the class of all subsets of A. We define the choice function F, such that

$$F : P(A) \to A. \tag{1}$$

For each non-empty subset D of A, $D \subset A$, $f(D) \in A$. By virtue of the axiom of choice, such a function exists. We shall start from the "top" and move "down":

$$F(A) = a_1$$

$$F(A - \{a_1\}) = a_2$$

$$F(A - \{a_1, a_2\}) = a_3$$

$$\vdots$$

$$F(A - \{a_1, a_2, \ldots, a_n\}) = a_{n+1}. \tag{2}$$

The set A is infinite, hence for every $n \in N$, the set

$$A - \{a_1, a_2, \ldots, a_n\} \neq \phi \tag{3}$$

is non-empty.

Since F is a choice function

$$a_n \neq a_k \quad \text{for} \quad n \neq k \tag{4}$$

Hence all a_n are distinct and the set

$$B = \{a_1, a_2, a_3, \ldots\} \tag{5}$$

is a denumerable subset of A, $B \subset A$. In the first step, the choice function F chooses one element from A, $F(A) = a_1$. Then from the set $A - \{a_1\}$, the choice function F chooses a_2, $F(A - \{a_1\}) = a_2$. The remaining set $A - \{a_1, a_2, \ldots, a_n\}$ is non-empty because the set A is infinite.

● **PROBLEM 5–13**

Prove the theorems:

1. A subset of a denumerable set is countable.

2. Every subset of a countable set is countable.

SOLUTION:

1. Let A denote any denumerable set

133

$$A = \{a_1, a_2, a_3, \dots\} \tag{1}$$

and let B denote a subset of A, $B \subset A$. When $B = \phi$, then of course B is finite. If $B \neq \phi$, then we can denote the least positive integer as k, such that

$$a_{k_1} \in B. \tag{2}$$

By k_2 we denote the least positive integer, such that

$$k_2 > k_1 \text{ and } a_{k_2} \in B. \tag{3}$$

By repeating the procedure, we obtain

$$B = \{a_{k_1}, a_{k_2}, a_{k_3}, \dots\}. \tag{4}$$

If the set of integers $\{k_1, k_2, \dots\}$ is finite, then B is finite, otherwise B is denumerable.

2. Suppose A is a countable set, then A is either finite or denumerable. In either case, its subsets are countable.

● **PROBLEM 5-14**

Let A and B denote disjoint sets, $A \cap B = \phi$, and let A denote an infinite set and B denote a denumerable set. Show that

$$A \cup B \sim A. \tag{1}$$

SOLUTION:

Since A is an infinite set, it must contain a denumerable subset $P = \{p_1, p_2, p_3, \dots\}$ (See Problem 5-12).
Let us write

$$A \cup B = (A - P) \cup (P \cup B) \text{ and } A = (A - P) \cup P. \tag{2}$$

We shall establish an equivalence relation between $A \cup B$ and A in the following manner:

$$f(x) = \begin{cases} x & \text{for } x \in A - P \\ p_{2n-1} & \text{for } x = p_n \\ p_{2n} & \text{for } x = b_n \end{cases} \tag{3}$$

where

$$B = \{b_1, b_2, b_3, \dots\} \tag{4}$$

Figure 1 illustrates function $f(x)$.

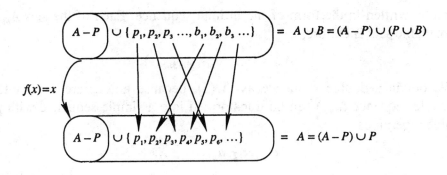

FIGURE 1

Function $f : A \cup B \rightarrow A$ is one-to-one and onto. Hence, the sets $A \cup B$ and A are equivalent

$$A \cup B \sim A. \tag{5}$$

● **PROBLEM 5-15**

Prove the following important theorem:

THEOREM

Let $\{A_1, A_2, A_3, ...\}$ denote a countable sequence of countable sets, then the union

$$A = A_1 \cup A_2 \cup A_3 \cup ... \tag{1}$$

is a countable set.

SOLUTION:

We proved (see Problem 5-9) that a Cartesian product of two countable sets is a countable set. From that, it follows that every double sequence can be transformed into a simple sequence. The elements of the array

$$a_{11}, a_{12}, ..., a_{1n}, ...$$

$$a_{21}, a_{22}, ..., a_{2n}, ...$$

$$\vdots$$

$$a_{m1}, a_{m2}, ..., a_{mn}, ... \tag{2}$$

can be written in the form of the infinite sequence. Each of the sets A_m is a countable set. Therefore,

$$A_m = \{a_{m1}, a_{m2}, a_{m3}, \ldots\}. \tag{3}$$

We obtain a double sequence as in (2), because all A_m are countable. A double sequence (a_{mn}) can be transformed into a simple sequence with possible repetitions

$$a_{11}, a_{12}, a_{21}, a_{31}, \ldots \tag{4}$$

We apply the axiom of choice, because the set of sequences consisting of the elements of the set A_m contains more than one element. We showed here that the elements of the set

$$A = A_1 \cup A_2 \cup A_3 \cup \ldots = \bigcup_{n=1}^{\infty} A_n \tag{5}$$

can be arranged in a sequence. Therefore, A is a countable set.

● **PROBLEM 5–16**

Prove that the set of all intervals on the real axis, with both endpoints rational numbers, is countable.

Is the set of all oriented intervals on the real axis, with both endpoints rational numbers, countable as well?

SOLUTION:

The set of rational numbers Q is countable. Hence the Cartesian product $Q \times Q$ is a countable set. Let P denote the set of intervals

$$P = \{[a, b] : a \in Q \wedge b \in Q\}$$

with endpoints rational numbers. We define mapping f, such that

$$f : P \to Q \times Q$$

$$f : [a, b] \to (a, b) \in Q \times Q$$

When P is the set of oriented intervals, then f is one-to-one and onto. Indeed, for example

$$f([0,1]) = (0, 1)$$

$$f([1, 0]) = (1, 0).$$

Therefore, the set of oriented intervals is countable. From that, we conclude that the set of intervals is countable.

Show that the power set 2^A of A is equivalent to the collection of characteristic functions on A, $C(A)$, i.e. prove that

$$2^A \sim C(A). \tag{1}$$

SOLUTION:

Let B represent any subset of A,

$$B \in 2^A. \tag{2}$$

We define function

$$f : 2^A \to C(A) \tag{3}$$

in such a way that

$$f(B) = X_B = \begin{cases} 1 & \text{if } x \in B \\ 0 & \text{if } x \notin B \end{cases}. \tag{4}$$

Then function f is one-to-one and onto. Therefore, the sets 2^A and $C(A)$ are equivalent.

$$2^A \sim C(A). \tag{5}$$

● **PROBLEM 5-18**

Prove that the set of spheres in a three-dimensional space, which have the rational coordinates of the center and rational radii, is countable.

SOLUTION:

The set Q of all rational numbers is countable.

Now, consider the set P of all points in a three-dimensional space, such that $p \in P$, if all three coordinates of p are rational numbers

$$P = \{p : p \in R^3; p = (p_1, p_2, p_3); p_1 \in Q \land p_2 \in Q \land p_3 \in Q \}.$$

It is easy to see that

$$P \sim Q \times Q \times Q.$$

Hence, the set of centers of the spheres is countable. The radii of the spheres are rational numbers.

Let Q^+ represent the set of all non-negative rational numbers. Then the set of all spheres, which have the rational coordinates of the center and rational radii, is equivalent to the set $Q \times Q \times Q \times Q^+$. The Cartesian product of a finite number of countable sets is a countable set.

Therefore, $Q \times Q \times Q \times Q^+$ is a countable set and the set of spheres is countable.

Show that every set of disjoint intervals is countable.

SOLUTION:

Let P denote the set of disjoint intervals, and let $p \in P$ represent an interval

$$p = [\alpha, \beta] \tag{1}$$

where α and β are real numbers. We can always find an interval p', such that

$$p' = [a, b] \tag{2}$$

and $p' \subset p$, that is

$$[a, b] \subset [\alpha, \beta] \tag{3}$$

where a and b are rational numbers.

The set of rational numbers is countable. Hence, the set of all intervals with rational endpoints is countable.

Let $F : P \to P'$ represent a one-to-one and onto function, such that to every interval $[\alpha, \beta] \in P$, F assigns an interval $[a, b]$, such that $[a, b] \subset [\alpha, \beta]$ and a and b are rational numbers. Here P' is the set of disjoint intervals with rational endpoints. Since P' is a subset of a countable set (the set of all intervals with rational endpoints), P' is countable.

Therefore, P is also countable.

Prove that the set of proper maxima of the function is countable.

SOLUTION:

Let f represent a function with real arguments and values.

DEFINITION

Function f has a proper maximum at the point a, if an interval $[b, c]$ exists

138

such that $a \in \,]b, c[$, and such that for each $x \in \,]b, c[$ and $x \neq a$,

$$f(x) < f(a). \tag{1}$$

For each point x', which is the proper maximum of function $f(x)$, an interval $[b, c]$ with the properties described in the definition exists.

It is possible that to the proper maximum x' there corresponds an interval $[b', c']$ which contains another proper maximum (or maxima) as shown in Figure 1.

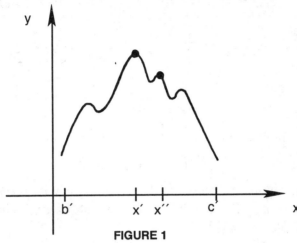

FIGURE 1

For each proper maximum, we can find an interval $[\alpha, \beta]$ with the rational endpoints which does not contain other proper maxima. Hence, we obtain a family of disjoint intervals, which is countable (see Problem 5-19).

Therefore, the set of proper maxima of the function is countable.

● PROBLEM 5-21

Prove that the set of points of discontinuity of a monotonic function in an interval $[a, b]$ is countable.

SOLUTION:

Suppose $f(x)$ is a monotonically increasing function. Function $f(x)$ is discontinuous at α, if and only if

$$p(\alpha) = f(\alpha + 0) - f(\alpha - 0) > 0 \tag{1}$$

where $f(\alpha + 0)$ denotes the right-hand limit, and $f(\alpha - 0)$ the left-hand limit. Of course, we should set

$$f(a - 0) = f(a) \quad \text{and} \quad f(b + 0) = f(b). \tag{2}$$

139

If

$$a < \alpha_1 < \alpha_2 < \dots < \alpha_s < b \tag{3}$$

and

$$\alpha_\mu < x_\mu < \alpha_{\mu+1} \tag{4}$$

then

$$f(x_\mu) - f(x_{\mu-1}) \geq f(\alpha_\mu + 0) - f(\alpha_\mu - 0) = p(\alpha_\mu). \tag{5}$$

Hence

$$f(b) - f(a) = \sum_{\mu=1}^{S} [f(x_\mu) - f(x_{\mu-1})] \geq \sum_{\mu=1}^{S} p(\alpha_\mu) \tag{6}$$

where

$$x_0 = a \quad \text{and} \quad x_s = b. \tag{7}$$

When α_μ are numbers with $p(\alpha_\mu) > 1/n$, it follows that

$$M < n\,[f(b) - f(a)]. \tag{8}$$

Therefore, the number of points of discontinuity α with $p(\alpha) > 1/n$ has a fixed upper bound. Thus, a finite number of points of discontinuity with $p(\alpha) > 1/n$ belong to the interval $[a, b]$.

Now we can write the finite number of points of discontinuity with $p(\alpha) > 1$, then $p(\alpha) > 1/2$, then $p(\alpha) > 1/3$, etc.

For each point of discontinuity α, n exists, such that $p(\alpha) > 1/n$. Therefore, each point of discontinuity appears in the sequence. The set of points of discontinuity is countable.

● **PROBLEM 5–22**

Prove this rather amazing result: that the set of points of a straight line and the set of points of a space are equivalent, that is, that R^1 and R^3 are equivalent.

SOLUTION:

We shall first prove that the set of points of a cube of unit length and the set of points of an interval [0, 1] are equivalent. The cube is shown in Figure 1.

The coordinates of a point P interior to the cube are x, y z. Each of these real numbers can be written as a decimal fraction

$$x = 0.a_1 a_2 a_3 \dots$$

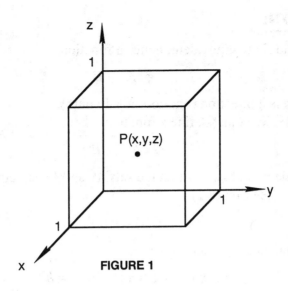

FIGURE 1

$$y = 0.b_1b_2b_3 \ldots$$

$$z = 0.c_1c_2c_3 \ldots$$

Let us form a real number p in the following manner:

$$p = 0.a_1b_1c_1a_2b_2c_2a_3b_3c_3\ldots$$

Hence, to every point P interior to the cube a unique point p can be assigned, where $0 < p < 1$.

This is a one-to-one and onto mapping. Therefore, the sets of points of an interval and of a cube are equivalent. Since the interval and the whole real axis are equivalent, the result follows.

It is surprising that the sets of different dimensions are equivalent. A one-to-one and onto correspondence between them exists.

Hilbert, Peano and Brouwer proved that this correspondence cannot be continuous.

THEOREM (HILBERT, PEANO, BROUWER)

No one-to-one correspondence that maintains continuity exists between two continuums of different order.

Correspondence is continuous when the neighboring points of one continuum can be mapped to the neighboring points of the other continuum.

● **PROBLEM 5–23**

Show that, if $A \sim B$, then

$$A^C \sim B^C. \qquad (1)$$

SOLUTION:

Sets A and B are equivalent, hence a function

$$g : A \rightarrow B \tag{2}$$

exists, which is a bijection (one-to-one and onto).

Let $f \in A^C$. We must define a function

$$G : A^C \rightarrow B^C \tag{3}$$

which is a bijection, to show that the sets A^C and B^C are equivalent.

Let us set (see Figure 1)

$$G(f) = g \circ f. \tag{4}$$

Function G is one-to-one, since

$$g \circ f = g \circ h \text{ implies } f = h.$$

Function G is also onto, because for $h \in B^C$, we obtain

$$G(g^{-1} \circ h) = g \circ g^{-1} \circ h = h \tag{5}$$

Since G is one-to-one and onto, it is bijective. Hence $A^C \sim B^C$.

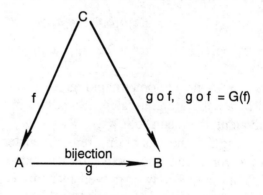

FIGURE 1

● **PROBLEM 5–24**

Show that

$$A^C \times B^C \sim (A \times B)^C. \tag{1}$$

SOLUTION:

We repeat

$$A^C = \{f : f : C \rightarrow A\}. \tag{2}$$

142

We define
$$G : A^C \times B^C \to (A \times B)^C \qquad (3)$$

in such a way that
$$G(f, g)(x) = (f(x), g(x)) \text{ for all } x \in C. \qquad (4)$$

Function G is one-to-one. Indeed let
$$G(f_1, g_1) = G(f_2, g_2). \qquad (5)$$

Then, for all $x \in C$
$$G(f_1, g_1)(x) = G(f_2, g_2)(x) \qquad (6)$$

and
$$(f_1(x), g_1(x)) = (f_2(x), g_2(x)). \qquad (7)$$

Hence
$$f_1(x) = f_2(x) \text{ for all } x \in C$$

and
$$g_1(x) = g_2(x) \text{ for all } x \in C.$$

Thus
$$f_1 = f_2 \quad \text{and} \quad g_1 = g_2,$$
$$(f_1, g_1) = (f_2, g_2). \qquad (8)$$

Now we shall prove that G is onto. Let $h \in (A \times B)^C$,
$$h(x) \in A \times B \text{ for all } x \in C.$$

Let
$$(h_1(x), h_2(x)) = h(x). \qquad (9)$$

Then
$$h_1 \in A^C \quad \text{and} \quad h_2 \in B^C$$
$$(h_1, h_2) \in A^C \times B^C. \qquad (10)$$

Thus
$$G(h_1, h_2) = h \quad \text{since} \qquad (11)$$
$$G(h_1, h_2)(x) = (h_1(x), h_2(x)) = h(x) \qquad (12)$$

for all $x \in C$.

Function $G : A^C \times B^C \to (A \times B)^C$ is bijective, hence, the sets $A^C \times B^C$ and $(A \times B)^C$ are equivalent.

CHAPTER 6

CARDINAL NUMBERS, CARDINAL ARITHMETIC

The power of the set of natural numbers is denoted by card N (or $\aleph$) and the power of the set of real numbers is denoted by card R or by C (continuum). These are the most important of the infinite cardinal numbers. The following important inequality holds:

$$\text{card } R \neq \text{card } N. \qquad (1)$$

Prove (1) by applying the results of Chapter 5.

SOLUTION:

In Chapter 5, we showed that for every sequence of real numbers a_1, a_2, a_3, ..., a_n, ..., we can define a real number b, which does not belong to this sequence. This can be formulated in the form of a theorem.

THEOREM

The set of all real numbers is non-countable.

Therefore, a cardinal number assigned to the set of natural numbers, card N, is different from the cardinal number assigned to the set of real numbers, card R.

$$\text{card } R \neq \text{card } N$$

The cardinal number assigned to the set A is denoted by card A.

In Problem 3-26, we discussed the axiomatic formulation of set theory. Describe how the concept of cardinal numbers can be established axiomatically.

SOLUTION:

The cardinal numbers describe the "size" of the sets. In this sense, they can be considered a generalization of natural numbers. To the set of, say, five books, we assign the natural number five, which describes how large the set is.

The axiomatic formulation given here does not define a cardinal number and is not related to counting. The concept of a cardinal number is related to "size." We want to determine if one of two given sets has more members than the other. Thus, we do not have to count; just pair off each member of one set with a member of the other and see if any elements are left over.

The following axioms introduce the concept of cardinal numbers:

I.   Each set X is associated with a cardinal number, denoted by card X, and for each cardinal number p, a set X exists such that

$$\text{card } X = p.$$

II.   Card $X = 0$, if and only if $X = \phi$.

III.   If X is a non-empty finite set, that is, $X \sim \{1, 2, ..., n\}$ for some $n \in N$, then card $X = n$.

IV.   For any two sets X and Y, card X = card Y, if and only if $X \sim Y$.

Axioms I and IV are usually called the axiom of cardinality.

From II and III, we see that the cardinal number of a finite set is the number of elements of that set.

● **PROBLEM 6–3**

Explain why

$$\text{card } N < \text{card } R \tag{1}$$

where N is the set of positive integers and R is the set of real numbers.

SOLUTION:

The cardinal number of a finite set is called a finite cardinal number. Finite cardinal numbers are actually non-negative integers. The cardinal number of an infinite set is called a transfinite cardinal number. Finite cardinal numbers have the inherited order of natural numbers

$$0 < 1 < 2 < ... < n < n + 1 < ... \tag{2}$$

The question is how to compare two transfinite cardinal numbers. By applying axiom IV (see Problem 6-2), we can determine whether two cardinal numbers are equal or not. To establish which of the two cardinal numbers is "larger," we shall use the following:

DEFINITION

Let A and B represent two sets and card A and card B denote their cardinal numbers, respectively. Then card A is said to be less than card B; we write

146

$$\text{card } A < \text{card } B \qquad (3)$$

when the set A is equipotent to a subset of B, but the set B is not equipotent to any subset of A.

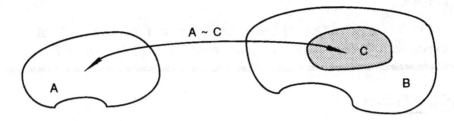

FIGURE 1

D does not exist, such that $D \subset A$ and $D \sim B$.

The set N is a subset of R, hence N is equipotent to a subset of R

$$N \sim N \subset R. \qquad (4)$$

The infinite set R is nondenumerable. Thus, R is not equipotent to any subset of N. Therefore,

$$\text{card } N < \text{card } R. \qquad (5)$$

● **PROBLEM 6-4**

Let m denote any finite cardinal number. Prove that

$$m < \text{card } N. \qquad (1)$$

SOLUTION:

Let A represent the subset of the set of natural numbers N.

$$A = \{1, 2, \ldots, m\} \qquad (2)$$

Then

$$\text{card } A = m \quad \text{and} \quad A \subset N. \qquad (3)$$

We have

$$m = \text{card } A < \text{card } N \qquad (4)$$

because the set A is equipotent to a subset of N, but N is not equipotent to any subset of A (see Problem 6-3).

Prove the following:

LEMMA

If B is a subset of A, and if an injection (one-to-one) $f : A \to B$ exists, then there is a bijection (one-to-one and onto) $g : A \to B$.

SOLUTION:

If $B = A$, then the identity function is one-to-one and onto (bijection). Suppose B is a proper subset of A. Let us denote the set

$$C = \overset{\infty}{\underset{n=0}{\cup}} f^n(A - B) \tag{1}$$

by C where f^0 is the identify function on A and

$$f^n(x) = f(f^{n-1}(x)) \tag{2}$$

for each $x \in A$ and for each positive integer n. Note that $A - B \subset C$ and $f(C) \subset C$. For any two distinct non-negative integers m,n, the sets $f^m(A - B)$ and $f^n(A - B)$ are disjoint.

Indeed, suppose $f^m(A - B) \cap f^n(A - B) \neq \phi$, then $x_1, x_2 \in A - B$ exists, such that

$$f^m(x_1) = f^n(x_2) \tag{3}$$

and

$$f^m(x_1) = f^{n-m}(f^m(x_2)) = f^m(f^{n-m}(x_2)). \tag{4}$$

Function f is one-to-one, hence

$$f^{n-m}(x_2) = x_1 \in (A - B) \cap B \tag{5}$$

is a contradiction.

We define function $g : A \to B$

$$g(x) = \begin{cases} f(x) & \text{for } x \in C \\ x & \text{for } x \in A - C \end{cases} \tag{6}$$

Function g is one-to-one, also:

$$g(A) = f(C) \cup (A - C) =$$

$$= f(\overset{\infty}{\underset{n=0}{\cup}} f^n(A - B)) \cup (A - \overset{\infty}{\underset{n=0}{\cup}} f^n(A - B)) =$$

$$= (\overset{\infty}{\underset{n=1}{\cup}} f^n(A - B)) \cup (A - \overset{\infty}{\underset{n=0}{\cup}} f^n(A - B)) =$$

$$= A - (A - B) = B. \tag{7}$$

Remember f^0 is the identity function. Thus, function g is a bijection. The diagram illustrates the proof.

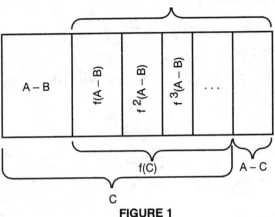

FIGURE 1

The set A is represented by the whole rectangle.

● PROBLEM 6-6

So far, we have considered the following cases:

1. If two sets A and B are equipotent, then

$$\text{card } A = \text{card } B. \qquad (1)$$

2. Set A is equipotent to a subset of B, but the set B is not equipotent to any subset of A, then

$$\text{card } A < \text{card } B. \qquad (2)$$

The question is how two cardinal numbers, card A and card B, compare when A is equipotent to a subset of B and B is equipotent to a subset of A. Georg Cantor conjectured that in such a case

$$\text{card } A = \text{card } B. \qquad (3)$$

It was later independently proven by both E. Schröder and F. Bernstein. Prove:

SCHRÖDER-BERNSTEIN THEOREM

If A and B are sets, such that A is equipotent to a subset of B and B is equipotent to a subset of A, then A and B are equipotent, i.e.

$$\text{card } A = \text{card } B.$$

SOLUTION:

Suppose $A_1 \subset A$ and $B_1 \subset B$ and

$$A_1 \sim B, \quad B_1 \sim A. \tag{4}$$

Then two bijections exist

$$f_1 : A \to B_1 \quad \text{and} \quad g_1 : B \to A_1. \tag{5}$$

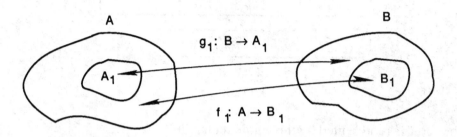

FIGURE 1

Let us define

$$f : A \to A_1 \text{ by } f(x) = g_1(f_1(x)). \tag{6}$$

Then f is one-to-one. By the lemma proved in Problem 6-5, a bijection $g : A \to A_1$ exists.

Therefore, since $g : A \to A_1$ is a bijection and $g_1^{-1} : A_1 \to B$ is a bijection, their composition

$$g_1^{-1} \circ g : A \to B \tag{7}$$

is a bijection. Hence, $A \sim B$ and

$$\text{card } A = \text{card } B. \tag{8}$$

● **PROBLEM 6-7**

Prove the following version of the Schröder-Bernstein theorem:

THEOREM

If card $A \leq$ card B and card $B \leq$ card A, then card $A =$ card B.

SOLUTION:

We shall write card $A \leq$ card B to mean card $A <$ card B or card $A =$ card B.

150

Suppose two sets are given, A and B. By applying the Schröder-Bernstein theorem, we conclude that one of three possibilities takes place:

$$\text{card } A < \text{card } B$$

or

$$\text{card } A > \text{card } B$$

or

$$\text{card } A = \text{card } B.$$

Hence, an immediate consequence of the Schröder-Bernstein theorem is

$$\begin{pmatrix} \text{card } A \leq \text{card } B \\ \text{card } B \leq \text{card } A \end{pmatrix} \Rightarrow \begin{pmatrix} \text{card } A = \text{card } B \end{pmatrix}$$

● **PROBLEM 6–8**

Prove that card N is the smallest transfinite cardinal number.

SOLUTION:

The cardinal number assigned to a finite set is called a finite cardinal number and the cardinal number assigned to an infinite set is called a transfinite cardinal number. We already know two transfinite cardinal numbers, card N and card R, where

$$\text{card } N < \text{card } R. \tag{1}$$

Thus

$$0 < 1 < 2 < \ldots \qquad\qquad < \ldots?\ldots < \text{card } N < \ldots?\ldots < \text{card } R \ldots?\ldots$$

Finite Cardinal Numbers   Transfinite Cardinal Numbers

FIGURE 1

Figure 1 illustrates the state of our knowledge about cardinal numbers. Each question mark indicates that we are not sure if the respective cardinal number exists.

Let A represent an infinite set. We shall apply the following:

THEOREM

Every infinite set contains a denumerable subset.

151

Therefore, a set B exists, such that

$$B \subset A \text{ and } B \sim N. \tag{2}$$

Thus

$$\text{card } B = \text{card } N \tag{3}$$

and

$$\text{card } B \leq \text{card } A. \tag{4}$$

We conclude, that for every infinite set A,

$$\text{card } N \leq \text{card } A. \tag{5}$$

The number, card N, is the smallest transfinite cardinal number. We can eliminate one of the question marks; see Figure 2.

$$0 < 1 < 2 < \ldots \qquad\qquad < \text{card } N < \ldots?\ldots < \text{card } R < \ldots?\ldots$$

Finite Cardinal Numbers   Transfinite Cardinal Numbers

FIGURE 2

We shall eliminate the remaining question marks later.

● **PROBLEM 6–9**

Prove the following:

$$\left(\begin{array}{c} A \subset B \subset C \\ A \sim C \end{array} \right) \Rightarrow (A \sim B).$$

SOLUTION:

We shall apply the Schröder-Bernstein theorem. Since $A \subset B \subset C$, we have

$$B \sim B \subset C \tag{1}$$

and

$$A \subset B. \tag{2}$$

But $A \sim C$, therefore

$$B \sim B \subset C \tag{3}$$

$$C \sim A \subset B \tag{4}$$

152

as shown in Figure 1.

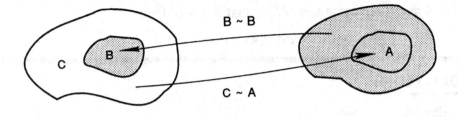

FIGURE 1

By the Schröder-Bernstein theorem we conclude that $C \sim B$. Therefore, since $A \sim C$ and $C \sim B$, we obtain $A \sim B$.

● PROBLEM 6–10

Show that the largest cardinal number does not exist.

SOLUTION:

Georg Cantor proved the following theorem:

CANTOR'S THEOREM

For any set X,

$$\text{card } X < \text{card } P(X). \tag{1}$$

Remember that the power set $P(X)$ of X is the set of all subsets of X.

Let X represent a set and $P(X)$ its power set. Then, by applying Cantor's theorem

$$\text{card } X < \text{card } P(X). \tag{2}$$

Since $P(X)$ is again the set, we have

$$\text{card } P(X) < \text{card } P(P(X)). \tag{3}$$

By applying Cantor's theorem, we obtain

$$\text{card } X < \text{card } P(X) < \text{card } P(P(X)) < \dots \tag{4}$$

In such a way, we can obtain a sequence of cardinal numbers.

By setting $X = R$, we obtain a sequence of transfinite cardinal numbers

$$\text{card } R < \text{card } P(R) < \text{card } P(P(R)) < \dots \tag{5}$$

Thus, there is no largest cardinal number.

Let A and B represent sets. Prove that if $A \sim B$, then

$$\text{card } P(A) = \text{card } P(B). \tag{1}$$

SOLUTION:

Since $A \sim B$, a bijection exists

$$f : A \to B. \tag{2}$$

We shall show that a bijection exists

$$F : P(A) \to P(B). \tag{3}$$

Let us define

$$F(X) = f(X) \text{ for all } X \in P(A). \tag{4}$$

Since f is a bijective function, F is also. Hence,

$$P(A) \sim P(B). \tag{5}$$

● **PROBLEM 6-12**

Let X represent a denumerable set. Prove that the power set $P(X)$ of X is nondenumerable.

SOLUTION:

We shall apply Cantor's theorem. Suppose, on the contrary, that a denumerable set X exists, such that its power set $P(X)$ is also denumerable. Then

$$\text{card } X = \text{card } P(X) \tag{1}$$

but this contradicts Cantor's theorem.
That completes the proof.

● **PROBLEM 6-13**

Prove the following:

CANTOR'S THEOREM

If A is a set, then

$$\text{card } A < \text{card } P(A). \tag{1}$$

SOLUTION:

If $A = \phi$, then card $\phi = 0$ and card $P(\phi) = 1$.
Suppose $A \neq \phi$. Let us define a function

$$f : A \rightarrow P(A) \tag{2}$$

such that

$$f(x) = \{x\} \in P(A) \tag{3}$$

for all $x \in A$.

Function f is one-to-one. Therefore, the sets A and $\{\{x\} : x \in A\}$ are equipotent. But

$$\{\{x\} : x \in A\} \subset P(A). \tag{4}$$

Therefore,

$$\text{card } A \leq \text{card } P(A). \tag{5}$$

Now we must show that the sets A and $P(A)$ are not equipotent, i.e. that card $A \neq$ card $P(A)$. Assume, on the contrary, that the sets A and $P(A)$ are equipotent, hence a bijection exists

$$g : A \rightarrow P(A). \tag{6}$$

Let us define the set

$$T = \{x \in A : x \notin g(x)\}. \tag{7}$$

Since $T \in P(A)$ and $g : A \rightarrow P(A)$, an element $a \in A$ exists, such that

$$g(a) = T. \tag{8}$$

The existence of such an element leads to a contradiction. Because if $a \in T$, then $a \notin g(a)$, according to the definition of T; on the other hand, $g(a) = T$ and $a \in T$.

If $a \notin T$, then since $g(a) = T$, we have $a \notin g(a)$ and $a \in T$ and consequently, $a \in g(a)$.

That completes the proof of Cantor's theorem.

● **PROBLEM 6-14**

Let α and β denote two cardinal numbers. Define their sum and show that the sum is uniquely defined, i.e. is independent of the choice of sets.

SOLUTION:

The sum $\alpha + \beta$ of two cardinal numbers α and β is defined to be the power of the union of two disjoint sets, which have the powers α and β, respectively.

We have

$$\text{card } A + \text{card } B = \text{card } (A \cup B), \text{ if } A \cap B = \phi. \qquad (1)$$

The cardinal number assigned to set X we denote by card X.

Sometimes the other notations are used, like $\overline{\overline{X}}$, $\#X$, etc.

In this definition, we employ two disjoint sets. Suppose A_1 and B_1 have powers, card A_1 and card B_1, respectively. Then, the sets A and B exist, which are disjoint and such that

$$\text{card } A_1 = \text{card } A \text{ and card } B_1 = \text{card } B. \qquad (2)$$

Indeed, suppose $a \neq b$ are two distinct elements, then

$$A = \{a\} \times A_1 \qquad B = \{b\} \times B_1$$

$$\text{and} \quad A \cap B = \phi. \qquad (3)$$

Since

$$\left(\begin{array}{c} A_1 \sim B_1 \\ A_2 \sim B_2 \\ A_1 \cap A_2 = B_1 \cap B_2 = \phi \end{array} \right) \Rightarrow (A_1 \cup A_2 \sim B_1 \cup B_2),$$

we conclude that for every two cardinal numbers, their sum is uniquely defined, that is, independent of the choice of sets A and B.

Remember that $A \sim B$ indicates that sets A and B have the same power (are equipotent).

● **PROBLEM 6-15**

Find the sum of cardinal numbers

$$\text{card } N + \text{card } N. \qquad (1)$$

SOLUTION:

Let N_O and N_E denote the set of odd natural numbers and the set of even natural numbers. We have

$$N_O \cap N_E = \phi \qquad (2)$$

156

$$N_O \subset N, \quad N_E \subset N \qquad (3)$$

$$N_O \cup N_E = N. \qquad (4)$$

The sets N_O and N_E are denumerable. Hence,

$$\text{card } N + \text{card } N = \text{card } N_O + \text{card } N_E =$$

$$= \text{card } (N_O \cup N_E) = \text{card } N. \qquad (5)$$

Similarly, card R + card R = card R. It is easy to show that card N + card R = card R. Indeed

$$(0,\, 1) \sim R \quad \text{and} \quad \text{card } (0,1) = \text{card } R. \qquad (6)$$

Let $P = (0,\, 1) \cup N$. Since N and $(0,1)$ are disjoint sets,

$$\text{card } P = \text{card } (0,1) + \text{card } N = \text{card } N + \text{card } R. \qquad (7)$$

On the other hand,

$$R \sim (0,1) \subset P \qquad (8)$$

and

$$P \sim P \subset R. \qquad (9)$$

According to the Schröder-Bernstein theorem,

$$P \sim R \quad \text{and} \quad \text{card } P = \text{card } N + \text{card } R = \text{card } R. \qquad (10)$$

The power of the set of natural numbers will be denoted throughout this book by card N or by $\aleph_0$ (aleph-null).

● **PROBLEM 6–16**

Define the product of two cardinal numbers and show that the product is uniquely defined.

SOLUTION:

Let α and β denote two cardinal numbers and let A and B represent two sets, such that

$$\text{card } A = \alpha \quad \text{and} \quad \text{card } B = \beta. \qquad (1)$$

Then, we define the product $\alpha \cdot \beta$ of α and β to be the power of the Cartesian product of sets A and B:

$$\text{card } A \cdot \text{card } B = \text{card } (A \times B). \qquad (2)$$

This definition is unique, in the sense that it does not depend on the choice of sets A and B. Suppose A_1 and B_1 are two sets, such that

$$A \sim A_1 \quad \text{and} \quad B \sim B_1 \tag{3}$$

Since the following holds

$$\left(\begin{matrix} A \sim A_1 \\ B \sim B_1 \end{matrix} \right) \Rightarrow (A \times B \sim A_1 \times B_1) \tag{4}$$

we conclude that the definition of the product of two cardinal numbers is unique.

● **PROBLEM 6–17**

Show that

$$\text{card } N + \text{card } N = \text{card } N \tag{1}$$

$$\text{card } N \cdot \text{card } N = \text{card } N \tag{2}$$

$$\text{card } N + n = \text{card } N \tag{3}$$

$$\text{card } N \cdot n = \text{card } N \tag{4}$$

where n is a natural number and card N is the power of the set of natural numbers.

SOLUTION:

In Chapter 5, we proved the following useful theorem:

THEOREM

The union $A \cup B$ of two countable sets A and B is countable.
Therefore, since the set of natural numbers is countable, we conclude that

$$\text{card } N + \text{card } N = \text{card } N \tag{5}$$

From this theorem, we conclude as well that

$$\text{card } N + n = \text{card } N \tag{6}$$

To prove (2) we shall apply:

THEOREM

The Cartesian product of two countable sets is a countable set.

Therefore,

$$\text{card } N \cdot \text{card } N = \text{card } N \tag{7}$$

and similarly,

$$\text{card } N \cdot n = \text{card } N. \tag{8}$$

● **PROBLEM 6–18**

Show that addition and multiplication satisfy the associative, commutative and distributive laws.

SOLUTION:

Since

$$A \cup (B \cup C) = (A \cup B) \cup C \quad \text{and} \quad A \cup B = B \cup A$$

addition is associative and commutative.

$$\alpha + (\beta + \gamma) = (\alpha + \beta) + \gamma$$

$$\alpha + \beta = \beta + \alpha \tag{1}$$

Similarly, since

$$A \times (B \times C) = (A \times B) \times C \tag{2}$$

multiplication is associative.
By applying the following formula

$$A \times B \sim B \times A \tag{3}$$

we conclude that multiplication is commutative

$$\alpha \cdot \beta = \beta \cdot \alpha. \tag{4}$$

To prove the distributive law

$$\alpha \cdot (\beta + \gamma) = \alpha \cdot \beta + \alpha \cdot \gamma \tag{5}$$

let us choose three sets A, B and C, such that

$$\alpha = \text{card } A, \ \beta = \text{card } B, \ \gamma = \text{card } C \tag{6}$$

where $B \cap C = \phi$. Then

$$A \times (B \cup C) = A \times B \cup A \times C \tag{7}$$

$$(A \times B) \cap (A \times C) = A \times (B \cap C) = \phi \tag{8}$$

and

$$\text{card } [A \times (B \cup C)] = \text{card } A \times B + \text{card } A \times C \qquad (9)$$

which proves (5).

Let x denote an arbitrary cardinal number. Compute the following cardinal numbers:

1. $0x$

2. $1x$.

SOLUTION:

1. Let X represent a set, such that

$$\text{card } X = x. \qquad (1)$$

Then

$$X \times \phi = \phi. \qquad (2)$$

Hence

$$0x = \text{card } \phi \cdot \text{card } X =$$

$$= \text{card } (\phi \times X) = \text{card } \phi = 0. \qquad (3)$$

2. Let $\{a\}$ represent any set consisting of one element. Then the sets X and $X \times \{a\}$ are equipotent

$$X \sim X \times \{a\}. \qquad (4)$$

We obtain

$$1x = \text{card } \{a\} \cdot \text{card } X =$$

$$= \text{card } \{a\} \times X = \text{card } X = x. \qquad (5)$$

Which of the following theorems is true?

1. If x, y, z are cardinal numbers, such that $x \leq y$, then $xz \leq yz$.

2. If x, y and z are cardinal numbers, such that $x < y$, $z \neq 0$, then $xz < yz$.

SOLUTION:

1. Let A, B, and C represent the sets, such that

$$\text{card } A = x, \text{ card } B = y, \text{ card } C = z. \tag{1}$$

Since

$$\text{card } A \leq \text{card } B \tag{2}$$

a one-to-one function exists, such that

$$f : A \rightarrow B. \tag{3}$$

Let $g : C \rightarrow C$, $g(c) = c$ for every $c \in C$. Then function

$$F : A \times C \rightarrow B \times C \tag{4}$$

defined by

$$F(a, c) = (f(a), g(c)) \tag{5}$$

is also one-to-one. Therefore,

$$\text{card } (A \times C) \leq \text{card } (B \times C) \tag{6}$$

or

$$\text{card } A \cdot \text{card } C = xz \leq \text{card } B \cdot \text{card } C = yz. \tag{7}$$

2. This theorem is not true. The following example proves why:
   Let

$$x = 1 \quad y = z = \text{card } N \tag{8}$$

then $x < y$ but

$$xz = 1 \cdot \text{card } N = \text{card } N \quad \text{and} \quad yz = \text{card } N \text{ card } N = \text{card } N.$$

Hence $xz = yz$.

● **PROBLEM 6–21**

Prove that

$$\text{card } R \cdot \text{card } R = \text{card } R. \tag{1}$$

SOLUTION:

In Chapter 5 we proved that the set of real numbers R and the open interval $(0, 1)$ are equipotent

$$R \sim (0, 1). \tag{2}$$

To evaluate card $R \cdot$ card R, let us note that

$$\text{card } R = \text{card } (0, 1) \tag{3}$$

$$\text{card } R \cdot \text{card } R =$$

$$= \text{card } (0,1) \cdot \text{card } (0,1) = \text{card } (0,1) \times (0,1). \tag{4}$$

Hence, we have to compute card $(0,1) \times (0,1)$. To do it, a function f will be defined, such that

$$f : (0,1) \times (0,1) \to (0,1) \tag{5}$$

and f is bijective (i.e. one-to-one and onto).

Each $x \in (0,1)$ can be expressed by its infinite decimal expansion. For example

$$^2/_7 = 0.28571428\ldots$$

$$^1/_2 = 0.4999999\ldots \tag{6}$$

We define f in the following manner:

$$x, y \in (0,1)$$

$$f(x, y) = f(.x_1x_2x_3\ldots, .y_1y_2y_3\ldots) =$$

$$= .x_1y_1x_2y_2x_3y_3\ldots \tag{7}$$

It is easy to verify that such a function is one-to-one and onto. Thus

$$(0,1) \times (0,1) \sim (0,1) \tag{8}$$

and

$$\text{card } (0,1) \times (0,1) = \text{card } (0,1). \tag{9}$$

From (4) and (9), we obtain

$$\text{card } R \cdot \text{card } R = \text{card } (0,1) = \text{card } R. \tag{10}$$

● **PROBLEM 6–22**

Let x and y denote cardinal numbers. Prove that

1. If $xy = 0$ then $x = 0$ or $y = 0$.

2. If $xy = 1$ then $x = 1$ or $y = 1$.

SOLUTION:

1. Let A and B represent two sets, such that

$$\text{card } A = x$$

$$\text{card } B = y \tag{1}$$

then

$$xy = \text{card } (A \times B) = 0. \tag{2}$$

Hence $A \times B$ is an empty set

$$A \times B = \phi \tag{3}$$

From (3), we conclude that

$$A = \phi \quad \text{and} \quad B = \phi. \tag{4}$$

Thus

$$\text{card } A = x = 0$$

$$\text{card } B = y = 0. \tag{5}$$

2.
$$xy = \text{card } A \cdot \text{card } B =$$

$$= \text{card } (A \times B) = 1. \tag{6}$$

Set $A \times B$ consists of one element. Thus, each of the sets A and B consists of one element. We have

$$\text{card } A = x = 1$$

$$\text{card } B = y = 1. \tag{7}$$

● PROBLEM 6–23

Define b^a (a^{th} power of b), where a and b are cardinal numbers, and justify this definition using an analogy with natural numbers.

SOLUTION:

Let us start with a definition:

163

DEFINITION

Let $a \neq 0$ and b denote cardinal numbers and A and B represent two sets such that

$$\text{card } A = a, \quad \text{card } B = b \tag{1}$$

Then, we define b^a by

$$b^a = \text{card } B^A \tag{2}$$

where B^A denotes the set of all functions from A to B.

Consider the finite case of two natural finite numbers m and n. Then

$$n^m = n \cdot n \cdot n \cdot \ldots \cdot n, \ m \text{ times.} \tag{3}$$

Let A represent a set with m elements and B a set with n elements.

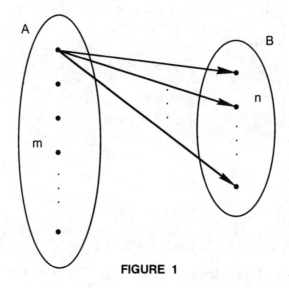

FIGURE 1

Each of the m elements of A has n choices for its image. The choices for each element of A are independently made and, since A consists of m elements, there are

$$n \cdot n \cdot n \cdot \ldots \cdot n = n^m$$

functions from A to B.

The definition of b^a is a generalization of this concept.

● **PROBLEM 6-24**

Show that the definition of b^a (see Problem 6-23) is independent of the choice of representatives, A and B.

164

SOLUTION:

We shall prove the following:

THEOREM

Let A, B, A', B' represent sets, such that $A \sim A'$ and $B \sim B'$. Then

$$B^A \sim B'^{A'} \tag{1}$$

Proof:

Since $A \sim A'$ and $B \sim B'$ are bijective functions, g and h exist

$$g : A \to A', \quad h : B \to B'. \tag{2}$$

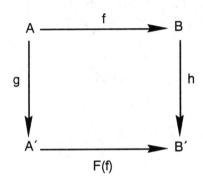

FIGURE 1

Suppose $f : A \to B$ is a function, $f \in B^A$. Let us define

$$F : B^A \to B'^{A'} \tag{3}$$

by $F(f) : A' \to B'$

$$F(f)(x) = h(f(g^{-1}(x))) = h \circ f \circ g^{-1}(x). \tag{4}$$

Functions g and h are bijective, hence $F(f)$ is bijective. Therefore,

$$F : B^A \to B'^{A'} \tag{5}$$

is bijective and

$$B^A \sim B'^{A'} \tag{6}$$

● **PROBLEM 6-25**

Show that for any set A

$$\operatorname{card} P(A) = 2^{\operatorname{card} A}. \tag{1}$$

165

SOLUTION:

Let $B = \{0, 1\}$. To each subset X of A, $X \subset A$, we assign the characteristic function

$$X_x : A \to B \tag{2}$$

as follows:

$$X_x(x) = \begin{cases} 1 & \text{if } x \in X \\ 0 & \text{if } x \in A - X \end{cases} \tag{3}$$

and a one-to-one and onto mapping exists between the sets $P(A)$ and B^A.

Remember that $P(A)$ is the set of all subsets of A, and B^A is the set of all functions from A to $\{0, 1\}$. To each subset $X \subset A$, $X \in P(A)$, we assign its characteristic function X_x, $X_x \in B^A$. Two different subsets cannot have the same characteristic function, hence mapping is one-to-one. Any characteristic function uniquely defines a subset, hence mapping is onto.

Bijection exists between $P(A)$ and B^A. Thus

$$\text{card } P(A) = \text{card } B^A \tag{4}$$

Since $B = \{0, 1\}$, card $B = 2$ and

$$\text{card } P(A) = 2^{\text{card } A}. \tag{5}$$

● **PROBLEM 6–26**

Let a, x, and y denote cardinal numbers. Show that

$$a^x \, a^y = a^{x+y}. \tag{1}$$

SOLUTION:

First of all, let us choose the sets A, X, Y, such that

$$\text{card } A = a$$

$$\text{card } X = x$$

$$\text{card } Y = y$$

$$X \cap Y = \phi. \tag{2}$$

From the definition of the sum of cardinal numbers, we have

$$\text{card } (X \cup Y) = x + y. \tag{3}$$

166

Now we shall show that the sets

$$A^X \times A^Y \text{ and } A^{X \cup Y}$$

are equipotent.

Suppose $f \in A^X$ and $g \in A^Y$, then $(f, g) \in A^X \times A^Y$. Then the function $f \cup g$ belongs to $A^{X \cup Y}$.

The union $f \cup g$ is the function

$$f \cup g : X \cup Y \to A \qquad (4)$$

defined by

$$f \cup g(x) = \begin{cases} f(x) & \text{if } x \in X \\ g(x) & \text{if } x \in Y \end{cases} \qquad (5)$$

Thus, we have established a one-to-one and onto relationship between the sets $A^X \times A^Y$ and $A^{X \cup Y}$.

Hence,

$$\text{card } A^X \times A^Y = a^x \, a^y =$$

$$= \text{card } A^{X \cup Y} = a^{x+y}. \qquad (6)$$

● **PROBLEM 6–27**

Let a, x, and y denote cardinal numbers. Show that

$$(a^x)^y = a^{xy}. \qquad (1)$$

SOLUTION:

Let A, X, and Y represent sets, such that

$$\text{card } A = a, \text{ card } X = x, \text{ card } Y = y. \qquad (2)$$

We shall prove that the sets $A^{X \times Y}$ and $(A^X)^Y$ are equipotent. Let $f : X \times Y \to A$ represent a given function and b a given element of Y, $b \in Y$.

Then a function exists

$$f^b : X \to A \qquad (3)$$

defined by

$$f^b(a) = f(a, b) \qquad (4)$$

for all $a \in X$.

Now, let us define the function

$$F : A^{X \times Y} \to (A^X)^Y \qquad (5)$$

such that to each $f \in A^{X \times Y}$, we assign the function

$$p_f \in (A^X)^Y \qquad (6)$$

defined by

$$p_f(b) = f^b \qquad (7)$$

for all $b \in Y$.

Function F is one-to-one and onto, hence the sets $A^{X \times Y}$ and $(A^X)^Y$ are equivalent. Thus

$$(a^x)^y = a^{xy}. \qquad (8)$$

● **PROBLEM 6-28**

Prove the following:

THEOREM

If a, b, and x are cardinal numbers, then
$$(ab)^x = a^x b^x. \qquad (1)$$

SOLUTION:

Let A, B, and X represent sets, such that

$$\text{card } A = a, \text{ card } B = b, \text{ card } X = x. \qquad (2)$$

We must prove that the sets $(A \times B)^X$ and $A^X \times B^X$ are equipotent.

Recall that the A-projection

$$p_A : A \times B \to A \qquad (3)$$

is a function which assigns element $a \in A$ to each ordered pair $(a, b) \in A \times B$. Let us define a bijection

$$F : (A \times B)^X \to A^X \times B^X. \qquad (4)$$

Function F assigns to each

$$f \in (A \times B)^X, f : X \to A \times B \qquad (5)$$

a function defined by

$$(p_A \circ f, p_B \circ f) \in A^X \times B^X. \qquad (6)$$

This assignment is one-to-one and onto (it can be easily verified), hence

$$(A \times B)^X \sim A^X \times B^X \qquad (7)$$

168

and

$$(ab)^x = a^x b^x. \tag{8}$$

● PROBLEM 6-29

Prove

$$2^{card\ N} = card\ R. \tag{1}$$

SOLUTION:

Recall that the symbols, card N and card R, denote the cardinal numbers of the sets N and R, respectively. Equation (1) will be proven in two steps: first we shall show that

$$2^{card\ N} \geq card\ R \tag{2}$$

and then we shall prove

$$card\ R \geq 2^{card\ N}. \tag{3}$$

Let Q represent the set of rational numbers; then

$$N \sim Q. \tag{4}$$

The function

$$f : R \rightarrow P(Q) \tag{5}$$

defined by

$$f(x) = \{y \in Q : y < x\} \tag{6}$$

is one-to-one. Indeed, if $x, x' \in R$ and $x < x'$, then a rational number $y \in Q$ exists, such that

$$x < y < x'. \tag{7}$$

The rational numbers are a dense subset of the set of real numbers.

Then $y \in f(x')$, but $y \notin f(x)$, thus, $f(x) \neq f(x')$ and f is one-to-one.

For any two sets, A and B, card $A \leq$ card B, if and only if an injection $f : A \rightarrow B$ exists. Hence,

$$card\ R \leq card\ P(Q) = 2^{card\ Q} = 2^{card\ N}. \tag{8}$$

To prove card $R \geq 2^{card\ N}$, define

$$F : \{0, 1\}^N \rightarrow R \tag{9}$$

such that

169

$$F(f) = 0. f(1) f(2) f(3) \ldots \qquad (10)$$

where $f : N \rightarrow \{0, 1\}$. $F(f)$ is a decimal number consisting of 0's and 1's. Function F is one-to-one. Suppose $f, g \in \{0, 1\}^N$ and $f \neq g$, then $F(f) \neq F(g)$ because decimals defining $F(f)$ and $F(g)$ are different. Therefore,

$$\text{card } \{0, 1\}^N \leq \text{card } R \qquad (11)$$

or

$$2^{\text{card } N} \leq \text{card } R \qquad (12)$$

$$Q.E.D.$$

● PROBLEM 6–30

Prove

$$\text{card } N < \text{card } R. \qquad (1)$$

SOLUTION:

According to Cantor's theorem (If X is a set, then card $X <$ card $P(X)$) we have

$$\text{card } N < \text{card } P(N). \qquad (2)$$

For any set A, the following equality holds:

$$\text{card } P(A) = 2^{\text{card } A} \qquad (3)$$

(see Problem 6-25). Thus

$$\text{card } N < \text{card } P(N) = 2^{\text{card } N} =$$

$$= \text{card } R. \qquad (4)$$

● PROBLEM 6–31

Show that for any finite $n \geq 2$

$$n^{\text{card } N} = (\text{card } N)^{\text{card } N} = \text{card } R. \qquad (1)$$

SOLUTION:

We proved (see Problem 6-29)

$$\text{card } R = 2^{\text{card } N}. \qquad (2)$$

Since $n \geq 2$,

$$2^{\text{card } N} \leq n^{\text{card } N} \tag{3}$$

and

$$n^{\text{card } N} \leq (\text{card } N)^{\text{card } N}. \tag{4}$$

In Problem 6-30, we proved

$$\text{card } N < 2^{\text{card } N}. \tag{5}$$

Combining (2), (3), (4), and (5)

$$\text{card } R = 2^{\text{card } N} \leq n^{\text{card } N} \leq (\text{card } N)^{\text{card } N} \leq (2^{\text{card } N})^{\text{card } N} =$$

$$= 2^{\text{card } N \cdot \text{card } N} = \tag{6}$$

but $\text{card } N \cdot \text{card } N = \text{card } N$ (see Problem 6-19). Hence

$$2^{\text{card } N \cdot \text{card } N} = 2^{\text{card } N} = \text{card } R. \tag{7}$$

From (7), we conclude that

$$\text{card } R = n^{\text{card } N} = (\text{card } N)^{\text{card } N}. \tag{8}$$

● PROBLEM 6–32

Prove that:

1. $\text{card } C = \text{card } R$ (1)
    where C is the set of complex numbers.

2. $\text{card } R \cdot \text{card } N = \text{card } R$.   (2)

SOLUTION:

1. A complex number

$$z = x + iy \tag{3}$$

can be represented by an ordered pair

$$(x, y)$$

where x and y are real numbers. Thus,

$$C \sim R \times R \tag{4}$$

and

$$\text{card } C = \text{card } R \cdot \text{card } R. \tag{5}$$

171

From Problem 6-21

$$\text{card } R \cdot \text{card } R = \text{card } R. \tag{6}$$

Hence

$$\text{card } C = \text{card } R. \tag{7}$$

2. It is easy to verify that

$$\text{card } R \le \text{card } R \cdot \text{card } N \le \text{card } R \cdot \text{card } R = \text{card } R. \tag{8}$$

Therefore

$$\text{card } R \cdot \text{card } N = \text{card } R. \tag{9}$$

● **PROBLEM 6–33**

In Problem 6-21, we proved

$$\text{card } R \cdot \text{card } R = \text{card } R. \tag{1}$$

Using

$$2^{\text{card } N} = \text{card } R \tag{2}$$

prove (1).

SOLUTION:

For three cardinal numbers a, b, and c

$$a^b \, a^c = a^{b+c} \tag{3}$$

(see Problem 6-26).
   Thus, from (2) and (3), we obtain

$$\text{card } R \cdot \text{card } R = 2^{\text{card } N} 2^{\text{card } N} = 2^{\text{card } N + \text{card } N} =$$

$$= 2^{\text{card } N} = \text{card } R. \tag{4}$$

Note that in Problem 6-15 we proved

$$\text{card } N + \text{card } N = \text{card } N. \tag{5}$$

Let
$$F = \{f : f : R \rightarrow R\} \tag{1}$$
represent a set of all functions from R to R. Compare the cardinal numbers, card F and card R.

SOLUTION:

According to the definition of exponentiation of cardinal numbers, we have
$$\text{card } F = \text{card } R^R = (\text{card } R)^{\text{card } R} \tag{2}$$
(see Problem 6-23).
Since
$$2^{\text{card } N} = \text{card } R \tag{3}$$
(2) leads to
$$\text{card } F = (\text{card } R)^{\text{card } R} = (2^{\text{card } N})^{\text{card } R} =$$
$$= 2^{\text{card } N \cdot \text{card } R}. \tag{4}$$
From Problem 6-32, we obtain
$$\text{card } N \cdot \text{card } R = \text{card } R \tag{5}$$
therefore
$$\text{card } F = 2^{\text{card } R}. \tag{6}$$
According to Cantor's theorem (for any set X, card $X <$ card $P(X)$) and
$$\text{card } P(X) = 2^{\text{card } X} \tag{7}$$
we find setting $X = R$
$$2^{\text{card } R} = \text{card } P(R) > \text{card } R. \tag{8}$$
From (6) and (8), we obtain
$$\text{card } F = 2^{\text{card } R} > \text{card } R. \tag{9}$$

Prove that
$$\text{card } C(R, R) = \text{card } C(Q, R) =$$

$$= \text{card } K(R, R) = \text{card } R \qquad (1)$$

where $C(R, R)$ and $C(Q, R)$ are the sets of continuous real-valued functions with domains R and Q, respectively. The set of all real-valued constant functions with the domain R is denoted by $K(R, R)$.

SOLUTION:

Note that the each function $f : R \rightarrow R$ a function corresponds $f \mid Q : Q \rightarrow R$, such that for all $x \in Q$, $f(x) = f \mid Q(x)$. Function $f \mid Q$ is the restriction of f to Q. We denote this correspondence by

$$F : C(R, R) \rightarrow C(Q, R). \qquad (2)$$

The restriction of a continuous function is a continuous function. We show that F is one-to-one. Indeed, let $f, g \in C(R, R)$ denote such that for all $x \in Q$

$$f(x) = g(x). \qquad (3)$$

Let x' be any real number, $x' \in R$. Then a sequence $(x_n) \in Q$ of rational numbers exists, such that

$$\lim_{n \to \infty} x_n = x' \qquad (4)$$

Since both f and g are continuous

$$f(x') = g(x'). \qquad (5)$$

Therefore, $f = g$. Function F is injective, thus

$$\text{card } C(R, R) \leq \text{card } C(Q, R). \qquad (6)$$

On the other hand,

$$\text{card } C(Q, R) \leq \text{card } R^Q = (\text{card } R)^{\text{card } N} =$$

$$= (2^{\text{card } N})^{\text{card } N} = 2^{\text{card } N} = \text{card } R. \qquad (7)$$

From (6) and (7),

$$\text{card } C(R, R) \leq \text{card } C(Q, R) \leq \text{card } R. \qquad (8)$$

Consider the set $K(R, R)$. To each real number $p \in R$, a constant function corresponds

$$f_p : R \rightarrow R \qquad (9)$$

such that for all $x \in R$, $f_p(x) = p$. Thus

$$\text{card } K(R, R) = \text{card } R. \qquad (10)$$

Each constant function is continuous

$$K(R, R) \subset C(R, R). \qquad (11)$$

Hence

$$\text{card } R = \text{card } K(R, R) \le \text{card } C(R, R). \tag{12}$$

We conclude from (8) and (12) that

$$\text{card } R = \text{card } C(R, R) = \text{card } C(Q, R) = \text{card } K(R, R). \tag{13}$$

Surprisingly, there are as many constant functions as continuous functions.

● **PROBLEM 6–36**

Let $D(R, R)$ represent the set of all differentiable real-valued functions. Show that

$$\text{card } D(R, R) = \text{card } R. \tag{1}$$

SOLUTION:

Each differentiable function is continuous

$$D(R, R) \subset C(R, R). \tag{2}$$

Hence

$$\text{card } D(R, R) \le \text{card } C(R, R) = \text{card } R \tag{3}$$

(see Problem 6-35).

Each constant function is differentiable

$$K(R, R) \subset D(R, R) \tag{4}$$

hence

$$\text{card } R = \text{card } K(R, R) \le \text{card } D(R, R) \tag{5}$$

(see Problem 6-35). Thus

$$\text{card } D(R, R) = \text{card } R. \tag{6}$$

● **PROBLEM 6–37**

The so-called classic Hilbert space, H, consists of all infinite sequences

$$(x_1, x_2, x_3, \ldots) ; x_k \in R ; k \in N \tag{1}$$

such that the series

$$x_1^2 + x_2^2 + x_3^2 + \ldots \tag{2}$$

converges. Find card H.

SOLUTION:

Define function

$$f : R \to H \qquad (3)$$

such that for every $x \in R$

$$f(x) = (x, 0, 0, 0, \ldots). \qquad (4)$$

The infinite sequence belongs to the Hilbert space, H,

$$(x, 0, 0, \ldots) \in H. \qquad (5)$$

Function f is an injection, hence

$$\text{card } R \leq \text{card } H. \qquad (6)$$

On the other hand,

$$\text{card } H \leq (\text{card } R)^{\text{card } N} = (2^{\text{card } N})^{\text{card } N} =$$

$$= 2^{\text{card } N \cdot \text{card } N} = 2^{\text{card } N} = \text{card } R. \qquad (7)$$

From (6) and (7), we conclude that

$$\text{card } H = \text{card } R. \qquad (8)$$

● **PROBLEM 6–38**

Compare the cardinal numbers of these two sets:

1. The set of all infinite sequences

$$(x_1, x_2, x_3, \ldots) \qquad (1)$$

of real numbers, denoted by $R^{\text{card } N}$.

2. The set of all infinite sequences of integers.

SOLUTION:

The cardinal number of the set of all infinite sequences of integers is

$$(\text{card } N)^{\text{card } N}. \qquad (2)$$

From Problem 6-31, we obtain

$$(\text{card } N)^{\text{card } N} = \text{card } R. \qquad (3)$$

The cardinal number of the set $R^{\text{card } N}$ is

$$\text{card } R^{\text{card } N} = (\text{card } R)^{\text{card } N} \tag{4}$$

Equation (4) leads to

$$(\text{card } R)^{\text{card } N} = (2^{\text{card } N})^{\text{card } N} = 2^{\text{card } N \cdot \text{card } N} =$$

$$= 2^{\text{card } N} = \text{card } R. \tag{5}$$

Therefore, both sets have the same cardinal numbers. Sometimes, points

$$(x_1, x_2, x_3, \ldots)$$

where all x_k's are integers are called lattice points.

● **PROBLEM 6–39**

Compare the cardinal numbers of the following sets:

1. The set of all functions of one real variable, which assume values of 0 or 1.

2. The set of all real-valued functions of n real variables.

SOLUTION:

Consider the set of all functions

$$f(x) = \begin{cases} 0 \\ 1 \end{cases} \qquad f : R \to \{0,1\}.$$

The cardinal number of this set is

$$\text{card } 2^R = 2^{\text{card } R}. \tag{1}$$

The cardinal number of the set of real-valued functions of n real variables

$$f : R^n \to R \tag{2}$$

is

$$\text{card } R^{R^n} = \text{card } R^{\text{card } R^n} \tag{3}$$

Since

$$(\text{card } R)^n = \text{card } R \tag{4}$$

we have

$$\text{card } R^{\text{card } R^n} = \text{card } R^{\text{card } R} = (2^{\text{card } N})^{\text{card } R} \tag{5}$$

but

$$\text{card } N \text{ card } R = \text{card } R \tag{6}$$

hence

$$\text{card } R^{\text{card } R^n} = 2^{\text{card } R}. \tag{7}$$

Both sets have the same cardinal number.

● **PROBLEM 6-40**

Explain the continuum hypothesis and the generalized continuum hypothesis.

SOLUTION:

We recall briefly the major facts concerning cardinal numbers. There are finite cardinal numbers (which are non-negative integers) and transfinite cardinal numbers. The smallest transfinite cardinal number is card N (denoted $\aleph_0$, aleph-null). There is no largest cardinal number. Let us denote card $R = c$

$$\underbrace{0 < 1 < 2 < \ldots}_{\text{finite cardinal numbers}} \qquad \underbrace{< \aleph_0 < \ldots < c < \ldots < 2^c < \ldots}_{\text{transfinite cardinal numbers}}$$

We proved that $\aleph_0 < 2^{\aleph_0} = c$.

The following question posed by Cantor is known as the continuum problem: Is there a cardinal number that lies between $\aleph_0$ and c? Many mathematicians have spent many hours trying to solve this problem. Nobody has found a set A, such that card $A = a$ and $\aleph_0 < a < c$. Hence, the hypothesis is that such a set does not exist.

CONTINUUM HYPOTHESIS

There is no cardinal number p, such that

$$\aleph_0 < p < c = 2^{\aleph_0}.$$

A similar question exists for any transfinite cardinal number: Is there a cardinal number that lies between a transfinite cardinal number p and 2^p? This question has not been answered.

For any transfinite cardinal number p, no cardinal number r has been found, such that $p < r < 2^p$. Thus:

GENERALIZED CONTINUUM HYPOTHESIS

For any transfinite cardinal number p, there is no cardinal number r, such that

$$p < r < 2^p.$$

● PROBLEM 6-41

Do we know the solution of the continuum problem?

SOLUTION:

The continuum problem was formulated by Cantor in 1880. It became one of the most famous unresolved mathematical problems. On Hilbert's list of 23 mathematical questions with no answers, it occupied first place. (This list was announced in 1900 at the International Congress of Mathematicians in Paris.)

In 1938, Kurt Gödel proved that the generalized continuum hypothesis is consistent with the axioms of set theory. What this means can be explained as follows: Consider two systems of axioms

A – axioms for set theory, and
B – all axioms of A with the generalized continuum hypothesis added.
Any contradiction that might be implied by system B could be formulated as a contradiction implied by system A.

Later, P.J. Cohen proved that the negation of the continuum hypothesis is also consistent with the axioms for set theory. Thus, both the continuum and generalized continuum hypotheses are independent of the other axioms for set theory. They cannot be proved or disproved on the basis of these axioms.

W. Sierpinski has shown that, if the generalized continuum hypothesis is added as an axiom to the axioms of set theory, then the axiom of choice becomes redundant. It can be derived from the generalized continuum hypothesis and remaining axioms.

The situation is that the generalized continuum hypothesis is not provable on the basis of the axioms of set theory. We may postulate the generalized continuum hypothesis or deny it and, in either case, obtain a consistent mathematical theory.

CHAPTER 7

AXIOM OF CHOICE AND ITS EQUIVALENT FORMS

Quote the axiom of choice and explain it for a finite family of sets.

SOLUTION:

AXIOM OF CHOICE

Let $\mathcal{F}$ be any non-empty family of non-empty sets. Then a function called a choice function exists

$$f : \mathcal{F} \to \bigcup_{A \in \mathcal{F}} A \qquad (1)$$

such that for all $A \in \mathcal{F}$,

$$f(A) \in A \qquad (2)$$

Suppose $\mathcal{F}$ is a set of phone books from different cities. It is possible to choose one and only one name from each book. The axiom of choice is trivial for $\mathcal{F}$ finite.

In this chapter we shall examine the axiom of choice and its three equivalent principles (the Hausdorff maximality principle, Zorn's lemma, and the well-ordering principle of Zermelo).

The axiom of choice is, indeed, an axiom that cannot be proven as a theorem using the classical axioms of mathematics.

● PROBLEM 7-2

Let

$$f : A \to B \qquad (1)$$

denote onto mapping. Apply the axiom of choice to prove that a subset C of A exists, such that C and B are equipotent and hence card $A \geq$ card $B =$ card C.

SOLUTION:

Suppose $B \neq \phi$ (if $B = \phi$, then $C = \phi$). Consider the family of sets

$$\{f^{-1}(y) : y \in B\} \qquad (2)$$

which forms a partition of A because sets (2) are disjoint and

$$\bigcup_{y \in B} \{f^{-1}(y) : y \in B\} = A. \qquad (3)$$

Note that f is onto.

According to the axiom of choice, the family of sets $\{f^{-1}(y) : y \in B\}$ has a set of representatives C, such that for each $y \in B$

$$\{f^{-1}(y) : y \in B\} \cap C \tag{4}$$

is a single element.

Restriction of f to C

$$f|C : C \to B \tag{5}$$

is one-to-one and onto. Therefore, sets C and B are equipotent. Hence

$$\text{card } A \geq \text{card } B = \text{card } C. \tag{6}$$

● PROBLEM 7-3

Let $\{A_i : i \in I\}$ represent a family of non-empty sets. The generalized Cartesian product $\underset{i \in I}{X} A_i$ of the family $\{A_i\}$ is the set of all functions

$$f : I \to \bigcup_{i \in I} A_i \tag{1}$$

such that

$$f(i) \in A_i \text{ for all } i \in I. \tag{2}$$

Prove that, if $I \neq \phi$, then $\underset{i \in I}{X} A_i$ is non-empty.

SOLUTION:

Set $\{A_i : i \in I\}$ is a non-empty set, whose elements are non-empty sets (i.e., for each $i \in I$, $A_i \neq \phi$). According to the axiom of choice, a function g, called a choice function, exists such that

$$g : \{A_i : i \in I\} \to \bigcup_{i \in I} A_i. \tag{3}$$

Then, we can define function f by

$$f(i) = g(A_i) \in A_i \tag{4}$$

for all $i \in I$, where

$$f : I \to \bigcup_{i \in I} A_i \tag{5}$$

Hence, if $I \neq \phi$, then

$$\underset{i \in I}{X} A_i \neq \phi. \tag{6}$$

1. Let A represent any set. Show that the relation $\{(a, b) : a = b\}$ is a preordering.

2. Show that in R, $\leq$ is a preordering and $<$ is not.

3. Is inclusion a preordering?

SOLUTION:

We shall start with

DEFINITION OF PREORDER

A relation ρ in a set A, $\rho \subset A \times A$, is called a preorder if it is reflexive and transitive, that is, if

1. $\forall a \in A : (a, a) \in \rho$

2. $\forall a, b \in A : (a, b) \in \rho \land (b, c) \in \rho \Rightarrow (a, c) \in \rho$

A set A with preorder is called a preordered set $(A, <)$. Preorder is usually denoted by $<$ ($a < b$ is read "a precedes b").

1. Relation $=$ is a preorder. It is reflexive because for each $a \in A$, $a = a$. It is transitive:

$$(a = b) \land (b = c) \Rightarrow a = c.$$

Hence $(A, =)$ is a preordered set.

2. For each $a \in R$, $a \leq a$. Also

$$(a \leq b) \land (b \leq c) \Rightarrow (a \leq c).$$

Therefore, $\leq$ is a preordering and $(R, \leq)$ is a preordered set. Of course, $a < a$ is not true and $<$ is not a preorder.

3. Consider the power set $P(X)$ (i.e., the set of all subsets of X). The relation $\subset$ is a preorder because for each $A, B, C \in P(X)$

$$A \subset A \text{ and } (A \subset B) \land (B \subset C) \Rightarrow A \subset C.$$

Preordering by inclusion is defined by

$$A < B \text{ iff } A \subset B.$$

183

A family of sets preordered in this manner is called preordered by inclusion.

Let C represent the set of complex numbers. Show that $|z_1| \le |z_1|$, $z_1, z_2 \in C$ defines a preordering.

SOLUTION:

Any complex number $z \in C$ can be represented by

$$z = x + iy$$

where x, y are real numbers. Then

$$|z| = \sqrt{x^2 + y^2}.$$

For every $z \in C$ $|z| \le |z|$ because

$$\sqrt{x^2 + y^2} \le \sqrt{x^2 + y^2}.$$

Hence relation $|\dots| \le |\dots|$ is reflexive. Let $z_1, z_2, z_3 \in C$ and $|z_1| \le |z_2|$ and $|z_2| \le |z_3|$. Then

$$\sqrt{x_1^2 + y_1^2} \le \sqrt{x_2^2 + y_2^2} \text{ and } \sqrt{x_2^2 + y_2^2} \le \sqrt{x_3^2 + y_3^2}$$

where $z_1 = x_1 + iy_1$, $z_2 = x_2 + iy_2$, $z_3 = x_3 + iy_3$. Hence

$$\sqrt{x_1^2 + y_1^2} \le \sqrt{x_3^2 + y_3^2}$$

or

$$|z_1| \le |z_3|.$$

Relation $|\dots| \le |\dots|$ is also transitive. Thus $|\dots| \le |\dots|$ is a preorder and $(C, |\dots| \le |\dots|)$ is a preordered set.

Let D denote the diagonal, $D \subset A \times A$. Show that $\rho \subset A \times A$ is a preorder, if and only if $D \subset \rho$ and $\rho \circ \rho = \rho$. The diagonal is defined by

$$D = \{(x, x) : x \in A\}.$$

SOLUTION:

We have to prove that

$$(\rho \text{ is a preorder}) \Leftrightarrow \left(\begin{matrix} D \subset \rho \\ \rho \circ \rho = \rho \end{matrix} \right).$$

$\Rightarrow$ ρ is a preorder. For each $a \in A$, $(a, a) \in \rho$. Hence $D \subset \rho$. Furthermore

$$(a, b) \in \rho \wedge (b, c) \in \rho \Rightarrow (a, c) \in \rho$$

which is equivalent to $\rho \circ \rho = \rho$.

$\Leftarrow$ Since $D \subset \rho$ we have

$$\forall a \in A : (a, a) \in \rho.$$

The set $\rho \circ \rho$ is

$$\rho \circ \rho = \{(a, c) : a, c \in A, \exists b \in A \text{ such that } (a, b) \in \rho \wedge (b, c) \in \rho\}.$$

By assumption

$$\rho \circ \rho = \rho.$$

Hence

$$(a, b) \in \rho \wedge (b, c) \in \rho \Rightarrow (a, c) \in \rho.$$

● **PROBLEM 7-7**

Show that the relation $\leq$ defined on R^2 by

$$(a_1, a_2) \leq (b_1, b_2) \qquad (1)$$

if and only if

$$a_1 \leq b_1 \text{ and } a_2 \leq b_2 \qquad (2)$$

is a partial order relation on R^2.

SOLUTION:

We shall apply the following:

DEFINITION OF PARTIAL ORDER

A relation $\leq$ on a set A is called a partial order, if and only if $\leq$ is reflexive, transitive and antisymmetric on A, that is, if $a \leq b$ and $b \leq a$, then $a = b$. Set A with a partial order relation on A is denoted by $(A, \leq)$ and is called a partially ordered set.

Relation $\leq$ defined in (1) is reflexive,

185

$$(a, b) \le (a, b) \text{ because } a \le a \text{ and } b \le b. \tag{3}$$

It is also transitive.

Suppose

$$(a_1, b_1) \le (a_2, b_2) \quad \text{and} \quad (a_2, b_2) \le (a_3, b_3) \tag{4}$$

then

$$a_1 \le a_2 \text{ and } a_2 \le a_3$$

$$b_1 \le b_2 \text{ and } b_2 \le b_3$$

Hence

$$a_1 \le a_3 \text{ and } b_1 \le b_3$$

or

$$(a_1, b_1) \le (a_3, b_3). \tag{5}$$

Finally we show that $\le$ is antisymmetric. Indeed, if

$$(a_1, b_1) \le (a_2, b_2) \quad \text{and} \quad (a_2, b_2) \le (a_1, b_1) \tag{6}$$

then

$$a_1 \le a_2 \text{ and } a_2 \le a_1$$

$$b_1 \le b_2 \text{ and } b_2 \le b_1.$$

Therefore,

$$a_1 = a_2 \text{ and } b_1 = b_2 \quad \text{or} \quad (a_1, b_1) = (a_2, b_2). \tag{7}$$

● **PROBLEM 7–8**

Show that the relation defined in Problem 7-7 is not a total order.

SOLUTION:

We shall use

DEFINITION OF TOTAL ORDER

A total order relation $\le$ on a set A is a partial order relation, such that for any pair $a, b \in A$, either $a \le b$ or $b \le a$. Set A with a total order relation, denoted by $(A, \le)$, is called a totally ordered set.

Total order relation is sometimes called linear order relation. According to the above definition, the relation described in Problem 7-7 is not a total order relation. Neither $(0, 1) \le (1, 0)$, nor $(1, 0) \le (0, 1)$.

This example shows that a partially ordered set need not be a totally ordered set.

● **PROBLEM 7–9**

Set $A = \{a, b, c\}$ is ordered as shown in the diagram.

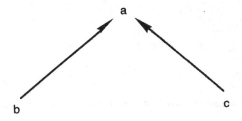

FIGURE 1

The order can be defined as follows: $x \le y$ iff $x = y$ or one can move from x to y in the direction of arrows. Let B denote the collection of all non-empty totally ordered subsets of A. Draw the diagram of the order of B, when B is partially ordered by set inclusion.

SOLUTION:

The set of all subsets of A consists of

$$\{a\}, \{b\}, \{c\}, \{a, b\}, \{a, c\}, \{b, c\}, \{a, b, c\}.$$

Only the subsets $\{a, b, c\}$ and $\{b, c\}$ are not totally ordered, because neither $b \le c$ nor $c \le b$.

Set B consists of

$$\{a\}, \{b\}, \{c\}, \{a, b\}, \{a, c\},$$

and is ordered by set inclusion. Hence

$$\{a\} \le \{a, b\}$$

$$\{b\} \le \{a, b\}$$

$$\{a\} \le \{a, c\}$$

$$\{c\} \le \{a, c\}$$

187

Set B is partially ordered as shown in the diagram

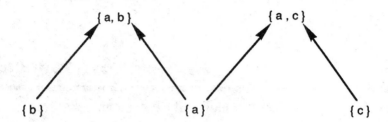

FIGURE 2

● **PROBLEM 7-10**

Consider the relation defined in Problem 7-7. Show that the diagonal of the plane R^2 is a chain.

SOLUTION:

Let A denote a partially ordered set $(A, \leq)$; then

$$P =: \{(x, y) : x, y \in A, x \leq y\}.$$

The sign =: will be used to denote "equals by definition."
    For any subset B of A, $B \subset A$, we define $\leq'$ to be the intersection

$$P \cap (B \times B).$$

The set $(B, \leq')$ is a partially ordered set. It may happen that $(B, \leq')$ is a totally ordered subset of a partially ordered set.

DEFINITION OF A CHAIN

A totally ordered subset of a partially ordered set is called a chain.

In Problem 7-7, we showed that a relation defined by (1) is a partial order. The diagonal

$$D = \{(x, x) : x \in R\}$$

of the plane R^2 is a chain because on D, a relation becomes a total order.

Prove that the relation "m divides n" defined on the set of natural numbers N is a partial order, but not a total order.

SOLUTION:

A relation is reflexive, because for any natural number n, n divides n is true.

This relation is also transitive. Suppose m divides n and n divides k, that is,

$$\frac{n}{m} \in N \,,\, \frac{k}{n} \in N$$

then

$$\frac{n}{m} \cdot \frac{k}{n} = \frac{k}{m} \in N \,.$$

That is, m divides k.

A relation is antisymmetric because, if m divides n and n divides m, then $m = n$. Hence, the relation is a partial order.

It is not a total order because

3 does not divide 2

and

2 does not divide 3.

● PROBLEM 7–12

Show that

1. The identity relation "=" is a partial order relation on any set.

2. The set $(R, \leq)$ is a totally ordered set, where R is the set of real numbers and "$\leq$" is understood in the ordinary sense.

SOLUTION:

1. Let A represent a set. Relation "=" is reflexive because for all $x \in A$

$$x = x. \tag{1}$$

It is also transitive. If

$$x = y \text{ and } y = z, \text{ then } x = z. \tag{2}$$

Furthermore, relation "=" is antisymmetric. Hence, $(A, =)$ is a partially ordered set.

If set A contains only one element, then $(\{a\}, =)$ is a totally ordered set.

2. Relation "$\leq$", understood in the ordinary sense and defined on the set of real numbers R, is a total order. For each $x \in R$

$$x \leq x.$$

If $x \leq y$ and $y \leq z$, then

$$x \leq z.$$

If $x \leq y$ and $y \leq x$, then

$$x = y.$$

For any pair of real numbers x, y, either $x \leq y$ or $y \leq x$.

● **PROBLEM 7-13**

Using the notion of temperature, define a total order relation on the set of all physical objects.

SOLUTION:

Temperature can be introduced in the following manner. Let a and b represent two objects. If there is a heat transfer from a to b, we say that the temperature of a is higher than the temperature of b. If there is no heat transfer, we say that a and b are of the same temperature. Here, we neglected all physical details.

Let us define the relationship $a \leq b$ iff the temperature of b is higher than or equal to the temperature of a. This relation is reflexive:

$$a \leq b.$$

It is transitive:
    If $a \leq b$ and $b \leq c$, then

$$a \leq c.$$

It is antisymmetric:
    If $a \leq b$ and $b \leq a$, then $a = b$ (that is, a and b have the same temperature).
    Note that "temperature" defines the total order, since, for any two a and b, we have either $a \leq b$ or $b \leq a$.

Consider the set F of all functions

$$f : R \to R \qquad (1)$$

with the relation S defined by

$$S = \{(f, g) \in F \times F : \forall x \in R, f(x) \le g(x)\}. \qquad (2)$$

Prove that (F, S) is a partially ordered set.

SOLUTION:

Relation S is reflexive since

$$\forall f \in F \quad \forall x \in R \quad f(x) \le g(x). \qquad (3)$$

S is transitive,

$$\begin{pmatrix} \forall f, g, h \in F, \forall x \in R \\ f(x) \le g(x) \\ g(x) \le h(x) \end{pmatrix} \Rightarrow \begin{pmatrix} f(x) \le h(x) \\ \forall x \in R \end{pmatrix}$$

S is antisymmetric,

$$\begin{pmatrix} f(x) \le g(x), \forall x \in R \\ g(x) \le f(x) \end{pmatrix} \Rightarrow \begin{pmatrix} f(x) = g(x) \\ \forall x \in R \end{pmatrix}$$

S is a partial order.

It is not a total order. For example, if $f(x) = x$ and $g(x) = x^2$, then neither $f(x) \le g(x)$ nor $g(x) \le f(x)$ for all $x \in R$.

Let A and B denote totally ordered sets $(A, \le')$, $(B, \le'')$. Show that $A \times B$ can be totally ordered.

SOLUTION:

We can define the order on $A \times B$ in this manner:

$$(a,b) \le (c,d) \quad \begin{cases} \text{if } a \le' c \\ \text{or, if } a = c \text{ and } b \le'' d \end{cases}$$

Since both $\le'$ and $\le''$ are reflexive, $\le$ is reflexive. Relation $\le$ is also transitive.

Suppose $(a, b) \leq (c, d)$ and $(c, d) \leq (a, b)$. Then $a \leq' c$ and $c \leq' a$ imply $a = c$. Also $b \leq'' d$ and $d \leq'' b$ imply $b = d$.

Thus, $(a, b) = (c, d)$ and relation $\leq$ is antisymmetric.

Since $\leq'$ and $\leq''$ are total orders, $\leq$ is a total order.

The order defined above is called the lexicographical order on $A \times B$ because it is similar to the arrangement of the words in a dictionary.

● **PROBLEM 7-16**

Give an example of the partially ordered set that has no upper bound.

SOLUTION:

We shall start with

DEFINITION OF UPPER (LOWER) BOUND

Let $(A, \leq)$ denote a partially ordered set and B its subset, $B \subset A$. An element, $x \in A$, is an upper bound (lower bound) of B, if and only if

$$x \geq a \ (x \leq a) \text{ for all } a \in B.$$

For example, let $(N, \leq)$ denote a set of natural numbers $N = \{1, 2, 3, 4, ...\}$ and $B = \{2, 4, 8\}$ its subset. Relation "$\leq$" is an ordinary relation. Element $10 \in N$ is an upper bound of B. Any element of the set $\{8, 9, 10, 11, ...\}$ is an upper bound of B.

Elements 1 and 2 are lower bounds of B. Now consider the set of even numbers

$$\{2, 4, 6, 8, ...\}.$$

This set has no upper bound.

● **PROBLEM 7-17**

Let S represent a family of sets partially ordered by the relation of inclusion, and let R represent a subfamily of S.

1. Show that

$$\cup \{A : A \in R\}$$

does not have to be an upper bound of R.

2. Show that if $B \in S$ is an upper bound of R, then

$$\cup \{A : A \in R\} \subset B. \tag{1}$$

192

SOLUTION:

1. Set S is a family of sets and R is its subfamily. It does not guarantee that the union of elements of R

$$\cup \{A : A \in R\}$$

is an element of S.

Hence, $\cup \{A : A \in R\}$ is an upper bound of R, if and only if it is an element of S.

2. Let

$$x \in \cup \{A : A \in R\}. \qquad (2)$$

Then set A_0 exists, such that $x \in A_0$ and $A_0 \in R$.
Set B is an upper bound of R, therefore

$$A_0 \subset B \quad \text{and} \quad x \in B$$

which proves (1).

● **PROBLEM 7-18**

The set

$$A = \{a, b, c, d, e, f, g, h\} \qquad (1)$$

is partially ordered as shown in the diagram.

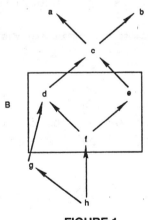

FIGURE 1

Two elements, x and y, are related, $x \le y$, iff $x = y$ or one can move along the arrows from x to y. Consider the subset B of A

$$B = \{d, e, f\}.$$

Find the least upper bound and the greatest lower bound of B.

SOLUTION:

The elements a, b, and c are upper bounds of B.

DEFINITION OF LEAST UPPER BOUND (GREATEST LOWER BOUND)

Let $(A, \leq)$ represent a partially ordered set and B its subset. An upper bound (lower bound) x_0 of B is the least upper bound (the greatest lower bound) of B, if and only if $x_0 \leq x$ ($x_0 \geq x$) for every upper (lower) bound x of B.

Element c is the least upper bound of B because

$$c \leq a \quad \text{and} \quad c \leq b.$$

Elements f and h are lower bounds of B.
    The greatest lower bound of B is f.
    Sometimes, we denote the least upper bound of A by

$$\sup (A) \quad \text{(or lub } A)$$

and the greatest lower bound of A by

$$\inf (A) \quad \text{(or glb } A).$$

● PROBLEM 7-19

Let B denote a subset of the partially ordered set $(A, \leq)$. Prove that if the least upper bound (greatest lower bound) of B exists, then it is unique.

SOLUTION:

Proof (reductio ad absurdum or r.a.a.)
    Suppose there are two least upper bounds of B, x and x', such that $x \neq x'$. Since x is the least upper bound of B and x' is an upper bound of B

$$x \leq x'. \tag{1}$$

Since x' is the least upper bound of B and x is an upper bound of B

$$x' \leq x. \tag{2}$$

Relation $\leq$ defined as A is a partial order. Hence, it is antisymmetric

$$\left(\begin{matrix} x \leq x' \\ x' \leq x \end{matrix} \right) \Rightarrow (x = x').$$

194

This is in contradiction with the assumption that $x \neq x'$. Thus, the least upper bound is unique.

In the same way, it can be shown that the greatest lower bound is unique.

● **PROBLEM 7-20**

Let X denote a non-empty set. The set $(P(X), \subset)$ is a partially ordered set. Show that, for any $Q \subset P(X)$, the least upper bound of Q is

$$\sup(Q) = \bigcup_{A \in Q} A \qquad (1)$$

and the greatest lower bound of Q is

$$\inf(Q) = \bigcap_{A \in Q} A. \qquad (2)$$

SOLUTION:

Let B represent any element of Q, $B \in Q$, then

$$B \subset \bigcup_{A \in Q} A. \qquad (3)$$

Hence, $\bigcup_{A \in Q} A$ is the upper bound of Q. Suppose D is the upper bound of Q, then for any $B \in Q$

$$B \subset D. \qquad (4)$$

Hence

$$\bigcup_{A \in Q} A \subset D \qquad (5)$$

and $\bigcup_{A \in Q} A$ is the least upper bound.

The set $\bigcap_{A \in Q} A$ is the lower bound of Q because for any $B \in Q$

$$\bigcap_{A \in Q} A \subset B. \qquad (6)$$

The set $\bigcap_{A \in Q} A$ is the greatest lower bound of Q. Indeed let D be the lower bound, then

$$D \subset B \text{ for all } B \in Q$$

and

$$D \subset \bigcap_{A \in Q} A. \qquad (7)$$

The set

$$A = \{2, 3, 4, 5, 6, 7, 8, 9, 10, 14\}$$

is ordered by

$$x \leq y \text{ iff } x \text{ is a multiple of } y.$$

Find the maximal and minimal elements of A.

SOLUTION:

The order of A is depicted in the diagram

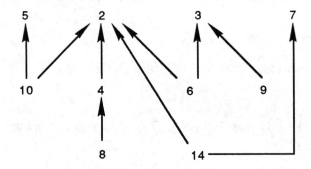

FIGURE 1

DEFINITION OF MAXIMAL (MINIMAL) ELEMENT

An element $x_0 \in A$ of a partially ordered set $(A, \leq)$ is said to be maximal (minimal), if and only if $x_0 \leq x$ ($x \leq x_0$) implies $x = x_0$ for all $x \in A$.

According to the definition, the maximal elements of A are 2, 3, 5 and 7. The minimal elements are 6, 8, 9, 10, and 14.

Consider the set

$$A = \{2, 3, 4, 5, \ldots\} = N - \{1\}$$

ordered by $m \leq n$, iff m divides n.

Find maximal and minimal elements of A.

SOLUTION:

All prime numbers are minimal elements. If $p \in A$ is a prime number, then only p divides p, because $1 \notin A$.

Only prime numbers are minimal elements. Suppose $a \in A$ is not a prime number, then $b \in A$ exists, such that

$$b \text{ divides } a \ (b \le a).$$

Hence, a is not a minimal element.

Set A has no maximal elements since, for every $m \in A$, $2m$ also belongs to A and

$$m \text{ divides } 2m \ (m \le 2m).$$

● **PROBLEM 7-23**

Give an example of a partially ordered set that has more than one maximal element and more than one minimal element.

SOLUTION:

Consider set A consisting of elements

$$A = \{a, b, c, d, e\}$$

which are partially ordered as shown in the diagram.

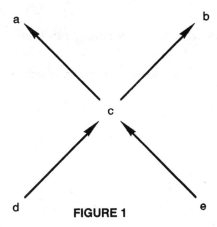

FIGURE 1

The maximal elements of A are a and b.

The minimal elements of A are d and e.

From this example, it is clear that maximal and minimal elements of a partially ordered set are not unique.

The situation changes for totally ordered sets (see Problem 7-24).

197

Show that if a totally ordered set has a maximal (or minimal) element, then this element is unique.

SOLUTION:

Set A is a totally ordered set $(A, \leq)$.

Suppose the set has two maximal elements, x_1 and x_2. We have $x_1 \in A$ and $x_2 \in A$. Since A is a totally ordered set for any pair $x_1, x_2 \in A$, either $x_1 \leq x_2$ or $x_2 \leq x_1$. Suppose $x_1 \leq x_2$. Then, since x_1 is a maximal element,

$$(x_1 \leq x_2) \Rightarrow (x_1 = x_2).$$

Hence, $x_1 = x_2$. The maximal element of a totally ordered set is unique.

In the same way, we can show that the minimal element of a totally ordered set is unique.

Draw the diagram for the ordered set (A, S) where

$$A = \{a, b, c, d, e, f, g\}$$

$$S = \{(b, c), (d, e), (e, f), (d, f), (g, a), (g, b), (g, c),$$

$$(g, d), (g, e), (g, f)\} \cup I_A.$$

Find the elements which are maximal, minimal, maximum, and minimum.

SOLUTION:

The diagram for the set (A, S) is shown in Figure 1.

The maximal elements are a, c, and f.

The minimal element is g.

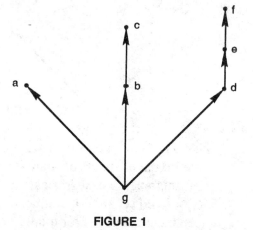

FIGURE 1

DEFINITION OF MAXIMUM (MINIMUM)

Element $a \in X$ of an ordered set X is a maximum (minimum) element of X or greatest (least) element of X, if and only if $x \leq a$ ($a \leq x$) for all $x \in X$.

Set (A, S) has no maximum. Element g is a minimum element of A.

● PROBLEM 7-26

Show that if an ordered set A has two distinct minimal elements, then it has no minimum element.

SOLUTION:

Let x_1 and x_2 denote two distinct minimal elements of A. Suppose a is a minimum element of A. Therefore

$$a \leq x_1 \quad a \leq x_2 \tag{1}$$

because a is a minimum element of A and $x_1, x_2 \in A$.

Since x_1 is a minimal element of A, there is no $x \in A$, such that

$$x_1 \neq x \leq x_1. \tag{2}$$

From (1) and (2), we conclude

$$x_1 = a. \tag{3}$$

Similarly, since x_2 is a minimal element of A

$$x_2 = a. \tag{4}$$

Hence

$$x_1 = x_2 \tag{5}$$

which is a contradiction. Set A with two distinct minimal elements has no minimum element.

● PROBLEM 7-27

Let X represent a linearly ordered set and A a finite totally ordered subset of X. Show that if

$$a = \sup A$$

then

$$a = \text{maximum } A$$

Show that the conclusion can be false when A is not totally ordered.

SOLUTION:

Let

$$A = \{a_1, a_2, \ldots, a_n\} \tag{1}$$

and let

$$b_1 = a_1 = \text{maximum } \{a_1\}. \tag{2}$$

Moving step by step, we define

$$b_2 = \text{maximum } \{b_1, a_2\}$$

$$b_3 = \text{maximum } \{b_2, a_3\} \tag{3}$$

In general

$$b_i = \text{maximum } \{b_{i-1}, a_i\}$$

and

$$b_n = \text{maximum } \{b_{n-1}, a_n\}.$$

Set A is totally ordered, hence we can always find the maximum. Since the maximum exists,

$$a = \sup A = \text{maximum } A. \tag{4}$$

Suppose $x = \{a, b, c\}$ and $A = \{a, b\}$. Set A is not totally ordered; see Figure 1.

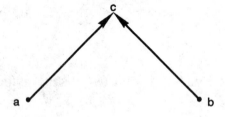

FIGURE 1

Then

$$c = \sup A \tag{5}$$

but c is not a maximum of A.

Prove that not every antisymmetric relation may be extended to an order.

SOLUTION:

Let X represent a set and T an antisymmetric relation on X. We have to choose T in such a way that no partial order Q on X exists, such that

$$T \subset Q. \tag{1}$$

Let

$$T = \{(a, b), (b, c), (c, d), (d, a)\} \tag{2}$$

denote an antisymmetric relation on X, where X is any set containing elements a, b, c, d.

Suppose a partial order Q exists, such that $T \subset Q$. Then

$$(a, b) \in Q \quad \text{and} \quad (b, c) \in Q. \tag{3}$$

Hence, since Q is transitive

$$(a, c) \in Q. \tag{4}$$

Furthermore

$$(a, d) \in Q \quad \text{and} \quad (d, a) \in Q \tag{5}$$

and, since Q is antisymmetric

$$a = d. \tag{6}$$

Similarly, we show that

$$a = b = c = d. \tag{7}$$

● PROBLEM 7-29

Let A and B represent ordered sets and let

$$f : A \to B \tag{1}$$

be onto. Prove that if for all $x, y \in A$

$$(x \le y) \Leftrightarrow (f(x) \le f(y)) \tag{2}$$

then f is one-to-one.

SOLUTION:

Suppose

$$f(x) = f(y) \tag{3}$$

Then

$$f(x) \geq f(y) \Rightarrow x \geq y$$

$$f(x) \leq f(y) \Rightarrow x \leq y. \tag{4}$$

Since order is a relation which is antisymmetric we conclude from (4) that

$$x = y. \tag{5}$$

● **PROBLEM 7–30**

We shall accept the following theorem without a proof.

THEOREM 1

Let $(A, \leq)$ represent a non-empty partially ordered set, such that every totally ordered subset of A has a least upper bound in A. If function $f : A \rightarrow A$ is such that for every $a \in A$, $f(a) \geq a$, then element $b \in A$ exists such that

$$f(b) = b.$$

Use Theorem 1 to prove:

THEOREM 2

Let $(A, \leq)$ represent a non-empty partially ordered set, such that every totally ordered subset of A has a greatest lower bound. If function $f : A \rightarrow A$ is such that for every $a \in A$, $f(a) \leq a$, then element $b \in A$ exists such that

$$f(b) = b.$$

SOLUTION:

Suppose (A, ρ) is a partially ordered set and ρ^{-1} is a relation inverse to ρ. Then (A, ρ^{-1}) is also a partially ordered set, where

$$(a \, \rho^{-1} \, b) \quad \text{iff} \quad (b \, \rho \, a). \tag{1}$$

Assume that (A, ρ) is such that every totally ordered subset of A has a least upper bound in A. Note that, if B is a totally ordered subset of (A, ρ), then (B, ρ^{-1}) is a totally ordered subset of (A, ρ^{-1}). If x_0 is a least upper bound of (B, ρ), then x_0 is a greatest lower bound of (B, ρ^{-1}).

Suppose function $f : A \rightarrow A$ is such that for every $a \in A$, $f(a) \geq a$. Then by applying Theorem 1, we conclude that element $b \in A$ exists, such that

$$f(b) = b.$$

If function $f : (A, \rho^{-1}) \rightarrow (A, \rho^{-1})$ is such that for every $a \in A$, $a \, \rho^{-1} \, f(a)$, then by virtue of Theorem 1, element $b \in A$ exists, such that

$$f(b) = b.$$

● **PROBLEM 7-31**

Show that the Hausdorff maximality principle and the axiom of choice are equivalent. Apply Theorem 1 of Problem 7-30.

SOLUTION:

HAUSDORFF MAXIMALITY PRINCIPLE

Let $(A, \leq)$ represent a partially ordered set and let S represent the set of all totally ordered subsets of A partially ordered by inclusion $(S, \subset)$. Then $(S, \subset)$ has a maximal element.

We shall use the reductio ad absurdum method.

Assume that S has no maximal element. Then, for every element s, $s \in S$, we can find a non-empty set

$$P = \{s' \in S : s \subset s', s \neq s'\}. \tag{1}$$

Elements s and s' of S are themselves ordered sets.

The axiom of choice guarantees the existence of a function f defined on $\{P : s \in S\}$ and such that

$$f(P) \in P. \tag{2}$$

Hence we can define a function

$$g : S \rightarrow S \tag{3}$$

such that

$$g(s) = f(P). \tag{4}$$

Note that since $g(s) = f(P)$, we have

$$s \subset g(s) = f(P), \quad g(s) \neq s \tag{5}$$

because $f(P) \in P$.

By applying Theorem 1 of Problem 7-30 [set $(S, \subset)$ and function g satisfy the hypotheses of the theorem], we conclude that an element $q \in S$ exists, such that

$$g(q) = q \tag{6}$$

which is a contradiction because for all $s \in S$

$$s \subset g(s) \quad (s \neq g(s)). \tag{7}$$

We showed that

(Axiom of Choice) $\Rightarrow$ (Hausdorff Maximality Principle).

Later we will carry out the proof in the opposite direction.

● **PROBLEM 7–32**

Use the Hausdorff maximality principle to prove the following:

Let $(A, \leq)$ represent a partially ordered set and B represent a totally ordered subset of A, $B \subset A$. Then A has a maximal totally ordered subset C such that

$$B \subset C. \tag{1}$$

SOLUTION:

Let S represent the set of all totally ordered subsets of A that contain B.

Note that the set $(S, \subset)$ is partially ordered.

The restriction that S consists of only these sets that contain B does not change the structure of the proof of theorem in Problem 7-31. Hence, we conclude that S has a maximal element, say C, such that

$$B \subset C.$$

● **PROBLEM 7–33**

Show that

(Axiom of Choice) $\Rightarrow$ (Zorn's Lemma).

SOLUTION:

ZORN'S LEMMA

If $(A, \leq)$ is a partially ordered set, such that every chain in A has an upper bound in A, then A contains a maximal element.

According to the Hausdorff maximality principle, the non-empty partially ordered set $(A, \leq)$ has a totally ordered subset $B \subset A$, which is maximal with respect to the relation of inclusion. By hypothesis, set B has an upper bound $x \in A$.

We shall show that element x is a maximal element of A.

Suppose, on the contrary, that an element $y \in A$ exists, such that

$$x \leq y.$$

Then $B \cup \{y\}$ is a totally ordered subset of A that contains B

$$B \subset B \cup \{y\}$$

where B is the maximal totally ordered subset of A. Hence

$$B \cup \{y\} = B$$

and

$$y \leq x.$$

Thus, x is a maximal element of $(A, \leq)$.

That completes the proof of

(Hausdorff Maximality Principle) $\Rightarrow$ (Zorn's Lemma).

From Problem 7-31, we obtain

(Axiom of Choice) $\Rightarrow$ (Hausdorff Maximality Principle) $\Rightarrow$

$$\Rightarrow \text{(Zorn's Lemma)}.$$

● PROBLEM 7–34

Prove that if set J of subsets of X has finite character, then $(J, \subset)$ has a maximal element.

SOLUTION:

DEFINITION OF FINITE CHARACTER

Set J of subsets of X has finite character, provided that $A \in J$, if and only if every finite subset of A belongs to J.

Set J is partially ordered by inclusion $(J, \subset)$.

Let $\mathcal{H}$ denote a chain in $(J, \subset)$. We shall show that $\mathcal{H}$ has an upper bound

$$B = \bigcup_{H \in \mathcal{H}} H$$

Let $\{x_1, x_2, \ldots, x_n\}$ denote any finite subset of B,

$$\{x_1, x_2, \ldots, x_n\} \subset \bigcup_{H \in \mathcal{H}} H.$$

Then

$$x_l \in H_l \text{ for some } H_l \in \mathcal{H}$$

$$l = 1, 2, \ldots, n.$$

Set $\mathcal{H}$ is a chain, therefore $H \in \mathcal{H}$ exists, such that for all $l = 1, 2, \ldots, n$

$$H_l \subset H.$$

Hence

$$\{x_1, x_2, \ldots, x_n\} \subset H$$

and

$$\{x_1, x_2, \ldots, x_n\} \in J.$$

Therefore $B \in J$ is an upper bound for $\mathcal{H}$.

According to Zorn's lemma, $(J, \subset)$ has a maximal element.

● **PROBLEM 7-35**

Apply Zorn's lemma to show that every vector space has a basis.

SOLUTION:

Let V denote a vector space, and let S denote the set of all linearly independent subsets of vectors in V.

Set S can be partially ordered by inclusion $(S, \subset)$. If H is a chain in $(S, \subset)$, then

$$T = \bigcup_{h \in H} h$$

is linearly independent. Hence T is an upper bound for H, $T \in S$.

According to Zorn's lemma, set $(S, \subset)$ has a maximal element. It can be easily shown that this element forms a basis for V.

Prove that every ring with an identity has a proper maximal ideal.

SOLUTION:

Let K denote a ring and let $1 \in K$ denote the identity.

Set $\mathcal{A}$ of all proper ideals of K is partially ordered by inclusion, $(\mathcal{A}, \subset)$. Let $(H, \subset)$ denote a chain in $(\mathcal{A}, \subset)$. Then

$$\bigcup_{h \in H} H$$

is a proper ideal of K.

Indeed

$$1 \notin \bigcup_{h \in H} h$$

Hence $\bigcup_{h \in H} h \in \mathcal{A}$ is an upper bound of H.

According to Zorn's lemma, we conclude that $(\mathcal{A}, \subset)$ has a maximal element. Hence, ring K with an identity has a proper maximal ideal.

● **PROBLEM 7-37**

Prove

$$\text{(Zorn's lemma)} \Rightarrow \text{(Kuratowski's lemma)}.$$

SOLUTION:

KURATOWSKI'S LEMMA

Each chain in a partially ordered set $(X, \leq)$ is included in a maximal chain.

Let $(X, \leq)$ denote a partially ordered set and h a given chain in X. Let H represent a set of all chains in X which include h, $h \in H$. Set H is ordered by inclusion $(H, \subset)$. Let a non-empty set $\mathcal{F} \neq \phi$ represent an inclusion chain of elements of H; then

$$\cup \, \mathcal{F} \in H. \tag{1}$$

Suppose $a, b \in \cup \, \mathcal{F}$; then for some $A, B \in \mathcal{F}$

$$a \in A \in \mathcal{F}$$

$$b \in B \in \mathcal{F}. \tag{2}$$

Since A and B are members of $\mathcal{F}$, we must have

$$A \subset B \quad \text{or} \quad B \subset A. \tag{3}$$

Hence

$$a \leq b \quad \text{or} \quad b \leq a$$

because $\cup \mathcal{F}$ is a chain. We have

$$h \subset A \subset \cup \mathcal{F} \in H. \tag{4}$$

Set $\mathcal{F}$ has an upper bound in H.

Therefore, according to Zorn's lemma, H has a maximal element.
Chain h is included in a maximal element.
One can prove that

$$\text{(Kuratowski's Lemma)} \Rightarrow \text{(Zorn's Lemma)}.$$

Therefore, Zorn's lemma and Kuratowski's lemma are equivalent.

● **PROBLEM 7–38**

Show that the set of natural numbers $(N, \leq)$ is well ordered when $\leq$ is understood in the usual sense.

SOLUTION:

DEFINITION OF A WELL ORDERED SET

A totally ordered set $(A, \leq)$ is a well ordered set iff every non-empty subset B of A contains a minimal element; that is, if

$$\underset{\phi \neq B \subset A}{\forall} \quad \underset{b \in B}{\exists} \quad \underset{x \in B}{\forall} \quad b \leq x$$

Element b is called the least element of B and relation $\leq$ is called a well ordered relation.

The set of natural numbers N is well ordered because every non-empty subset of N contains the least element.

On the other hand, the set of rational numbers with the ordinary "less than or equal" $\leq$ relation is not well ordered. Not all subsets of R (for example, $\{-1, -2, -3, -4, \ldots\}$, contain the least element.

● **PROBLEM 7–39**

Show that the set Q of rational numbers can be well ordered.

SOLUTION:

Set $(N, \le)$ is well ordered. Set Q is denumerable, hence a bijection

$$f: Q \to N \qquad (1)$$

exists.

Relation $\le$ on Q can be introduced in the following manner:

$$\left(\underset{a,b \in Q}{\forall} \quad a \le b \right) \Leftrightarrow \left(\begin{array}{c} f(a) \le f(b) \\ f(a), f(b) \in N \end{array} \right).$$

Now, since N is well ordered, so is Q.

In this proof, Q can be replaced by any denumerable set.

● **PROBLEM 7-40**

Let (X, T) denote a totally ordered set and $Y \subset X$. Prove that a is the minimum element of Y, if and only if

$$\text{Im}(T \cap (\{a\} \times Y)) = Y. \qquad (1)$$

SOLUTION:

Let $y \in Y$, then

$$y \in \text{Im}(T \cap (\{a\} \times Y)). \qquad (2)$$

Hence

$$(x, y) \in T \cap (\{a\} \times Y) \qquad (3)$$

for some $x \in X$ if $(a, y) \in T$. Furthermore

$$(x, y) \in (T \cap (\{a\} \times Y)) = Y \qquad (4)$$

if and only if

$$(a, y) \in T$$

for all $y \in Y$ and if and only if a is the minimum element of Y.

In 1904 Ernst Zermelo used the axiom of choice to prove the amazing well ordering principle.

WELL ORDERING PRINCIPLE

Every set can be well ordered.

Prove

$$(\text{Zorn's Lemma}) \Rightarrow (\text{Well Ordering Principle}).$$

SOLUTION:

Let A represent any set and $\tilde{A}$ a family of all well ordered sets $(A_0, \leq_0)$ which are subsets of A. Set $\tilde{A}$ can be partially ordered by the relation $\tilde{\leq}$ defined by:

$$(A_1, \leq_1) \;\tilde{\leq}\; (A_2, \leq_2) \tag{1}$$

iff

1. $A_1 \subset A_2$, and

2. $x, y \in A_1$ and $x \leq_1 y$ imply $x \leq_2 y$, and

3. $x \in A_2 - A_1$ implies $y \leq_2 x$ for all $y \in A_1$. Set $(\tilde{A}, \tilde{\leq})$ is partially ordered.

Now, let $\tilde{B}$ represent a totally ordered subset of $\tilde{A}$. To apply Zorn's lemma, we have to show that $\tilde{B}$ has an upper bound. Consider $\left(\bigcup_{A \in \tilde{B}} A, \leq^* \right)$ where $\leq^*$ is defined by:

$$x \leq^* y \text{ iff both } x, y \text{ belong to some } (A_0, \leq_0)$$

and $(A_0, \leq_0) \in \tilde{B}$, and $x \leq_0 y$.

It is easy to verify that $\leq^*$ is a total order relation on $\bigcup_{A \in \tilde{B}} A$. Furthermore, we will show that

$$\left(\bigcup_{A \in \tilde{B}} A, \leq^* \right) \tag{2}$$

is well ordered.

Let

$$\phi \neq C \subset \bigcup_{A \in \tilde{B}} A, \tag{3}$$

then $(A_0, \leq_0) \in \tilde{B}$ exists, such that

$$A_0 \cap C \neq \phi.$$

Set $(A_0, \leq_0)$ is well ordered, so $A_0 \cap C$ contains the unique least element of $A_0 \cap C$ denoted by a_0.

For any $x \in C$, $(A_1, \leq_1) \in \tilde{B}$ exists, such that

$$(A_0, \leq_0) \tilde{\leq} (A_1, \leq_1) \tag{4}$$

and $a_0, x \in A_1$.

We have $a_0 \leq x$ and $a_0 \leq^* x$. Hence a_0 is the least element for $(C, \leq^*)$.

Thus $\left(\bigcup_{A \in \tilde{B}} A, \leq^*\right)$ is well ordered.

According to Zorn's lemma, $(\tilde{A}, \tilde{\leq})$ has a maximal element $(A_1, \leq_1)$. We will show that $A = A_1$.

Indeed, suppose $A \neq A_1$; then $y \in A - A_1$ and relation $\leq_1$ on A_1 can be extended to $A_1 \cup \{y\}$ by

$$x \leq_1 y \text{ for all } x \in A_1.$$

We conclude that

$$(A_1, \leq_1) \tilde{\leq} (A_1 \cup \{y\}, \leq_1). \tag{5}$$

But $(A_1, \leq_1)$ is a maximal element, hence, a contradiction.

● **PROBLEM 7-42**

Prove that

(Well Ordering Principle) $\Rightarrow$ (Axiom of Choice).

SOLUTION:

Let $\mathcal{A}$ represent a non-empty family of non-empty sets. According to the well ordering principle (see Problem 7-41) a total order relation $\leq$ exists, such that

$$\left(\bigcup_{A \in \mathcal{A}} A, \leq\right) \tag{1}$$

is a well ordered set.

Since (1) is well ordered, each $A \in \mathcal{A}$ contains a least element. Thus, we can define function f

$$f : \mathcal{A} \to \bigcup_{A \in \mathcal{A}} A \tag{2}$$

by

$$f(A) = \text{least element of } A. \tag{3}$$

Function f is the choice function of the axiom of choice.

Show that the following statements are equivalent:

1. The axiom of choice

2. Hausdorff maximality principle

3. Zorn's lemma

4. Kuratowski's lemma

5. The well ordering principle.

SOLUTION:

There is nothing left to prove. We will merely summarize the results of this chapter.

In Problem 7-31, we showed that

$$\text{(Axiom of Choice)} \Rightarrow \text{(Hausdorff Maximality Principle)}.$$

In Problem 7-33 we proved

$$\text{(Hausdorff Maximality Principle)} \Rightarrow \text{(Zorn's Lemma)}.$$

In Problem 7-41 we proved

$$\text{(Zorn's Lemma)} \Rightarrow \text{(Well Ordering Principle)}.$$

To make the circle complete, we proved in Problem 7-42

$$\text{(Well Ordering Principle)} \Rightarrow \text{(Axiom of Choice)}.$$

Hence

$$\text{(Axiom of Choice)} \equiv \text{(Zorn's Lemma)} \equiv$$

$$\equiv \text{(Well Ordering Principle)} \equiv \text{(Hausdorff Maximality Principle)} \equiv$$

$$\equiv \text{(Kuratowski's Lemma)}.$$

For Kuratowski's lemma, see Problem 7-37.

Note that all these propositions are purely existential. They state the fact that "something" exists but do not offer the means of finding "it." For example, the well ordering principle states that the set of real numbers can be well ordered. So far it is not known *how* the set of real numbers can be well ordered.

Give an example of a segment of $(N, \le)$, where $\le$ is the usual "less than or equal."

SOLUTION:

DEFINITION OF A SEGMENT

A segment of a totally ordered set $(A, \le)$ is a subset S of A such that, if

$$x \in S, \text{ and } y \in A, \text{ and } y \le x,$$

then $y \in S$. A proper segment of A is a segment which is a proper subset of A.

Set $\{1\}$ is a segment of N, where $N = \{1, 2, 3, 4, ...\}$ is the set of natural numbers.

Each set of the form

$$\{1\}, \{1, 2\}, \{1, 2, 3\}, \{1, 2, 3, 4\}, ...$$

is a segment.

Also the set of natural numbers is a segment.

● **PROBLEM 7-45**

Show that any union or intersection of segments is a segment.

SOLUTION:

Let $(A, \le)$ represent a totally ordered set, and let $\mathcal{F}$ represent a family of segments of A.

Suppose

$$x \in \bigcup_{S \in \mathcal{F}} S \tag{1}$$

and $y \in A$ where

$$y \le x. \tag{2}$$

Element x belongs to some segment, say S_0 of A, $x \in S_0$. Since S_0 is a segment and $x \in S_0$, $y \in A$, $y \le x$, we have

$$y \in S_0. \tag{3}$$

Hence,

$$y \in \bigcup_{S \in \mathcal{F}} S \qquad (4)$$

and $\bigcup_{S \in \mathcal{F}} S$ is a segment of A.

Suppose

$$x \in \bigcap_{S \in \mathcal{F}} S \qquad (5)$$

and $y \in A$, and $y \le x$. Then for every $S \in \mathcal{F}$, $x \in S$ and since S is a segment, $y \in S$.

Hence, $y \in \bigcap_{S \in \mathcal{F}} S$.

Therefore, $\bigcap_{S \in \mathcal{F}} S$ is a segment.

● **PROBLEM 7–46**

Prove that if $(A, \le)$ is a well ordered set, then:

1. All segments of a segment are again segments.

2. For each segment S of A, $S \ne A$, an element $x \in A$ exists, such that $S = A_x$, where

$$A_x = \{a \in A : a < x\}. \qquad (1)$$

We use $a < x$ to indicate $a \le x$ and $a \ne x$.

SOLUTION:

1. Let S represent a segment of A and P represent a segment of S. Suppose $x \in P$ and $y \in A$, and

$$y \le x. \qquad (2)$$

Since x belongs to the segment S, we have

$$y \in S.$$

Now from $x \in P$, $y \in S$ and $y \le x$ we conclude that $y \in P$ because P is a segment of S. Thus, P is a segment of A.

2. Let S represent a segment of A, such that $S \ne A$. Then set $A - S$ is non-empty. This set has a least element, which we will denote by x.

Thus, as can be easily verified

$$A_x = S \quad \text{where} \quad A_x = \{a \in A : a < x\}. \qquad (3)$$

Prove that if $(A, \leq)$ is a well ordered set and $\mathcal{F}$ a family of segments of A, such that

1. any union of elements of $\mathcal{F}$ belongs to $\mathcal{F}$.

2. if $A_x \in \mathcal{F}$, then $A_x \cup \{x\} \in \mathcal{F}$,

$$A_x = \{a \in A : a < x\}$$

then $\mathcal{F}$ contains all segments of A.

SOLUTION:

Suppose $x \in A$ exists, such that $A_x \notin \mathcal{F}$. Then, non-empty set
$$B = \{x \in A : A_x \notin \mathcal{F}\} \subset A \tag{1}$$
has a least element b, because $(A, \leq)$ is well ordered. We have
$$b \in B \quad \text{and} \quad A_b \notin \mathcal{F}. \tag{2}$$
If $y \in A$ and $y < b$, then
$$y \notin B \text{ and hence } A_y \notin \mathcal{F}.$$
By number 1, of the hypothesis,
$$\bigcup_{y<b} A_y \in \mathcal{F} \tag{3}$$
An element $a \in A$ exists (see Problem 7-46), such that
$$\bigcup_{y<b} A_y = A_a \in \mathcal{F} \tag{4}$$
hence $a < b$ and
$$A_a \cup \{a\} = A_c \text{ for some } c \in A. \tag{5}$$
Thus
$$a < c < b \quad \text{and}$$
$$a \in A_c \subset \bigcup_{y<b} A_y = A_a. \tag{6}$$
But $a \notin A_a$, hence a contradiction. Thus
$$\text{for all } x \in A, A_x \in \mathcal{F}. \tag{7}$$
Now, we shall prove that $A \in \mathcal{F}$. Indeed
$$\bigcup_{x \in A} A_x \in \mathcal{F} \tag{8}$$

if

$$A = \bigcup_{x \in A} A_x \,;$$

(9)

then everything is all right. Suppose

$$A \neq \bigcup_{x \in A} A_x.$$

(10)

Then $y \in A$ exists, such that

$$\bigcup_{x \in A} A_x = A_y.$$

(11)

Furthermore

$$A_y \cup \{y\} \in \mathcal{F}.$$

(12)

Then

$$x \leq y \quad \text{for all} \quad x \in A.$$

Thus

$$A = A_y \cup \{y\} \in \mathcal{F}.$$

(13)

● **PROBLEM 7–48**

Prove the following important principle:

PRINCIPLE OF TRANSFINITE INDUCTION

Let $(A, \leq)$ denote a well ordered set and let $p(x)$ represent a sentence for each $x \in A$.

If, for each $x \in A$,

$$(p(y) \equiv 1 \ \text{for every} \ y < x) \ \Rightarrow \ (p(x) \equiv 1)$$

(1)

then for every $x \in A$, $p(x) \equiv 1$.

Note that $p(x) \equiv 1$ denotes that the sentence $p(x)$ is true and $p(x) \equiv 0$ denotes that the sentence $p(x)$ is false.

SOLUTION:

Suppose $x \in A$ exists, such that

$$p(x) \equiv 0.$$

(2)

Let B denote a non-empty subset of A, such that

$$B = \{x \in A : p(x) \equiv 0\},$$

(3)

and let b denote the least element of B. Remember that $(A, \leq)$ is well ordered.

Since $b \in B$, $p(b) \equiv 0$. Suppose $y \in A$, such that $y < b$; then $y \notin B$ and $p(y) \equiv 1$. Hence for every $y < b$

$$p(y) \equiv 1. \tag{4}$$

From (1) and (4), we conclude that

$$p(b) \equiv 1$$

which is a contradiction.

Thus

$$p(x) \equiv 1 \tag{5}$$

for every $x \in A$.

● **PROBLEM 7–49**

Apply the principle of transfinite induction to prove the following:

THEOREM

Let $(A, \leq)$ and $(A_0, \leq_0)$ denote well ordered sets. If $f : A \to A_0$ is increasing, $f(A)$ is a segment of A_0 and $F : A \to A_0$ is strictly increasing, then

$$\text{for all } x \in A, \quad f(x) \leq_0 F(x). \tag{1}$$

SOLUTION:

We shall start with

DEFINITION OF INCREASING FUNCTION

Let $(A, \leq)$ and $(A_0, \leq_0)$ denote well ordered sets. A function $f : A \to A_0$ is increasing iff

$$\begin{pmatrix} a \leq b \\ a, b \in A \end{pmatrix} \Rightarrow \begin{pmatrix} f(a) \leq_0 f(b) \\ f(a), f(b) \in A_0 \end{pmatrix}.$$

Function $f : A \to A_0$ is strictly increasing iff

$$\begin{pmatrix} a < b \\ a, b \in A \end{pmatrix} \Rightarrow \begin{pmatrix} f(a) < f(b) \\ f(a), f(b) \in A_0 \end{pmatrix}.$$

Let $p(x)$ represent the sentence

$$f(x) \leq_0 F(x).$$

217

Suppose an element $a \in A$ exists, such that for all $x < a$, $p(x) \equiv 1$ but $p(a) \equiv 0$. That is, for all $x < a$,

$$f(x) \leq_0 F(x) \quad \text{and} \quad F(a) <_0 f(a). \tag{2}$$

Function f is increasing and F is strictly increasing, thus for all $x < a$ and for all $a \leq y$

$$f(x) \leq_0 F(x) <_0 F(a) <_0 f(a) \leq_0 f(y). \tag{3}$$

Hence

$$F(a) <_0 f(a) \quad \text{and} \quad F(a) \notin f(A),$$

which is a contradiction because $f(A)$ is a segment of A_0. Hence, for each $x \in A$

$$\text{if } f(y) \leq_0 F(y) \text{ for all } y < x, \text{ then } f(x) \leq_0 F(x).$$

Now we can apply the principle of transfinite induction:

$$f(x) \leq_0 F(x) \text{ for all } x \in A.$$

CHAPTER 8

ORDINAL NUMBERS AND ORDINAL ARITHMETIC

Show that if the functions

$$f : A \rightarrow B$$

$$g : B \rightarrow C \tag{1}$$

are order isomorphic, then f^{-1} and $g \circ f$ are also order isomorphic.

SOLUTION:

DEFINITION OF ORDER ISOMORPHISM

Two well ordered sets $(A, \leq)$ and $(B, \leq_0)$ are order isomorphic if a bijection $f : A \rightarrow B$ exists, such that

$$\begin{pmatrix} a_1, a_2 \in A \\ a_1 \leq a_2 \end{pmatrix} \Rightarrow (f(a_1) \leq_0 f(a_2)). \tag{2}$$

Function $f : A \rightarrow B$ is called an order isomorphism.

∎

If $f : A \rightarrow B$ is an order isomorphism, then so is $f^{-1} : B \rightarrow A$. Indeed, both sets A and B are well ordered. If f is a bijection, then f^{-1} is a bijection. Furthermore, if f has property (2), so does f^{-1}.

If functions f and g of (1) are bijections, then $g \circ f : A \rightarrow C$ is a bijection. Also if f and g have property (2), then $g \circ f$ has property (2).

Let $(A, \leq)$ and $(B, \leq_0)$ represent well ordered sets each consisting of the same finite number of elements. Show that $(A, \leq)$ and $(B, \leq_0)$ are order-isomorphic.

SOLUTION:

Sets $(A, \leq)$ and $(B, \leq_0)$ are equivalent, finite, and well ordered. Hence, their elements can be arranged as follows:

$$A = \{\alpha_1, \alpha_2, ..., \alpha_n\}$$

$$B = \{\beta_1, \beta_2, ..., \beta_n\} \tag{1}$$

The arrangement $\alpha_1, \alpha_2, ..., \alpha_n$ is such that we denote the least element of the set A by α_1; then we denote the least element of $A - \{\alpha_1\}$ by α_2, etc. The same procedure is repeated for the set B.

Bijection

$$f : A \to B$$

is defined by

$$f(\alpha_k) = \beta_k$$

for $k = 1, ..., n$.

Function f is an order isomorphism because if $\alpha_l \le \alpha_k$, then $f(\alpha_l) \le f(\alpha_k)$, where $\alpha_l \le \alpha_k$ indicates that α_k appears after α_l in the sequence $\alpha_1, ..., \alpha_n$. We shall write

$$(A, \le) \approx (B, \le_0) \quad \text{or} \quad A \approx B$$

to indicate that the sets $(A, \le)$ and $(B, \le_0)$ are order isomorphic.

● **PROBLEM 8-3**

The set of natural numbers N is well ordered as follows.

$$(N, \le) = \{1, 2, 3, 4, ...\} \tag{1}$$

and as

$$(N, \le_0) = (1, 3, 5, ..., 2, 4, 6, ...). \tag{2}$$

Show that both sets are not order isomorphic.

SOLUTION:

It is easy to verify that both sets are well ordered. Suppose

$$f : (N, \le) \to (N, \le_0) \tag{3}$$

exists, such that f is order isomorphic. Then, to ensure that

$$\left(\begin{matrix} a_1, a_2 \in (N, \le) \\ a_1 \le a_2 \end{matrix} \right) \Rightarrow (f(a_1) \le f(a_2))$$

we must have

$$f(1) = 1, f(2) = 3, f(3) = 5, ...$$

or in general

$$f(n) = 2n - 1. \tag{4}$$

Function (4) maps set N onto the set of odd numbers. Hence, (3) is not a bijection.

Thus, $(N, \le)$ and $(N, \le_0)$ are not order isomorphic.

Describe the concept of ordinal numbers.

SOLUTION:

1. To each well ordered set $(A, \leq)$, an ordinal number denoted by $\text{ord}(A, \leq)$, is assigned. Also, if α is an ordinal number, then a well ordered set $(A, \leq)$ exists, such that

$$\alpha = \text{ord}(A, \leq). \qquad (1)$$

2. $\text{ord}(A, \leq) = \text{ord}(B, \leq_0)$ iff $(A, \leq) \approx (B, \leq_0)$

i.e., iff the sets $(A, \leq)$ and $(B, \leq_0)$ are order isomorphic.

3. $\text{ord}(A, \leq) = 0$ iff $A = \phi$.

4. If $(A, \leq)$ is a well ordered set consisting of n elements, then

$$\text{ord}(A, \leq) = n.$$

The ordinal number of the set N with the usual 1, 2, 3, ... order is denoted by ω. Thus

$$\text{ord}(N, \leq) = \omega.$$

A given set can have only one cardinal number, but depending on its order set, it can have many different ordinal numbers (for example see Problem 8-3).

Show that if $(A, \leq)$ is a well ordered set, then the only order isomorphism of $(A, \leq)$ onto a segment of $(A, \leq)$ is the identity function of A onto A.

SOLUTION:

Suppose, on the contrary, that an order isomorphism exists

$$f : A \rightarrow A_a \qquad (1)$$

where A_a is a segment

$$A_a = \{x \in A : x < a\} \tag{2}$$

then $f(a) \in A_a$ and $f(a) < a$.

Let

$$B = \{x \in A : f(x) < x\} \neq \phi \tag{3}$$

and let b denote the least element of B. Then, since $b \in B$

$$f(b) < b \tag{4}$$

hence $f(b) \in B$, which is a contradiction because $f(b) < b$ and b is the least element of B. We conclude that a well ordered set cannot be order isomorphic to any of its proper segments. Now, suppose

$$f : A \rightarrow A \tag{5}$$

is order isomorphism.

Both functions f and the identity function $1_A : A \rightarrow A$ are strictly increasing. Hence,

$$1_A(x) \leq f(x) \leq 1_A(x) \tag{6}$$

for all $x \in A$ and $f = 1_A$.

● **PROBLEM 8-6**

Let α and β denote two ordinal numbers. Show that the relation $\alpha \leq \beta$ (α is less than or equal to β) is reflexive and transitive.

SOLUTION:

We shall begin with the following:

DEFINITION

Let α and β denote ordinal numbers and let $(A, \leq)$ and $(B, \leq_0)$ denote well ordered sets, such that

$$\alpha = \mathrm{ord}(A, \leq), \quad \beta = \mathrm{ord}(B, \leq_0). \tag{1}$$

Number α is less than or equal to β we denote it by $\alpha \leq \beta$, if and only if $(A, \leq)$ is order isomorphic to a segment of $(B, \leq_0)$. If $\alpha \leq \beta$ and $\alpha \neq \beta$, then we write $\alpha < \beta$.

■

Relation is reflexive if for every ordinal number α, $\alpha \leq \alpha$. Let $(A, \leq)$ represent a well ordered set, such that

$$\alpha = \mathrm{ord}(A, \leq). \tag{2}$$

Set A is a segment of $(A, \le)$. Identity function on A is an order isomorphism. Hence, $\alpha \le \alpha$.

To prove that relation $\le$ is transitive, we must show that for any ordinal numbers

$$(\alpha \le \beta, \beta \le \rho) \Rightarrow (\alpha \le \rho). \tag{3}$$

Three well ordered sets exist, such that

$$\alpha = \text{ord}(A, \le), \ \beta = \text{ord}(B, \le_1), \ \rho = \text{ord}(C, \le_2).$$

If $(A, \le)$ is order isomorphic to a segment of $(B, \le_1)$ and $(B, \le_1)$ is order isomorphic to a segment of $(C, \le_2)$, then $(A, \le)$ is order isomorphic to a segment of $(C, \le_2)$.

● **PROBLEM 8-7**

Show that the relation $\le$ for the ordinal numbers is a partial order relation.

SOLUTION:

A partial order relation is reflexive, transitive, and antisymmetric. We proved in Problem 8-5, that $\le$ is reflexive and transitive. Now, we shall prove

THEOREM

$$\left(\begin{array}{c} \alpha, \beta \text{ are ordinal numbers} \\ \alpha \le \beta \wedge \beta \le \alpha \end{array} \right) \Rightarrow (\alpha = \beta). \tag{1}$$

■

Let $(A, \le)$ and $(B, \le_0)$ represent well ordered sets, such that

$$\alpha = \text{ord}(A, \le), \ \beta = \text{ord}(B \le_0). \tag{2}$$

Since $\alpha \le \beta$ and $\beta \le \alpha$, two order isomorphisms exist, such that

$$f : A \to B_b \quad g : B \to A_a \tag{3}$$

where A_a and B_b are segments of A and B respectively. Function

$$h : A \to A_c \tag{4}$$

defined by

$$h(x) = g(f(x)), \text{ for all } x \in A \tag{5}$$

is an order isomorphism of A to a segment A_c of A_a. Since A_a is a segment of A, so is A_c. By Problem 8-5, $A_c = A$.

Thus, $A_a = A$ and since g is the order isomorphism $g : B \rightarrow A$, we conclude that

$$\alpha = \beta. \tag{6}$$

In Problem 8-3 we proved that

$$\mathrm{ord}(N, \leq) \neq \mathrm{ord}(N, \leq_0) \tag{1}$$

where

$$(N, \leq) = \{1, 2, 3, 4, \ldots\}$$

$$(N, \leq_0) = \{1, 3, 5, \ldots, 2, 4, 6, \ldots\}. \tag{2}$$

Show that

$$\mathrm{ord}(N, \leq) < \mathrm{ord}(N, \leq_0). \tag{3}$$

SOLUTION:

Consider the function

$$f(n) = 2n - 1$$

$$f : (N, \leq) \rightarrow (N, \leq_0). \tag{4}$$

Let A denote the segment of $(N, \leq_0)$

$$A = \{1, 3, 5, 7, \ldots\}$$

Function $f(n) = 2n - 1$

$$f : (N, \leq) \rightarrow A \tag{5}$$

is a bijection, which preserves order. Thus, f defined by (5) is an order isomorphism. Therefore,

$$\mathrm{ord}(N, \leq) \leq \mathrm{ord}(N, \leq_0). \tag{6}$$

From (6) and (1), we obtain

$$\mathrm{ord}(N, \leq) < \mathrm{ord}(N, \leq_0). \tag{7}$$

Is the relation ≤, defined on ordinal numbers, a total order relation?

SOLUTION:

In Problem 8-7 we proved that relation ≤ is reflexive, transitive, and antisymmetric. That is, ≤ is a partial order.

We shall quote the following without a proof.

THEOREM

For any two cardinal numbers α and β, either

$$\alpha \leq \beta \quad \text{or} \quad \beta \leq \alpha.$$

∎

Thus, for any two ordinal numbers, their order is determined.

We conclude that ≤ is a total order and ordinal numbers are totally ordered.

The above results can be summarized in the following theorem.

THEOREM

If α and β are ordinal numbers, then one and only one of the following is true:

1. $\alpha < \beta$

2. $\alpha = \beta$

3. $\beta < \alpha$.

∎

Let $(A, \leq)$ and $(B, \leq_0)$ represent well ordered sets. Prove that

$$\left(\text{ord}(A,\leq) < \text{ord}(B,\leq_0)\right) \Leftrightarrow \left(\begin{array}{l} A \text{ is order isomorphic} \\ \text{to a proper segment of } B \end{array}\right).$$

SOLUTION:

⇐ If A is order isomorphic to a segment of B, then

$$\text{ord}(A, \leq) \leq \text{ord}(B, \leq_0). \tag{1}$$

But A is order isomorphic to a proper segment of B. From (1) and the theorem of Problem 8-5, we conclude that

$$\text{ord}(A, \leq) < \text{ord}(B, \leq_0). \tag{2}$$

$\Rightarrow$ Condition (2) can be written as

$$\text{ord}(A, \leq) \leq \text{ord}(B, \leq_0) \tag{3}$$

and

$$\text{ord}(A, \leq) \neq \text{ord}(B, \leq_0). \tag{4}$$

From (3), we conclude that A must be order isomorphic to a segment of B; and from (4) and theorem of Problem 8-5, we conclude that A is order isomorphic to a proper segment of B.

● PROBLEM 8-11

Show that

$$\text{ord}\ \{k, k + 1, k + 2, \ldots\} = \text{ord}\ \{1, 2, \ldots\ \} \tag{1}$$

where k is any natural number.

SOLUTION:

Let us define function

$$f : \{1, 2, 3, \ldots\} \rightarrow \{k, k + 1, k + 2, \ldots\} \tag{2}$$

by

$$f(n) = n + k - 1 \quad \text{for}\ \ n = 1, 2, 3, \ldots \tag{3}$$

Function f is a bijection and such that if $n_1 \leq n_2$ where $n_1, n_2 \in \{1, 2, 3, \ldots\}$, then

$$f(n_1) \leq f(n_2)\ \text{where}\ f(n_1), f(n_2) \in \{k, k + 1, \ldots\}.$$

Thus, f is an order isomorphism and

$$\{k, k + 1, k + 2, \ldots\} \approx \{1, 2, 3, 4, \ \ldots\ \} \tag{4}$$

which proves (1).

Prove that

$$\text{ord } \{1, 2, 3, \ldots\} < \text{ord } \{k, k+1, k+2, \ldots, 1, 2, 3, \ldots, k-1\}. \quad (1)$$

SOLUTION:

In other words, we have to prove that $\{1, 2, 3, \ldots\}$ is order isomorphic to a proper segment of $\{k, k+1, \ldots, 1, 2, 3, \ldots, k-1\}$.

The subset

$$\{k, k+1, k+2, \ldots\} \quad (2)$$

is a proper segment of

$$\{k, k+1, k+2, \ldots, 1, 2, \ldots, k-1\}. \quad (3)$$

In Problem 8-11, we proved that (2) is order isomorphic to

$$\{1, 2, 3, 4, \ldots\}.$$

That completes proof of (1).

● **PROBLEM 8-13**

What is the smallest transfinite ordinal number?

SOLUTION:

The ordinal number of the set consisting of n elements is n, assuming the set is well ordered. The infinite set with the lowest cardinal number is the set of natural numbers or any set equivalent to it.

A given set has only one cardinal number, but a set may have different ordinal numbers under distinct well orderings. The ordinal number of the set of natural numbers ordered as

$$\{1, 2, 3, 4, \ldots\}$$

is denoted by ω. Hence,

$$\omega = \text{ord } (N, \leq).$$

Any other ordering of N leads to the same or larger ordinal number. See, for example, Problem 8-12.

Hence, the smallest transfinite ordinal number is ω.

Find the ordinal sum

$$3 + 4$$

of the two finite ordinal numbers 3 and 4.

SOLUTION:

DEFINITION OF SUM

The ordinal sum $\alpha + \beta$ of ordinal numbers α and β is the ordinal number of the set $(A \cup B, \leq')$, where $(A, \leq)$ and $(B, \leq_0)$ are disjoint well ordered sets, such that

$$\alpha = \mathrm{ord}(A, \leq) \quad \text{and} \quad \beta = \mathrm{ord}(B, \leq_0).$$

Relation $\leq'$ is defined by

1. If $a, b \in A$ (or $a, b \in B$), then we write $a \leq' b$ iff $a \leq b$ (or $a \leq_0 b$).

2. If $a \in A$ and $b \in B$, we write $a \leq' b$.

Since

$$3 = \mathrm{ord}\{1, 2, 3\}$$

and

$$4 = \mathrm{ord}\ \{4, 5, 6, 7\}$$

we have

$$3 + 4 = \mathrm{ord}\ \{1, 2, 3, 4, 5, 6, 7\} = 7.$$

Consider the definition of the sum of ordinal numbers given in Problem 8-14. Explain why the sets

$$(A \cup B, \leq') \text{ and } (B \cup A, \leq')$$

should be considered distinct.

SOLUTION:

To find the sum of ordinal numbers α and β, we take two well ordered disjoint sets A and B, such that

$$\alpha = \text{ord}(A, \leq), \quad \beta = \text{ord}(B, \leq_0).$$

If the sets are not disjoint we can replace them by $A \times \{x\}$ and $B \times \{y\}$. Then

$$A \approx A \times \{x\} \quad \text{and} \quad B \approx B \times \{y\}.$$

In the next step, we form the sum of the sets A and B and define well ordering, $(A \cup B, \leq')$, such that:

$$\text{if } a \in A \quad \text{and} \quad b \in B, \text{ we write } a \leq' b.$$

Hence, any element $a \in A$ is less than any element $b \in B$. While, for the sum $(B \cup A, \leq')$, any element $b \in B$ is less than any element $a \in A$, $b \leq' a$.

To avoid this whole problem, one can use different notation, such as $(A \cup B, \leq')_{A,B}$ meaning "A is less than B." Then

$$(A \cup B, \leq')_{A,B} = (B \cup A, \leq')_{A,B}.$$

● **PROBLEM 8-16**

Let $k \neq 0$ denote any finite ordinal number. Show that

$$k + \omega = \omega \tag{1}$$

$$\omega + k \neq \omega. \tag{2}$$

SOLUTION:

We have

$$k = \text{ord}\{1, 2, \ldots, k\} \tag{3}$$

$$\omega = \text{ord}\{k + 1, k + 2, \ldots\} \tag{4}$$

(see Problem 8-11.) Thus,

$$k + \omega = \text{ord}\{1, 2, \ldots, k, k + 1, \ldots\} = \omega \tag{5}$$

Similarly, we obtain

$$\omega + k = \text{ord}\{k + 1, k + 2, \ldots, 1, 2, 3, \ldots\} > \omega \tag{6}$$

(see Problem 8-12.) Thus,

$$\omega + k \neq \omega. \tag{7}$$

In general, addition of ordinal numbers is not communicative.
It is rather amazing that

$$\omega + k \neq k + \omega. \tag{8}$$

230

● **PROBLEM 8-17**

Let $(A1, \leq_1)$, $(A_2, \leq_2)$, $(B_1, \leq_1')$, and $(B_2, \leq_2')$ represent well ordered sets, such that

$$A_1 \cap B_1 = A_2 \cap B_2 = \phi \text{ and}$$

$$A_1 \approx A_2, \quad B_1 \approx B_2. \tag{1}$$

Prove that

$$(A_1 \cup B_1, \leq_1^*) \approx (A_2 \cup B_2, \leq_2^*). \tag{2}$$

SOLUTION:

Since $A_1 \approx A_2$ and $B_1 \approx B_2$, we have: bijection $f: A_1 \to A_2$ exists, such that

$$(a_1 \leq_1 a_2) \Rightarrow (f(a_1) \leq_2 f(a_2)) \tag{3}$$

and bijection $g: B_1 \to B_2$ exists, such that

$$(b_1 \leq_1' b_2) \Rightarrow (g(b_1) \leq_2' g(b_2)). \tag{4}$$

Functions f and g are order isomorphisms. We define

$$F: A_1 \cup B_1 \to A_2 \cup B_2 \tag{5}$$

by

$$F(x) = \begin{cases} f(x) \text{ if } x \in A_1 \\ g(x) \text{ if } x \in B_1. \end{cases} \tag{6}$$

Remember that $A_1 \cap B_1 = A_2 \cap B_2 = \phi$. It is easy to verify that F is a bijection. We define well orderings $\leq_1^*$ and $\leq_2^*$ according to the definition of Problem 8-14.

$$x_1, x_2 \in A_1, \quad f(x_1) = F(x_1)$$
$$f(x_2) = F(x_2)$$
or ↓
if $x_1 \leq_1^* x_2$, then $\quad x_1, x_2 \in B_1, \quad g(x_1) = F(x_1) \quad - \quad F(x_1) \leq_2^* F(x_2)$
$$g(x_2) = F(x_2)$$
or ↓
$$x_1 \in A_1, x_2 \in B_1$$
$$F(x_1) = f(x_1) \in A_2$$
$$F(x_2) = g(x_2) \in B_2 \tag{7}$$

Thus, F defined in (4) and (5) is an order isomorphism and

$$(A_1 \cup B_1, \leq_1^*) \approx (A_2 \cup B_2, \leq_2^*). \tag{8}$$

231

Prove that ordinal sum is associative; that is, for any ordinals α, β, and ρ

$$(\alpha + \beta) + \rho = \alpha + (\beta + \rho). \tag{1}$$

SOLUTION:

It is possible to find three pair-wise disjoint sets A, B, and C, such that

$$\alpha = \text{ord}(A, \leq_1)$$

$$\beta = \text{ord}(B, \leq_2)$$

$$\rho = \text{ord}(C, \leq_3). \tag{2}$$

According to the definition of Problem 8-14, we obtain

$$\alpha + \beta = \text{ord}(A \cup B, \leq_1{}^*) \tag{3}$$

and

$$\beta + \rho = \text{ord}(B \cup C, \leq_2{}^*) \tag{4}$$

where well ordering $\leq_1{}^*$ means "A is less than B" and well ordering $\leq_2{}^*$ means "B is less than C." We obtain

$$(\alpha + \beta) + \rho = \text{ord}(A \cup B \cup C, \leq_3{}^*)$$

$$\alpha + (\beta + \rho) = \text{ord}(A \cup B \cup C, \leq_4{}^*) \tag{5}$$

but $\leq_3{}^*$ and $\leq_4{}^*$ are exactly the same orders; i.e., "A is less than B" and "B is less than C." Thus,

$$(\alpha + \beta) + \rho = \alpha + (\beta + \rho) = \alpha + \beta + \rho =$$

$$= \text{ord}(A \cup B \cup C, \leq^*). \tag{6}$$

If $x_1, x_2 \in A$, we write $x_1 \leq^* x_2$ iff $x_1 \leq_1 x_2$.

If $x_1, x_2 \in B$, we write $x_1 \leq^* x_2$ iff $x_1 \leq_2 x_2$.

If $x_1, x_2 \in C$, we write $x_1 \leq^* x_2$ iff $x_1 \leq_3 x_2$.

We write $x_1 \leq^* x_2$ if $x_1 \in A$ and $x_2 \in B \cup C$, or if $x_1 \in B$ and $x_2 \in C$.

Prove that

$$(\beta < \rho) \Rightarrow (\alpha + \beta < \alpha + \rho) \qquad (1)$$

for any ordinal numbers α, β, and ρ.

SOLUTION:

Let $(A, \leq)$, $(B, \leq_1)$, and $(C, \leq_2)$ represent well ordered sets, such that

$$\alpha = \mathrm{ord}(A, \leq)$$

$$\beta = \mathrm{ord}(B, \leq_1)$$

$$\rho = \mathrm{ord}(C, \leq_2) \qquad (2)$$

and

$$A \cap B = \phi$$

$$A \cap C = \phi. \qquad (3)$$

Since $\beta < \rho$, the set $(B, \leq_1)$ is order isomorphic to some segment of $(C, \leq_2)$, which we will denote by C_t. Hence

$$f : B \to C_t. \qquad (4)$$

We form the sets $(A \cup B, \leq_1{}^*)$ and $(A \cup C, \leq_2{}^*)$ with the orders defined in Problem 8-14. Then

$$F : A \cup B \to A \cup C_t \qquad (5)$$

where

$$F(x) = \begin{cases} x & \text{if } x \in A \\ f(x) & \text{if } x \in B \end{cases} \qquad (6)$$

is an order isomorphism to $A \cup B$ onto a proper segment $A \cup C_t$ of $A \cup C$. Therefore,

$$\alpha + \beta < \alpha + \rho. \qquad (7)$$

Prove that for any ordinal numbers α, β, and ρ

$$(\alpha + \beta = \alpha + \rho) \Rightarrow (\beta = \rho). \qquad (1)$$

SOLUTION:

r.a.a. (reductio ad absurdum)
Suppose ordinal numbers exist, such that

$$\alpha + \beta = \alpha + \rho \quad \text{and} \quad \beta \neq \rho.$$

When $\beta \neq \rho$, then either $\beta < \rho$ or $\rho < \beta$.
Suppose $\beta < \rho$. From Problem 8-19, we have

$$(\beta < \rho) \Rightarrow (\alpha + \beta < \alpha + \rho), \tag{2}$$

which is a contradiction.

Rule (1) is called left cancellation. In Problem 8-21, we will show that right cancellation is not true for ordinal numbers.

● **PROBLEM 8-21**

Give a counter-example to show that the right cancellation rule is not true for ordinal numbers.

SOLUTION:

We have to find three ordinal numbers, such that

$$\beta + \alpha = \rho + \alpha \quad \text{and} \quad \beta \neq \rho.$$

In Problem 8-16, we proved

$$n + \omega = \omega. \tag{1}$$

Hence,

$$1 + \omega = \omega$$

and

$$2 + \omega = \omega. \tag{2}$$

We obtain

$$1 + \omega = 2 + \omega \quad \text{but} \quad 1 \neq 2.$$

● **PROBLEM 8-22**

Prove that for any ordinal numbers α, β, and ρ

$$(\beta < \rho) \Rightarrow (\beta + \alpha \leq \rho + \alpha). \tag{1}$$

SOLUTION:

Let $(A, \leq)$, $(B, \leq')$, and $(C, \leq'')$ represent well ordered sets, such that

$$\alpha = \text{ord}(A, \leq), \quad \beta = \text{ord}(B, \leq'), \quad \rho = \text{ord}(C, \leq'') \qquad (2)$$

and

$$B \cap A = \phi \quad C \cap A = \phi. \qquad (3)$$

Since $\beta < \rho$, the set B is order isomorphic to some proper segment C' of C. Hence,

$$f : B \to C' \qquad (4)$$

is an order isomorphism.

We shall use the reductio ad absurdum method.

Suppose that

$$\beta + \alpha > \rho + \alpha. \qquad (5)$$

Hence, $C \cup A$ is order isomorphic to some proper segment $(B \cup A)'$ of $B \cup A$. Hence,

$$g : C \cup A \to (B \cup A)' \qquad (6)$$

is order-isomorphic, which leads to contradiction.

● PROBLEM 8–23

Show that relation $\leq$ cannot be replaced by $<$ in Problem 8-22.

SOLUTION:

We have to find three ordinal numbers α, β, and ρ, such that

$$\beta < \rho \qquad (1)$$

and

$$\beta + \alpha < \rho + \alpha. \qquad (2)$$

We shall use the helpful formula,

$$k + \omega = \omega. \qquad (3)$$

Let us substitute $\beta = 1$, $\rho = 2$, and $\alpha = \omega$ to obtain

$$1 < 2 \quad \text{and} \quad 1 + \omega = \omega = 2 + \omega. \qquad (4)$$

Hence, (2) is not true.

235

Prove that for any ordinal number α, $\alpha + 1$ is an ordinal number and

$$\alpha < \alpha + 1. \tag{1}$$

SOLUTION:

It is easy to show that for any ordinal number α,

$$\alpha + 0 = 0 + \alpha = \alpha. \tag{2}$$

Substituting $\beta = 0$ and $\rho = 1$ in equation (1) of Problem 8-22, we obtain

$$(0 < 1) \Rightarrow (\alpha + 0 < \alpha + 1). \tag{3}$$

From (2) and (3), we conclude that for any ordinal number α,

$$\alpha < \alpha + 1. \tag{4}$$

Show that the greatest ordinal number does not exist.

SOLUTION:

Suppose, on the contrary, that the greatest ordinal number exists, and denote it by δ. From Problem 8-24, we know that for any ordinal number

$$\alpha < \alpha + 1.$$

Hence,

$$\delta < \delta + 1.$$

Therefore, $\delta + 1$ is an ordinal number which is larger than δ.
Contradiction. Conclusion: the greatest ordinal number does not exist.

Let α and β denote ordinal numbers, such that $\alpha \leq \beta$. Prove that a unique ordinal number ρ exists, such that

$$\alpha + \rho = \beta. \tag{1}$$

Number ρ is sometimes denoted by $(-\alpha) + \beta$.

SOLUTION:

First we shall show that if ρ exists, then it is unique. Suppose, on the contrary, that ρ and ρ' exist, such that

$$\alpha + \rho = \beta \quad \text{and} \quad \alpha + \rho' = \beta. \tag{2}$$

Then

$$\alpha + \rho = \alpha + \rho' \tag{3}$$

and

$$\rho = \rho'. \tag{4}$$

If $\alpha = \beta$, then $\rho = 0$. Suppose $\alpha < \beta$; then sets A, B, B_t exist, such that $A \cap B = \phi$ and

$$\alpha = \text{ord}(A, \leq), \quad \beta = \text{ord}(B, \leq_1)$$

and an order isomorphism exists

$$f : A \to B_t$$

where B_t is a proper segment of B.

Let

$$\rho = \text{ord}(B - B_t, \leq_1). \tag{5}$$

Then

$$\alpha + \rho = \beta \tag{6}$$

where $\alpha + \gamma = \text{ord}(A \cup (B - B_t), \leq^*)$.

Order isomorphism can be defined as follows:

$$F(x) = \begin{cases} f(x) & \text{if } x \in A \\ x & \text{if } x \in B - B_t \end{cases} \tag{7}$$

● PROBLEM 8-27

Prove that for any ordinal numbers α, β, and ρ

$$(\beta < \rho) \Leftrightarrow (\alpha + \beta < \alpha + \rho). \tag{1}$$

SOLUTION:

$\Rightarrow$ Demonstrated in Problem 8-19.

$\Leftarrow$ Since $\alpha + \beta < \alpha + \rho$, an ordinal number δ exists, such that (see Problem 8-26)

$$\alpha + \beta + \delta = \alpha + \rho. \tag{2}$$

By applying left cancellation, we obtain

$$\beta + \delta = \rho$$

which leads to

$$\beta < \rho \tag{3}$$

because $\delta \neq 0$.

● **PROBLEM 8–28**

Let

$$A = \{a, b, c, d\} \tag{1}$$

and

$$B = \{1, 2, 3\} \tag{2}$$

represent well ordered sets with orderings defined by:

for A $a \leq b \leq c \leq d$

and for B $1 \leq 2 \leq 3$.

Find any well ordering of the Cartesian product of A and B.

SOLUTION:

Let us denote the sets by

$$(A, \leq) \text{ and } (B, \leq_1).$$

We have to determine when

$$(a, b) \leq^* (c, d) \tag{3}$$

where $a, c \in A$ and $(b, d) \in B$.

The elements of $A \times B$ can be well ordered as follows:

$$(a, 1) \leq^* (a, 2) \leq^* (a, 3) \leq^* (b, 1) \leq^* (b, 2) \leq^*$$

$$\leq^* (b, 3) \leq^* (c, 1) \leq^* (c, 2) \leq^* (c, 3) \leq^* (d, 1) \leq^*$$

$$\leq^* (d, 2) \leq^* (d, 3).$$

It is obvious that the set $(A \times B, \leq^*)$ is well ordered.

There are many ways (exactly 12!) of well ordering the set $A \times B$. We have just chosen one possibility.

● **PROBLEM 8–29**

Let $(A, \leq)$ and $(B, \leq_1)$ denote well ordered sets. Define the lexicographic (dictionary type) ordering of $A \times B$.

SOLUTION:

We construct the Cartesian product of the sets A and B. The lexicographic ordering is defined as follows:

$$(a,b) \leq *(c,d) \text{ iff } \begin{cases} a < c \\ \text{or} \\ a = c \text{ and } b \leq_1 d \end{cases}$$

where $\qquad\qquad a, c \in A \quad b, d \in B.$ $\hfill(1)$

If both sets $(A, \leq)$ and $(B, \leq_1)$ are well ordered, then $(A \times B, \leq^*)$ is a well ordered set.

In a similar manner, we can impose the lexicographic ordering on the Cartesian product of n well ordered sets $(A_1, \leq_1), (A_2, \leq_2), \ldots, (A_n, \leq_n)$.

$$(x_1, x_2, \ldots x_n) \leq *(y_1, \ldots, y_n) \text{ iff } \begin{cases} x_1 <_1 y_1 \\ x_1 = y_1 \text{ and } x_2 <_2 y_2 \\ x_1 = y_1 \text{ and } x_2 = y_2 \text{ and } x_3 <_3 y_3 \\ \vdots \\ x_1 = y_1 \text{ and } \ldots x_{n-1} = y_{n-1} \text{ and } x_n <_n y_n \end{cases}$$

where $x_1, y_1, \in A_1, x_2, y_2 \in A_2, \ldots, x_n, y_n \in A_n.$

● **PROBLEM 8–30**

Prove the following:

THEOREM

If $(A, \leq)$ and $(B, \leq_1)$ are well ordered sets, then the set $(A \times B, \leq^*)$ where $\leq^*$ is the lexicographic ordering is a well ordered set. ∎

SOLUTION:

Since both sets A and B are totally ordered for any two pairs $(a,b) \in A \times B$ and $(c, d) \in A \times B$, we have either $(a, b) \leq^* (c, d)$ or $(c, d) \leq^* (a, b)$.

The set $(A \times B, \leq^*)$ is totally ordered.

To show that $(A \times B, \leq^*)$ is well ordered, we must show that every subset $\phi \neq T \subset A \times B$ contains the least element. Consider the following projection:

$$p_A(T) = \{x \in A : \text{for some } b \in B, (x, b) \in T\}. \tag{1}$$

The set $p_A(T)$ is a non-empty subset of a well ordered set $(A, \leq)$. Hence, it contains the least element; i.e., t_1. By the same token,

$$p_B(T) = \{y \in B : \text{for some } a \in A, (a, y) \in T\}$$

contains the least element; i.e., t_2. The element $(t_1, t_2) \in T$ is the least element of T.

Therefore, $(A \times B, \leq^*)$ is a well ordered set.

● **PROBLEM 8-31**

Prove the following:

THEOREM

Let $(A_1, \leq_1)$, $(A_2, \leq_2)$, $(B_1, \leq_1')$, and $(B_2, \leq_2')$ denote well ordered sets, such that

$$(A_1, \leq_1) \approx (A_2, \leq_2) \tag{1}$$

and

$$(B_1, \leq_1') \approx (B_2, \leq_2') \tag{2}$$

then

$$(A_1 \times B_1, \leq_1^*) \approx (A_2 \times B_2, \leq_2^*) \tag{3}$$

where $\leq_1^*$ and $\leq_2^*$ are lexicographic orderings.

■

SOLUTION:

Since $A_1 \approx A_2$ and $B_1 \approx B_2$, order isomorphisms exist, such that

$$f : A_1 \to A_2 \quad \text{and} \quad g : B_1 \to B_2. \tag{4}$$

We will show that the function

240

$$F : A_1 \times B_1 \rightarrow A_2 \times B_2 \qquad (5)$$

defined by

$$F(x, y) = (f(x), g(y)) \text{ for all } (x, y) \in A_1 \times B_1$$

is an order isomorphism.

Both functions f and g are bijections, thus F is a bijection. To complete the proof, we must show that

$$((a, b) \leq_1{}^* (c, d)) \Rightarrow (F(a, b) \leq_2{}^* F(c, d)) \qquad (6)$$

where $(a, b), (c, d) \in A_1 \times B_1$.

If

$$(a, b) \leq_1{}^* (c, d) \qquad (7)$$

then two possibilities exist:

1. $a <_1 c$ and hence, f is an order isomorphism

$$f(a) <_2 f(c). \qquad (8)$$

Then

$$F(a, b) = (f(a), g(b)) \leq_2{}^* (f(c), g(d)) = F(c, d). \qquad (9)$$

2. $a = c$ and $b \leq_1{}' d$. Functions f and g are order isomorphisms. Thus $f(a) = f(c)$ and $g(b) \leq_2{}' g(d)$, and

$$F(a, b) = (f(a), g(b)) \leq_2{}^* (f(a), g(d)) = F(c, d). \qquad (10)$$

Thus, (6) is proven and F is an order isomorphism.

● PROBLEM 8–32

Define the ordinal product $\alpha \beta$ of ordinal numbers α and β, and explain why the ordinal product is not commutative.

SOLUTION:

DEFINITION

Let α and β denote ordinal numbers, the ordinal product $\alpha \beta$ is defined by

$$\alpha \beta = \text{ord}(B \times A, \leq^*) \qquad (1)$$

where $(A, \leq)$ and $(B, \leq_1)$ are well ordered sets, such that

$$\alpha = \text{ord}(A, \leq) \qquad \beta = \text{ord}(B, \leq_1)$$

and $\leq^*$ is the lexicographic ordering of $B \times A$.

■

According to the definition above, the product $\alpha\,\beta$ is different from $\beta\,\alpha$. In general, the Cartesian product

$$A \times B \neq B \times A. \tag{2}$$

Furthermore, the lexicographic ordering of $A \times B$ is different from the lexicographic ordering of $B \times A$.

● **PROBLEM 8–33**

Show that

$$2\omega \neq \omega 2. \tag{1}$$

SOLUTION:

Let $(N, \leq) = \{1, 2, 3, \ldots\}$ and $(B, \leq_1) = \{a, b\}$, so that

$$\text{ord}(N, \leq) = \omega \quad \text{ord}(B, \leq_1) = 2 \tag{2}$$

Hence,

$$2\omega = \text{ord}(N \times B, \leq^*) \tag{3}$$

where $\leq^*$ is the lexicographic ordering.

Let us define the function

$$f : N \times B \rightarrow N \tag{4}$$

$$f(n, a) = 2n - 1$$

$$f(n, b) = 2n.$$

It is easy to see that f is an order isomorphism. Thus

$$2\omega = \omega. \tag{5}$$

Now we shall form the set $(B \times N, \leq^*)$ with the lexicographic ordering.

$$\text{ord}(B \times N, \leq^*) = \omega 2 \tag{6}$$

$$(B \times N, \leq^*) = \{(a, 1), (a, 2), \ldots, (b, 1), (b, 2), \ldots\} \tag{7}$$

We have

$$\text{ord}(B \times N, \leq^*) = \text{ord}((\{a\} \times N) \cup (\{b\} \times N), \leq^*) =$$

$$= \text{ord}(\{a\} \times N, \leq_1) + \text{ord}(\{b\} \times N, \leq_2) =$$

$$= \omega + \omega. \tag{8}$$

Thus,

$$2\omega \neq \omega 2. \tag{9}$$

● **PROBLEM 8–34**

Prove that for any finite ordinal k

$$k\omega = \omega. \tag{1}$$

SOLUTION:

Let $(A, \leq_1)$ denote a well ordered set

$$(A, \leq_1) = \{a_1, a_2, ..., a_k\}$$

and $(N, \leq_2) = \{1, 2, 3, ...\}$

$$\omega = \text{ord}(N, \leq_2) \tag{2}$$

Then

$$k\omega = \text{ord}(N \times A, \leq^*) \tag{3}$$

where $\leq^*$ is the lexicographic ordering. Consider the function

$$f : N \times A \to N \tag{4}$$

defined by

$$f(m, a_n) = k(m - 1) + n \tag{5}$$

where m is any natural number and $n = 1, 2, ..., k$.

Function f, defined by (4) and (5), is one-to-one and onto, hence it is a bijection.

Furthermore,

$$\left((n_1, a_{l_1}) \leq {}^*(n_2, a_{l_2})\right) \Rightarrow \left(f(n_1, a_{l_1}) \leq f(n_2, a_{l_2})\right). \tag{6}$$

Function f is order isomorphic. Hence,

$$k\omega = \omega. \tag{7}$$

Show that

1. $\alpha 0 = 0\alpha = 0$ for any ordinal number α.

2. $\alpha 1 = 1\alpha = \alpha$ for any ordinal number α.

3. $\alpha\beta = 0$, if and only if either $\alpha = 0$ or $\beta = 0$.

SOLUTION:

1. $\text{ord}(A, \leq_1) = 0$, if and only if $A = \phi$. Let

$$\alpha = \text{ord}(B, \leq_2). \tag{1}$$

Then

$$\alpha 0 = \text{ord}(A \times B, \leq^*) = \text{ord}(\phi \times B, \leq^*) =$$

$$= \text{ord}(\phi, \leq^*) = 0. \tag{2}$$

Similarly,

$$\alpha 0 = \text{ord}(B \times A, \leq_1{}^*) = \text{ord}(\phi, \leq_1{}^*) = 0 \tag{3}$$

2. Let

$$1 = \text{ord}(\{a\}, \leq_1) \quad \alpha = \text{ord}(B, \leq_2). \tag{4}$$

Then

$$\alpha 1 = \text{ord}(\{a\} \times B, \leq_1{}^*) = \text{ord}(B, \leq_2) = \alpha. \tag{5}$$

Similarly,

$$1\alpha = \text{ord}(B \times \{a\}, \leq_2{}^*) = \text{ord}(B, \leq_2) = \alpha. \tag{6}$$

3. We shall prove that

$$(\alpha\beta = 0) \Leftrightarrow (\alpha = 0 \text{ or } \beta = 0). \tag{7}$$

$\Leftarrow$ was proven in part 1 of this problem.

$\Rightarrow$ Suppose $\alpha\beta = 0$. But

$$\text{ord}(A, \leq) = 0, \text{ if and only if } A = \phi.$$

On the other hand,

$$\alpha\beta = \text{ord}(B \times A, \leq^*).\tag{8}$$

Hence,

$$B \times A = \phi.$$

Therefore, either $A = \phi$ or $B = \phi$.

Prove the following:

THEOREM

Let α, β, and ρ denote ordinal numbers, such that $0 < \rho$. Then

$$(\alpha < \beta) \Rightarrow (\rho\alpha < \rho\beta).\tag{1}$$

∎

SOLUTION:

Let $(A, \leq_1)$, $(B, \leq_2)$, and $(C, \leq_3)$ denote well ordered sets, such that

$$\alpha = \text{ord}(A, \leq_1), \quad \beta = \text{ord}(B, \leq_2), \quad \rho = \text{ord}(C, \leq_3).\tag{2}$$

Since $\alpha < \beta$, $b \in B$ exists, such that

$$A \approx B_b.\tag{3}$$

We form the sets $(A \times C, \leq_1^*)$ and $(B \times C, \leq_2^*)$ with the lexicographic orderings. Hence,

$$(A \times C, \leq_1^*) \approx (B_b \times C, \leq_2')\tag{4}$$

where $\leq_2'$ is ordering inherited from $(B \times C, \leq_2^*)$. Let $c \in C$ denote the least element of C.

The set $B_b \times C$ is the segment of $B \times C$

$$B_b \times C = \{(x, y) \in B \times C : (x, y) \leq_2'' (b, c)\}.\tag{5}$$

Hence,

$$\rho\alpha < \rho\beta.\tag{6}$$

Show that, for any ordinal numbers α, β, and ρ, such that $0 < \rho$

$$(\rho\alpha = \rho\beta) \Rightarrow (\alpha = \beta).\tag{1}$$

SOLUTION:

Assume, on the contrary, that the ordinal numbers α, β, and ρ exist, such that $\rho > 0$ and $\rho\alpha = \rho\beta$ and $\alpha \neq \beta$.

Suppose $\alpha < \beta$.

Then by Problem 8-36, we obtain

$$\rho\alpha < \rho\beta \tag{2}$$

which is a contradiction.

It should be noted that ordinal multiplication is not right cancellative. The ordinal numbers α, β, and $0 < \rho$ exist, such that

$$\alpha\rho = \beta\rho, \quad \text{but } \alpha \neq \beta.$$

For example,

$$2\omega = 1\omega \text{ but } 1 \neq 2.$$

● **PROBLEM 8–38**

Prove the following important theorem.

THEOREM

Let α denote any ordinal number. Then the set

$$\{\beta : \beta \text{ is an ordinal}, \beta < \alpha\} \tag{1}$$

is a well ordered set whose ordinal number is α.

∎

SOLUTION:

Let

$$\text{ord}(A, \leq) = \alpha. \tag{2}$$

Then for any ordinal number $\beta < \alpha$ and any well ordered set $(B, \leq_1)$, such that

$$\beta = \text{ord}(B, \leq_1) \tag{3}$$

the set $(B, \leq_1)$ is order isomorphic to a proper segment A_b, $b \in A$, of A. Element $b \in A$ is uniquely determined by the ordinal number β. This defines a function.

$$f : \{\beta : \beta < \alpha\} \to A \tag{4}$$

defined by

$$f(\beta) = b \tag{5}$$

where

$$\beta = \operatorname{ord}(B, \leq_1) \text{ and } A_b \approx B. \qquad (6)$$

Function f is a bijection. Furthermore,

$$(\beta_1 < \beta_2) \Rightarrow (f(\beta_1) \leq f(\beta_2)). \qquad (7)$$

Hence, f is an order isomorphism and the set $\{\beta : \beta < \alpha\}$ is well ordered.

$$\operatorname{ord}\{\beta : \beta < \alpha\} = \operatorname{ord} A = \alpha. \qquad (8)$$

● **PROBLEM 8–39**

Prove the following theorem:

THEOREM

Any set of ordinal numbers is well ordered.

■

SOLUTION:

r.a.a. (reductio ad absurdum)

Suppose a set of ordinal numbers A exists which is not well ordered. Then, at least one subset $B \subset A$ exists that does not have a least element. Hence, set B must contain a strictly decreasing infinite sequence of ordinal numbers

$$\beta_1 > \beta_2 > \beta_3 > \beta_4 \ldots$$

This sequence is contained in

$$\{\beta : \beta < \beta_1\}.$$

Hence, the set $\{\beta : \beta < \beta_1\}$ is not well ordered because it contains a strictly decreasing infinite sequence. That is a contradiction.

By theorem of Problem 8-38, this set is well ordered.

● **PROBLEM 8–40**

Use the theorem of Problem 8-38 to explain how each ordinal number α can be identified with the set

$$\{\beta : \beta < \alpha\}.$$

SOLUTION:

Let α denote an arbitrary ordinal number. Then the set

$$\{\beta : \beta < \alpha\} \tag{1}$$

is well ordered and

$$\alpha = \text{ord}\{\beta : \beta < \alpha\}. \tag{2}$$

Hence, we can identify the ordinal number α with the set $\{\beta : \beta < \alpha\}$. Each ordinal number can be regarded as a well ordered set of ordinal numbers. We have

$0 \equiv \phi$

$1 \equiv \{0\}$

$2 \equiv \{0, 1\}$

$3 \equiv \{0, 1, 2\}$

$4 \equiv \{0, 1, 2, 3\}$

$\vdots$

$\omega \equiv \{0, 1, 2, 3, \ldots\}$

$\omega + 1 \equiv \{0, 1, 2, \ldots, \omega\}$

$\omega + 2 \equiv \{0, 1, 2, \ldots, \omega, \omega + 1\}$

$\omega 2 \equiv \{0, 1, 2, \ldots, \omega, \omega + 1, \ldots\}$

$\omega 2 + 1 \equiv \{0, 1, 2, \ldots, \omega, \omega + 1, \ldots, \omega 2\}$

$\vdots$

● PROBLEM 8-41

Russell's paradox leads to the conclusion that a set of all sets does not exist. Similar reasoning for the ordinal numbers is called the Burali-Forti paradox. Prove that:

THEOREM

A set of all ordinal numbers does not exist. ∎

SOLUTION:

Suppose a set S of all ordinal numbers exists. By theorem of Problem 8-39, S is a well ordered set. Let s denote the ordinal number of S, then s

must be an element of S. We have
$$S = \mathrm{ord}\ \{\alpha \in S : \alpha < s\} = \mathrm{ord}\ S_s$$
$$< \mathrm{ord}\ S\ = s$$
which is a contradiction.

CHAPTER 9

FUNDAMENTAL CONCEPTS OF TOPOLOGY

1. Which of the figures are congruent?

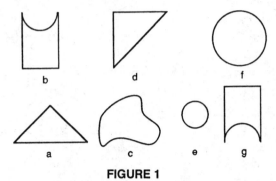

FIGURE 1

2. List a few properties of the triangle shown in Figure 2 which are geometric and a few which are not.

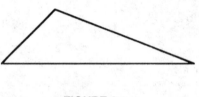

FIGURE 2

SOLUTION:

Geometry deals with figures and certain properties of figures in Euclidean space. To determine which properties are geometric and which are not, we shall introduce the notion of geometric equivalence, often called congruence.

Two figures are congruent if and only if one can be placed on top of the other, so as to fit perfectly.

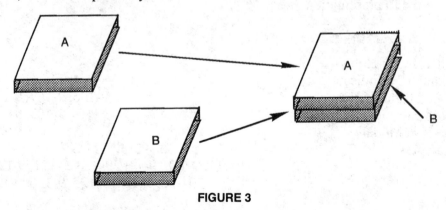

FIGURE 3

A and B shown in Figure 3 are congruent. The properties which are common for the congruent class of figures are called geometric properties.

1. From Figure 1, we see that a and d are congruent, b and g are congruent.

2. The geometric properties of the triangle are:
    the number of sides
    the length of its sides
    the number of angles
    the value of its angles
    the area enclosed by its perimeter.

The properties which are not geometric are:
    its color
    its orientation with respect to some given axes in the plane.

● **PROBLEM 9–2**

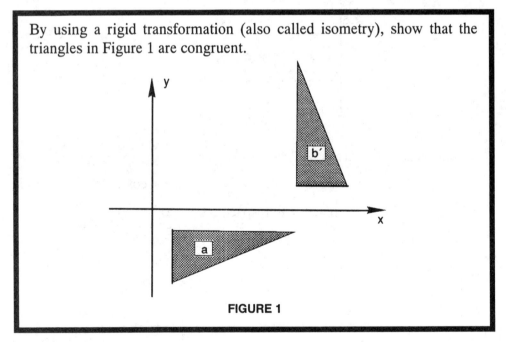

By using a rigid transformation (also called isometry), show that the triangles in Figure 1 are congruent.

FIGURE 1

SOLUTION:

There are three fundamental rigid transformations: translation, rotation, and reflection. Figure 2 shows translation of a point P into P'.

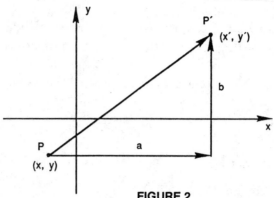

FIGURE 2

$$x' = x + a \qquad y' = y + b \qquad (1)$$

or

$$(x', y') = (x, y) + (a, b) \qquad (2)$$

Rotation of a point P about the origin is illustrated in Figure 3. The coordinates of P are (x, y).

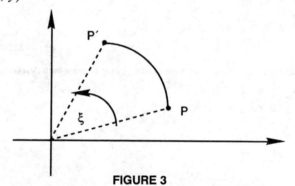

FIGURE 3

The coordinates of P' are (x', y') where

$$x' = x \cos \zeta - y \sin \zeta \qquad y' = x \sin \zeta + y \cos \zeta \qquad (3)$$

Suppose reflection takes place with respect to the x axis, then

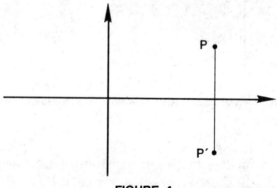

FIGURE 4

254

$$x' = x \qquad y' = -y \tag{4}$$

Now, back to the main problem (Figure 1).

Transformation yields a'. Rotation yields a'' and reflection yields a''', which is triangle b.

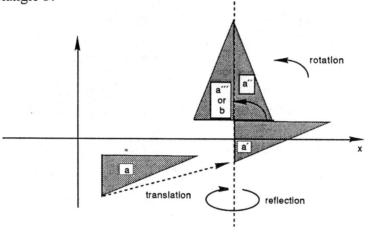

FIGURE 5

Thus, triangles a and b are congruent.

● **PROBLEM 9-3**

Why are rigid transformations isometric?

SOLUTION:

Isometric means preserving the distance.

Consider translation by a vector (a, b). Points A and B become A' and B'. The distance between A and B is

$$d_{AB} = \sqrt{(x_2 - x_1)^2 + (y_2 - y_1)^2} \tag{1}$$

FIGURE 1

while the distance between A' and B' is

$$d_{A'B'} = \sqrt{(x_2'-x_1')^2 + (y_2'-y_1')^2} =$$

$$= \sqrt{[(x_2 + a) - (x_1 + a)]^2 + [(y_2 + b) - (y_1 + b)]^2} =$$

$$= \sqrt{(x_2 - x_1)^2 + (y_2 - y_1)^2} = d_{AB} \qquad (2)$$

Translation preserves the distances. Similarly, rotation preserves the distances. Consider rotation by angle ζ. From Problem 9-2, Equation (3), we obtain:

$$\sqrt{\begin{array}{c} [(x_2 \cos\zeta - y_2 \sin\zeta) - (x_1 \cos\zeta - y_1 \sin\zeta)]^2 + \\ + [(x_2 \sin\zeta - y_2 \cos\zeta) - (x_1 \sin\zeta - y_1 \cos\zeta)]^2 \end{array}} =$$

$$= \sqrt{(x_2 - x_1)^2 + (y_2 - y_1)^2}. \qquad (3)$$

As for reflection, we can always assume that reflection takes place with respect to the x axis. Then

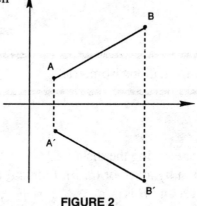

FIGURE 2

$$\sqrt{(x_2' - x_1')^2 + (y_2' - y_1')^2} = \sqrt{(x_2 - x_1)^2 + ((-y_2) - (-y_1))^2} =$$

$$= \sqrt{(x_2 - x_1)^2 + (y_2 - y_1)^2} \qquad (4)$$

Conclusion: rigid transformations are isometric.

Show that the relation of congruence divides all geometric figures into disjoint classes. Explain the role that these classes play in Euclidean geometry.

SOLUTION:

Instead of writing

$$a \text{ is congruent to } b$$

we shall write

$$a \approx b. \tag{1}$$

Note that since

$$a \approx a \tag{2}$$

each geometric figure belongs to some class. Furthermore, if

$$a \approx b \text{ and } b \approx c, \text{ then } a \approx c. \tag{3}$$

Also, if $a \approx b$, then $b \approx a$.

Congruence is an equivalence relation. It separates the set of figures in Euclidean space into disjoint equivalence classes. Geometry deals with the equivalence classes and not with the particular elements of some class. Within each equivalence (congruence) class, all elements share the same geometric properties.

Suppose we are given two figures a and b. If we can point out one geometric property which these figures do not have in common, then a and b belong to two different congruence classes.

It is quite possible that these figures share many other geometric properties.

● **PROBLEM 9-5**

Suppose the congruence classes are given as described in Problem 9-1. Can you give an example (or two) of how to subdivide further each class?

SOLUTION:

We can take into account the location of figures in the space. For example, a class of squares of a certain size can be subdivided according to the distance of their centers from the origin of the coordinate system. All the squares (of the same size) whose centers are located at distance p from the origin, belong to the same subclass as shown in Figure 1.

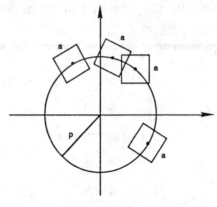

FIGURE 1

Here is another example. Suppose a line is given. The figures of some congruence class belong to the same subclass when their sides make the same angles with the line (see Figure 2).

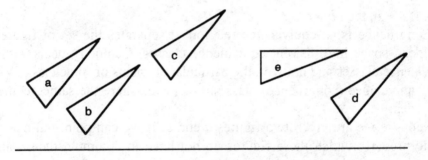

FIGURE 2

Note that triangles a, b, c, and d belong to the same subclass, while e does not. Often, vectors are subdivided into the classes in this manner; that is, vectors with equal length and orientation belong to the same class.

● **PROBLEM 9–6**

Remember the geometry described in Problem 9-4. We shall move a step further and define what might be called geometry of the magnifying glass or similarity geometry. Now, within each equivalence class, we permit right transformations and proportional magnification or contraction.

1. What are the conditions for two rectangles to belong to the same equivalence class?

2. Name a few geometric properties which remain invariant under the permitted transformations.

SOLUTION:

1. Note that all straight line segments belong to the same equivalence class in similarity geometry. Also, all squares are equivalent, and all circles are equivalent. Obviously, area is no longer an invariant.

The rectangles with the same ratio of side lengths are equivalent.

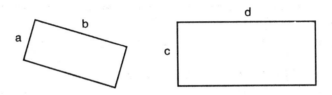

FIGURE 1

Rectangles of Figure 1 are equivalent, if and only if $b/a = d/c$.

2. The values of angles remain the same under the new transformations. Also, straight line segments continue to be straight line segments. In similarity geometry, the overall shape of the figures is preserved. Certain congruence classes of ordinary geometry have been combined to form equivalence classes of similarity geometry (see Figure 2).

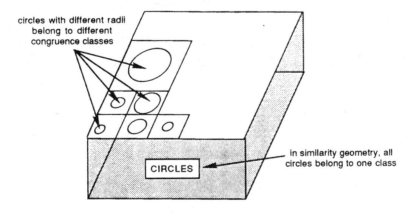

FIGURE 2

259

As a next step on our way to more and more general equivalence classes, we shall define affine geometry. The permitted transformations in the plane are defined now by

$$(x, y) \rightarrow (x', y') \tag{1}$$

where

$$x' = ax + by + c \quad y' = dx + ey + f. \tag{2}$$

The numbers a, b, c, d, e, and f are real and, such that

$$ae - bd \neq 0. \tag{3}$$

Show that in affine geometry the lines which were originally parallel, remain parallel although the angles are not invariant.

SOLUTION:

Suppose two parallel lines AB and CD are given, as shown in Figure 1 below:

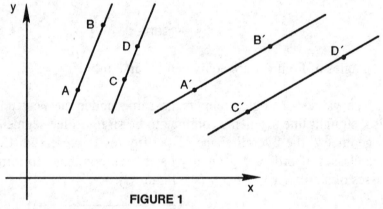

FIGURE 1

We denote the coordinates of the points A, B, C, and D by (x_1, y_1), (x_2, y_2), (x_3, y_3), and (x_4, y_4). Since the lines are parallel, we have

$$\frac{y_2 - y_1}{x_2 - x_1} = \frac{y_4 - y_3}{x_4 - x_3}. \tag{4}$$

Under transformation (2), the points A, B, C, D are transformed to A', B', C', and D' to correspond with coordinates

$$A' \ (ax_1 + by_1 + c, dx_1 + ey_1 + f) = (x_1', y_1')$$

$$B' \ (ax_2 + by_2 + c, dx_2 + ey_2 + f) = (x_2', y_2')$$

$$C' \ (ax_3 + by_3 + c, \ dx_3 + ey_3 + f) = (x_3', y_3')$$

$$D' \ (ax_4 + by_4 + c, \ dx_4 + ey_4 + f) = (x_4', y_4'). \tag{5}$$

To show that the lines $A'B'$ and $C'D'$ are parallel, we must prove that

$$\frac{y_2' - y_1'}{x_2' - x_1'} = \frac{y_4' - y_3'}{x_4' - x_3'}. \tag{6}$$

Indeed,

$$\frac{d(x_2 - x_1) + e(y_2 - y_1)}{a(x_2 - x_1) + b(y_2 - y_1)} = \frac{d + e\dfrac{y_2 - y_1}{x_2 - x_1}}{a + b\dfrac{y_2 - y_1}{x_2 - x_1}} =$$

$$= \frac{d + e\dfrac{y_4 - y_3}{x_4 - x_3}}{a + b\dfrac{y_4 - y_3}{x_4 - x_3}} = \frac{d(x_4 - x_3) + e(y_4 - y_3)}{a(x_4 - x_3) + b(y_4 - y_3)}. \tag{7}$$

Hence, lines $A'B'$ and $C'D'$ are parallel.

● **PROBLEM 9-8**

1. Using drawings, explain shear and strain.

2. Show that all triangles are equivalent in affine geometry and not all four-sided polygons are equivalent.

SOLUTION:

1. Shear is illustrated in Figure 1.

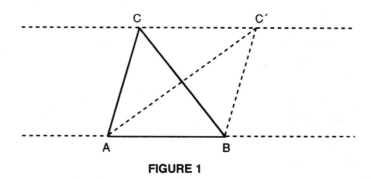

FIGURE 1

Here, the base of the triangles remains the same, while the vertex C is moved along a line parallel to the base.

Another transformation allowed in affine geometry is strain. Here the vertex is moved along a line which is not parallel to the base. See Figure 2.

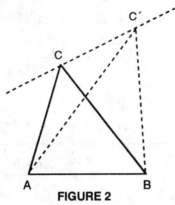

FIGURE 2

2. Suppose we are given any two triangles. Then by using only affine transformations, we can transform one triangle into another. Hence, all triangles belong to one equivalence class in affine geometry. Consider the two four-sided polygons, shown in Figure 3.

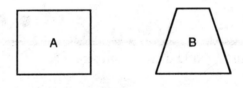

FIGURE 3

Since affine transformations transform parallel lines into parallel lines, A cannot be transformed into B. Hence, A and B belong to two different equivalence classes in affine geometry.

● **PROBLEM 9–9**

Explain why the equivalence classes of affine geometry are combinations of equivalence classes of similarity geometry which are, in fact, combinations of equivalence (congruence) classes of ordinary geometry.

SOLUTION:

Each geometry class is determined by the set of transformations it allows.

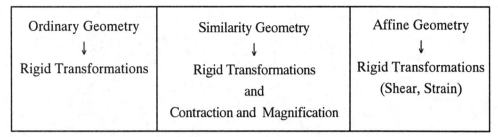

Ordinary Geometry	Similarity Geometry	Affine Geometry
↓	↓	↓
Rigid Transformations	Rigid Transformations and Contraction and Magnification	Rigid Transformations (Shear, Strain)

FIGURE 1

Each transformation allowed in ordinary geometry is allowed in similarity geometry, and each transformation allowed in similarity geometry is also allowed in affine geometry. The situation is illustrated in Figure 2.

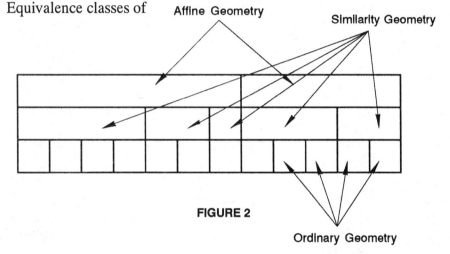

Equivalence classes of

Affine Geometry

Similarity Geometry

FIGURE 2

Ordinary Geometry

● PROBLEM 9–10

One of the invariants of affine geometry is that the parallel lines are transformed into parallel lines. Show that this is not true for projective geometry.

SOLUTION:

In projective geometry, the projective transformations are permitted. Such transformations are perspective projections of a figure.

In Figure 1, the triangle ABC in one plane is transformed into $A'B'C'$ in the other plane by projection from an exterior point. We shall also consider parallel projections, as shown in Figure 2.

Obviously, two parallel lines can be projected into two lines which are not parallel.

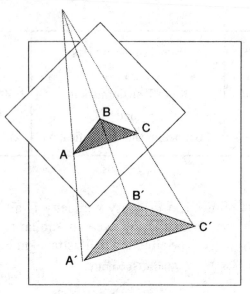

FIGURE 1

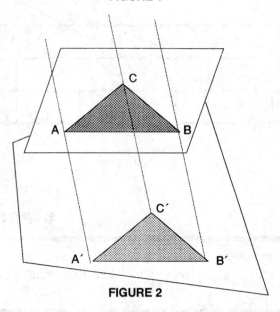

FIGURE 2

In projective geometry, straight lines remain straight lines. Show that cross-ratio is also an invariant; that is, prove that

$$\frac{AC:BC}{AD:BD} = \frac{A'C':B'C'}{A'D':B'D'}. \tag{1}$$

SOLUTION:

Here we shall prove (1) for projection from an exterior point. The proof

264

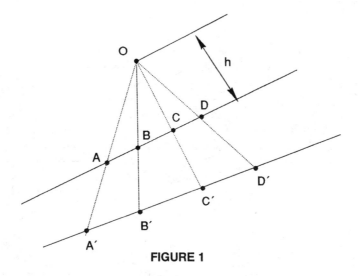

FIGURE 1

for the parallel projection is similar. Consider the triangle AOC. Its area is equal to

$$S_{AOC} = \tfrac{1}{2}\,h{\cdot}AC = AO{\cdot}OC\,\sin AOC. \qquad (2)$$

We obtain similar equations for all involved triangles. Thus,

$$\frac{AC{:}BC}{AD{:}BD} = \frac{S_{AOC}{:}S_{BOC}}{S_{AOD}{:}S_{BOD}} =$$

$$= \frac{AO{\cdot}OC{\cdot}\sin AOC}{BO{\cdot}OC{\cdot}\sin BOC} \cdot \frac{BO{\cdot}OD{\cdot}\sin BOD}{AO{\cdot}OD{\cdot}\sin AOD} =$$

$$= \frac{\sin AOC{\cdot}\sin BOD}{\sin BOC{\cdot}\sin AOD}. \qquad (3)$$

The final result in (3) does not depend on the choice of points A', B', C', and D'; hence, cross-ratio is an invariant.

● PROBLEM 9-12

All the geometries considered, so far, have had one important invariant in common: a straight line remained a straight line. This is not true in topology, where a square can be transformed into a circle, since we allow all one-to-one bi-continuous transformations. The new transformations, now permitted, are called elastic deformations.

Bending, stretching, and twisting is now allowed. Cutting is not allowed, unless the cuts are repaired in such a way that the points which were "near" remain "near."

Which figures shown in Figure 2 and Figure 3 belong to the same topological equivalence classes?

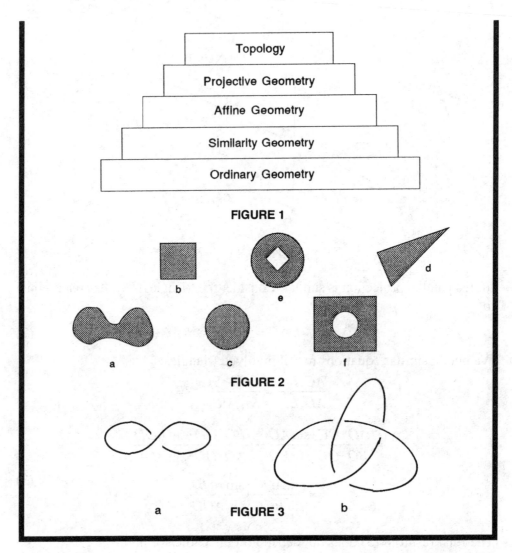

FIGURE 1

FIGURE 2

FIGURE 3

SOLUTION:

Figures a, b, c, and d of Figure 2 belong to the same topological equivalence class. Each may be transformed into any of the others by elastic deformations.

Figures e and f belong to the same class.

Figure 3 shows two plane curves which belong to the same topological class. In three-dimensional space, a cannot be transformed into b without cutting and subsequent re-connecting.

● **PROBLEM 9-13**

The plane closed curve C is deformed into C^I, then into C^{II}, then into C^{III} and finally into C^{IV}, see Figure 1. Which of the transformations are

266

topological?

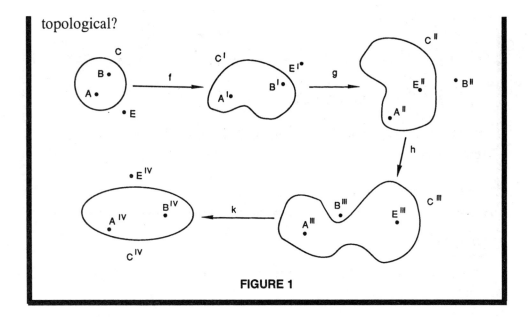

FIGURE 1

SOLUTION:

In topology the distances are much less important than in the geometries defined so far. Here we investigate non-metric spatial relationships. What is important is the preservation of the "nearness" of the points. We understand "nearness" in the topological sense. The property of separating a plane surface into a region outside and a region inside the plane closed curve is a topological invariant.

We conclude that f (of Figure 1) is a topological transformation. Points A and B inside C are transformed into A^I and B^I inside C^I while the point E outside C is transformed into E^I outside C^I. Transformations g and k are not topological while h is a topological transformation.

● **PROBLEM 9–14**

Consider the curves C_1 and C_2.

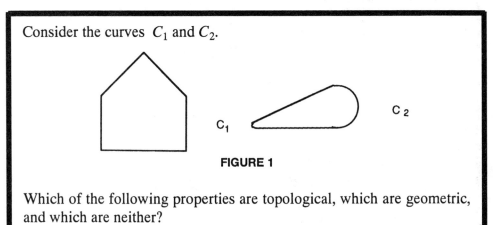

FIGURE 1

Which of the following properties are topological, which are geometric, and which are neither?

267

1. C_1 is above C_2.

2. The area enclosed by C_1 is larger than the area enclosed by C_2.

3. C_1 and C_2 have no common points.

4. C_1 consists of straight lines, while C_2 consists of straight lines and a semicircle.

5. C_1 has more vertices than C_2.

SOLUTION:

1. This property is neither geometric nor topological. Rotation can place C_2 above C_1.

2. The area is an invariant in ordinary geometry. But under affine transformations, the area is not preserved. This is not a topological property.

3. This is a geometric and topological property.

4. This is not a topological property. It is a geometric property in ordinary geometry.

5. It is a geometric but not topological property.

● **PROBLEM 9-15**

There is a sphere with a continuous closed non-self-intersecting curve C drawn on its surface. The curve C divides the surface into two disjoint regions A_1 and A_2. One cannot move from any point in A_1 to any point in A_2 without crossing C, see Figure 1.

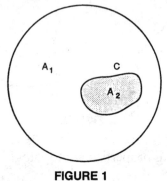

FIGURE 1

1. How many times does a path which joins a point belonging to one region with a point in the other region cross the curve C? What happens when both points belong to the same region?

2. Is it possible, by continuous transformations, to contract C to a point?

SOLUTION:

1. Figure 2 depicts a sphere with a curve C.

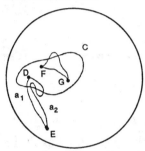

FIGURE 2

Points D and E belong to different regions; path a_1 crosses C once, a_2 three times. Hence, a point belonging to one region may be joined to a point belonging to the other by paths which cross C an odd number of times.

When both points belong to the same region (points F and G), the path from F to G will cross C an even number of times.

2. Curve C, which is closed, non-self-intersecting, and continuous may gradually be contracted on the surface of a sphere to a point. This is an important property, which we shall discuss later in detail.

● PROBLEM 9-16

The surface of a one-fold torus is shown in Figure 1. It has the form of a well-known doughnut. (If you are on a diet, compare it to a bicycle tire).

1. Curve C separates the surface into two disjoint regions, and it may be continuously contracted on the surface into a point. Is this true for any continuous non-self-intersecting closed curve?

2. What is the maximum number of continuous non-self-intersecting closed curves which may be drawn on the surface of a torus without dividing it?

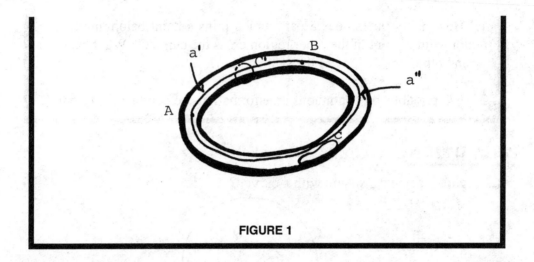

FIGURE 1

SOLUTION:

1. Consider curve C' of Figure 1. It cannot be continuously contracted into a single point. Furthermore, C' does not divide the surface of the torus into two regions. Points A and B can be joined by path a'' which does not cross C'.

The sphere and the torus belong to different topological equivalence classes. Topologically equivalent surfaces are called homeomorphic.

2. One. If we draw any other closed curve besides C (as in Figure 2), the surface of the torus will be divided. Any path AB crosses either C or C' at least once.´

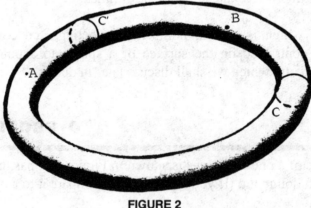

FIGURE 2

● **PROBLEM 9–17**

We shall move a step further and investigate the curves on the surface of a two-fold torus, see Figure 1.

270

FIGURE 1

How many continuous, closed curves can be drawn on the surface of this torus without dividing it into distinct regions?

SOLUTION:

Curve C, which can be continuously contracted into a point, divides the torus into two regions. Hence, we shall investigate only the curves which cannot be contracted into a point. Draw curves C' and C''. Any two points A and B can be joined without crossing C' or C''. Adding one more curve C''' (see Figure 2) changes the situation. Any path AB has to cross at least one of the curves at least once. The maximum number of curves which do not divide a two-fold torus is two.

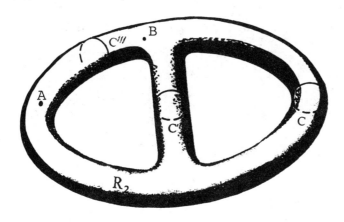

FIGURE 2

What is the genus of the surface of:

1. a sphere

2. a one-fold torus

3. an *n*-fold torus

4. a fork with a hot-dog on it

5. a ladder with five rungs.

SOLUTION:

The greatest number of continuous non-self-intersecting closed curves which may be drawn on a surface without dividing it into distinct regions defines the genus of the surface.

1. The genus of the surface of a sphere is 0.

2. The genus of the surface of a one-fold torus is 1.

3. Generally, the genus of an *n*-fold torus is *n*.

4. The genus of a fork with a hot dog on it is three, as shown in Figure 1.

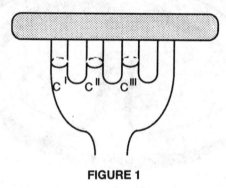

FIGURE 1

5. A ladder with five rungs has a genus of 4.

The genus of a surface is invariant under all one-to-one bicontinuous transformations; hence, it is a topological property. All surfaces belonging to the same topological equivalence class have the same genus.

Which of the following surfaces are closed and which are open?

1. a sphere

2. an *n*-fold torus

3. a hollow cylinder

4. a cube.

SOLUTION:

A surface is closed if it has no boundary curves. Otherwise, the surface is called open. A sphere, an *n*-fold torus, and a cube are closed surfaces.

The edges of a cube are not boundary curves. A hollow cylinder is an open surface. Similarly, a disc is open. Boundary curves of two-sided surfaces are curves which separate one side of a surface from the other.

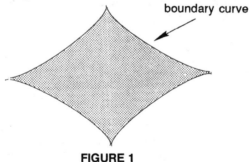

boundary curve

FIGURE 1

The edges of a piece of infinitely thin paper are boundary curves of this surface.

Describe the Möbius band, which is a one-sided surface. What is the boundary curve of the Möbius band? (See following figure.)

SOLUTION:

To obtain the Möbius band, take a rectangular strip of paper *ABCD*, make a 180° twist with the end *DC* and join up the ends. The obtained surface does not have two sides; it cannot be painted in different colors. An ant walking

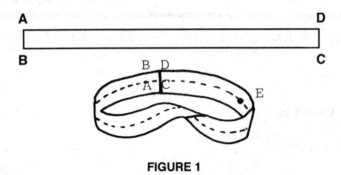

FIGURE 1

along a dotted line from a point E after making one round will end up on the other side of E without crossing the boundary line.

The Möbius band has only one boundary curve, which is topologically equivalent to a circle.

● **PROBLEM 9-21**

Why is the surface of a Klein bottle closed and one-sided?

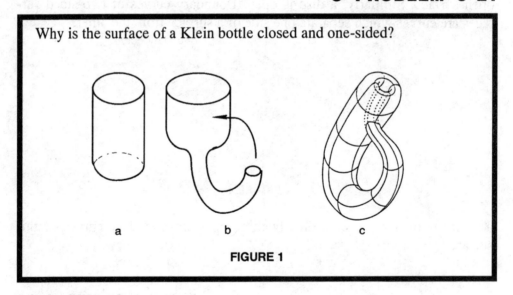

FIGURE 1

SOLUTION:

We start with an open cylinder, stretch out and bend one end, then pass through the side surface without intersecting or breaking it. At the final stage, we join up the ends.

This bottle cannot be physically constructed in the three-dimensional space. Similarly, the Möbius band, which is a two-dimensional structure, cannot be constructed in two dimensions, because of the twist. Any two points on the surface of a Klein bottle may be joined by a continuous path not crossing any boundary curve. The surface of a Klein bottle has no boundary curve. Note that two open ends of the original cylinder were joined together in such a way that the outside of one was connected to the inside of the other. Hence, the outside and inside became impossible to distinguish.

The surface of a Klein bottle is one-sided and closed.

● **PROBLEM 9-22**

Why is a Möbius band a non-orientable surface?

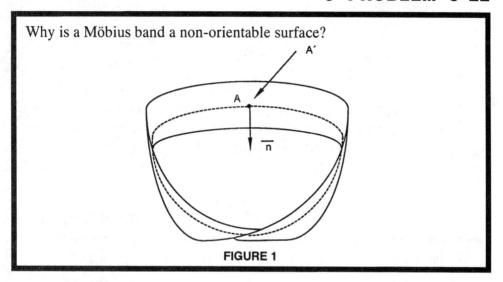

FIGURE 1

SOLUTION:

Consider a point A on the surface of a Möbius band. Let $\bar{n}$ be a normal to the surface at A. Suppose $\bar{n}$ moves along a dotted line from A to A'. The normal of A' has the opposite direction of that at A. Note that $\bar{n}$ is continuously defined as it moves from A to A'. Similarly, the surface of a Klein bottle is nonorientable.

● **PROBLEM 9-23**

A one-sided surface is connected if it is possible to join any two points of this surface with a continuous path. A two-sided surface is connected if its sides, taken separately, are both connected. Which of the surfaces depicted in Figure 1 is connected?

SOLUTION:

It is important to keep in mind the distinction between a solid and the surface of a solid. In most cases of a sphere, a torus, etc., we understand a surface is not a solid.

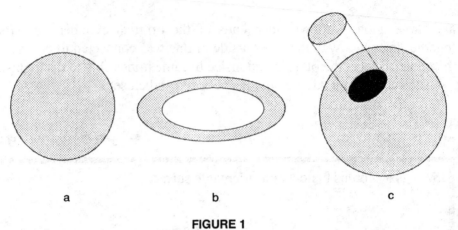

a   b   c

FIGURE 1

Hence, a sphere and a torus (Figure 1) are two-sided connected surfaces.

In Figure 1 c a disc is separated from a sphere. The total surface is disconnected. Now the total surface consists of two surfaces (each connected in itself) which are homeomorphic to a disc.

● **PROBLEM 9–24**

Which of these surfaces are simply connected?

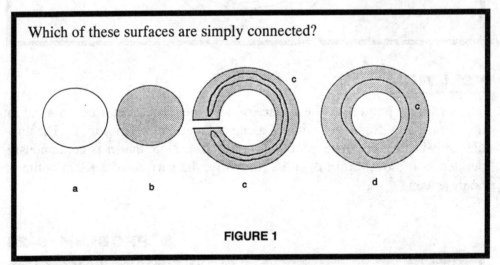

a   b   c   d

FIGURE 1

SOLUTION:

A surface is said to be simply connected if every continuous non-self-intersecting closed curve drawn on it may be continuously contracted on the surface into a point. The disc (a) and the sphere (b) are simply connected.

The surface shown in (c) is topologically equivalent to (a); hence, it is simply connected. The annulus (d) is not simply connected. Curve C cannot be continuously contracted to a point.

Annulus is doubly connected, since it requires only one cut to make it homeomorphic to a disc.

Which of the surfaces is quadruply connected?

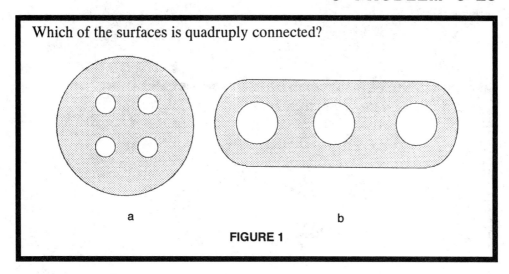

a

b

FIGURE 1

SOLUTION:

The surface is n-tuply connected if it requires $n - 1$ cuts in order to make it homeomorphic to a disc. Thus, annulus is doubly connected; it requires one cut to render it homeomorphic to a disc. Surface (a) of Figure 1 requires four cuts; hence, it is quintuply connected ($n = 5$).

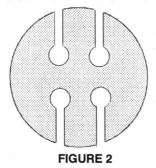

FIGURE 2

Surface (b) of Figure 1 is quadruply connected. It requires three cuts which can be made in many different ways (See Figure 3 (a) and (b)).

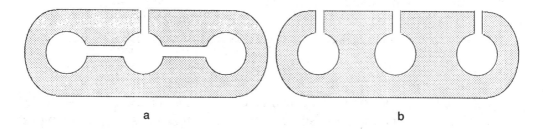

a

b

FIGURE 3

Note that each new cut forms a boundary of the surface, and the cuts cannot intersect.

● **PROBLEM 9–26**

Determine n for these n-tuply connected surfaces:

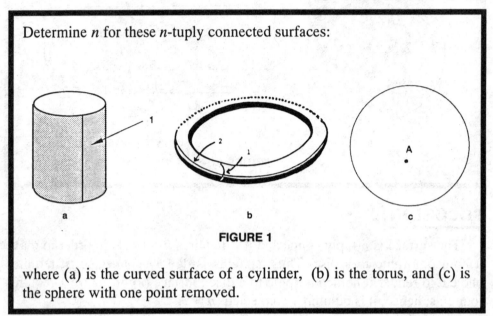

FIGURE 1

where (a) is the curved surface of a cylinder, (b) is the torus, and (c) is the sphere with one point removed.

SOLUTION:

We can cut the cylinder along line 1 and obtain a rectangle, which is homeomorphic to a disc; hence, $n = 2$.

We can cut the torus along line 1 to obtain a cylinder and then along 2. Thus, $n = 3$ and the surface of the torus is triply connected.

A sphere with a single hole is homeomorphic to a disc, $n = 1$.

● **PROBLEM 9–27**

Show that any two continuous non-self-intersecting closed curves on the sphere are homotopic. (Figures shown on following page.)

How many homotopy classes does the torus have?

SOLUTION:

Any continuous closed non-self-intersecting figure on a simply connected surface can be continuously contracted to a point. It follows that, on a simply connected surface, any curve may be continuously deformed into any other curve. C in Figure 1 can be continuously deformed into C'. All continuous

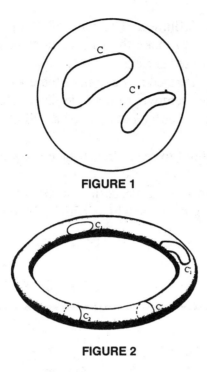

FIGURE 1

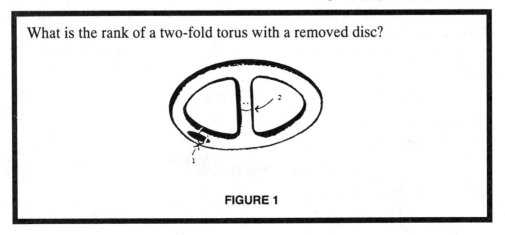

FIGURE 2

closed non-self-intersecting curves on a simply connected surface are homotopic to each other.

On the torus, C_1 can be transformed to C_1' and C_2 can be transformed to C_2'. But it is impossible to continuously transform C_1 into C_2. Curves C_1 and C_1' are homotopic, and curves C_2 and C_2' are homotopic. There are two homotopy classes for the surface of the torus.

Curves like C_1 and C_1', which may be continuously contracted to a point, belong to the null homotopy class.

● **PROBLEM 9-28**

What is the rank of a two-fold torus with a removed disc?

FIGURE 1

279

SOLUTION:

Let us start with a definition: The rank of an open surface is the least number of cuts required to make the surface homeomorphic to a disc. The rank of an n-tuply connected surface is $n - 1$.

The rank of a closed surface is the rank of the open surface obtained from the closed surface by the removal of a single disc. A disc and a sphere have rank zero. An annulus, the curved surface of a cylinder, and a Möbius band have rank 1.

Three cuts are required to make the surface shown in the figure homeomorphic to a disc.

We can also use an alternative definition of rank. Rank is the greatest number of non-intersecting cuts which can be made without making the surface disconnected.

● **PROBLEM 9-29**

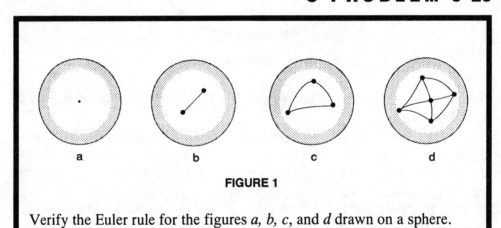

a   b   c   d

FIGURE 1

Verify the Euler rule for the figures *a, b, c,* and *d* drawn on a sphere.

SOLUTION:

Let us draw a figure having the number of vertices V of regions (or faces) F and of edges (or arcs) E, on a sphere.

Then, any figure drawn on a surface of a sphere satisfies the following rule:

$$V + F - E = 2 \tag{1}$$

discovered by Euler.

For *a*, we have:

$$V = 1, \quad F = 1, \quad E = 0 \tag{2}$$

hence $1 + 1 - 0 = 2$.

For b, we have

$$V = 2, \quad F = 1, \quad E = 1 \tag{3}$$

hence $2 + 1 - 1 = 2$.
   For c, we have

$$V = 3, \quad F = 2, \quad E = 3 \tag{4}$$

hence $3 + 2 - 3 = 2$.
   For d, we have

$$V = 5, \quad F = 5, \quad E = 8 \tag{5}$$

hence $5 + 5 - 8 = 2$.
   In the next problem, we shall prove Euler's formula.

● **PROBLEM   9–30**

Prove Euler's formula:

$$V + F - E = 2 \tag{1}$$

for the figures drawn on a sphere.

SOLUTION:

Suppose (1) is true for some figure depicted in Figure 1.
   Let us draw a line connecting a vertex to another vertex (which does not cross any existing edges), as shown in Figure 2 *a* and *b*:

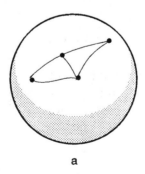

a

FIGURE 1

281

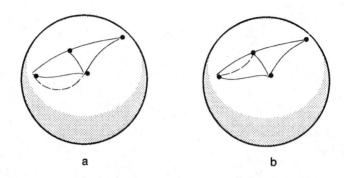

a   b

FIGURE 2

Let Δ denote the increase, then

$$\Delta V = 0, \quad \Delta F = 1, \quad \Delta E = 1. \tag{2}$$

Then

$$V + \Delta V + F + \Delta F - E - \Delta E = V + F - E = 2 \tag{3}$$

and (1) holds.

Suppose we draw a line which ends with a new vertex (Figure 3).

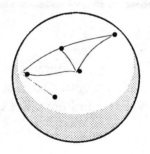

FIGURE 3

Then, $\Delta V = 1$, $\Delta F = 0$, $\Delta E = 1$ and (1) holds again.

Any figure on the sphere can be drawn, beginning with a vertex and then adding lines one after another. Hence, Euler's rule is proved.

Find one or two simple regular divisions of a sphere. A division is called regular when every face has the same number of lines f as a boundary and the same number of lines v meet at every vertex.

SOLUTION:

Euler's formula states that

$$V + F - E = 2. \tag{1}$$

Since there are F faces on the sphere, the total number of lines (edges) is

$$\frac{Ff}{2} = E \tag{2}$$

On the other hand,

$$\frac{Vv}{2} = E \tag{3}$$

From (1), (2), and (3), we obtain:

$$\frac{1}{v} + \frac{1}{f} = \frac{1}{E} + \frac{1}{2} \tag{4}$$

The simple solutions of (4) are

	v	f	E	V	F
I	$v = E$	2		2	$F = E$
II	2	$f = E$		$V = E$	2

Division I has 2 vertices and as many faces as edges. Each face is bounded by two lines, $f = 2$. The number of edges is arbitrary $E = 1, 2, 3, \ldots$ (see Figure 1).

SOLUTION I

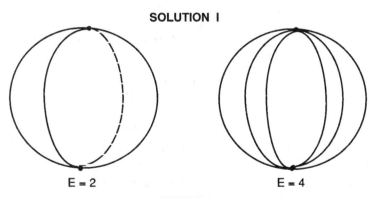

E = 2   E = 4

FIGURE 1

Division II has two faces and as many vertices as edges, see Figure 2. The number of edges is arbitrary $E = 1, 2, 3, \ldots$

SOLUTION II

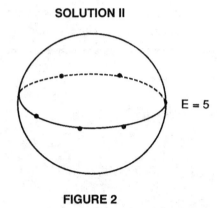

$E = 5$

FIGURE 2

Here we have a sphere divided into two hemispheres separated by an E-gon.

● **PROBLEM 9–32**

Consider the equation:

$$\frac{1}{v} + \frac{1}{f} = \frac{1}{E} + \frac{1}{2} \tag{1}$$

which describes the regular divisions of a sphere. Find all possible non-trivial solutions of (1).

SOLUTION:

For $E = 1$, we obtain trivial solutions

$$v = 1, \quad f = 2 \quad \text{or} \quad v = 2, \quad f = 1.$$

For $E = 2$, the only solution is $f = v = 2$, discussed in Problem 9-31. We are left with $E = 3, 4, 5, \ldots$ Hence,

$$\frac{1}{2} < \frac{1}{v} + \frac{1}{f} < 1 \tag{2}$$

(2) excludes $f = 1$. Case $f = 2$ was already discussed.

The possible values of f are $f = 3, 4, 5$. Note that $f = 6, 7, 8, \ldots$ are incompatible with (2).

By the same token, the admissible values of v are $v = 3, 4, 5$.

v / f	3	4	5
3	OK	OK	OK
4	OK		
5	OK		

The crossed-out values do not satisfy (2). Hence, we are left with five cases.

● **PROBLEM 9–33**

Find the number of faces F, edges E and vertices V for each of five regular polyhedra found in Problem 9-32.

SOLUTION:

In Problem 9-32, we found possible values of f and v:

v	f	E	F	V	Name
3	3	6	4	4	tetrahedron
4	3	12	8	6	octahedron
3	4	12	6	8	cube
5	3	30	20	12	icosahedron
3	5	30	12	20	dodecahedron

The values of E are computed from

$$\frac{1}{E} = \frac{1}{f} + \frac{1}{v} - \frac{1}{2} \tag{1}$$

The values of F and V are obtained from

$$F = \frac{2E}{f}, \qquad V = \frac{2E}{v} \tag{2}$$

Such five regular polyhedra, known as Platonic polyhedra, are depicted in the following figure.

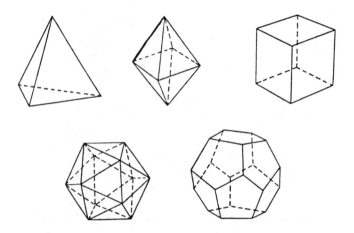

A separation of a surface consisting of vertices linked together by edges is called a map. For any given surface, the expression

$$V - E + F = \chi \tag{1}$$

is invariant. Hence, it remains invariant for all surfaces homeomorphic to the given surface.

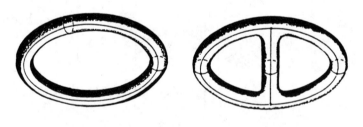

FIGURE 1   FIGURE 2

This invariant number χ is called the Euler characteristic of the surface, and it is a topological property of the surface independent of the actual values of V, E, and F. For the sphere $\chi = 2$, find χ for a torus and for a two-fold torus, see Figures 1 and 2.

SOLUTION:

The Euler characteristic remains the same for any map on a torus, so we can use the map of Figure 1.

$$V = 2 \quad E = 4 \quad F = 2. \tag{2}$$

Thus,

$$V - E + F = 2 - 4 + 2 = 0. \tag{3}$$

For torus, $\chi = 0$.

From Figure 2, for a two-fold torus

$$V = 5 \quad E = 9 \quad F = 2. \tag{4}$$

Hence, for a two-fold torus

$$\chi = 5 - 9 + 2 = -2. \tag{5}$$

● **PROBLEM 9–35**

1. Figure 1 depicts regular polyhedra. Find the Euler characteristic for each one.

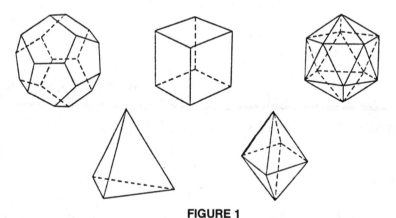

FIGURE 1

2. Is it possible

   a) to draw a map on a sphere having seven vertices linked by ten arcs and defining four regions?

   b) to draw a map on a two-fold torus having five vertices linked by eight arcs and defining two regions?

SOLUTION:

1. Consider the tetrahedron

$$V = 4 \quad E = 6 \quad F = 4.$$

Hence,

$$V - E + F = 4 - 6 + 4 = 2. \quad (1)$$

We can perform calculations for all polyhedra and in each case, obtain $\chi = 2$. This result is obvious and does not require any computing. Any simple polyhedron can be continuously deformed into a sphere; hence, $\chi = 2$.

2. a) For a sphere

$$V - E + F = 2$$

$$7 - 10 + 4 = 1 \neq 2 \quad (2)$$

Not possible.

b) For a two-fold torus

$$V - E + F = -2$$

$$5 - 8 + 2 = -1 \neq -2 \quad (3)$$

It is not possible to draw such a map.

● PROBLEM 9–36

Use the triangulation method to prove Euler's formula

$$V - E + F = 2 \quad (1)$$

for a cube.

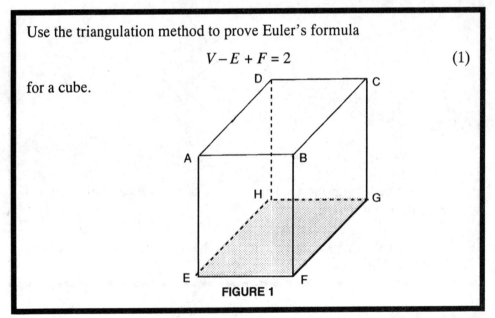

FIGURE 1

SOLUTION:

From the cube $ABCDEFGH$, we have to remove any face, say $EFGH$. The Euler characteristic $\chi = V - E + F$ for a cube is now decreased by 1 since V

and E remain constant and F decreases by 1. The cube with the removed face is transformed until all vertices lie in a plane (see Figure 2).

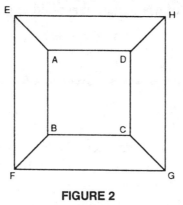

FIGURE 2

This deformation does not change χ. Now we shall carry out the triangulation as shown in Figure 3.

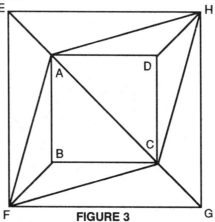

FIGURE 3

We have $V = 8$, $E = 17$, $F = 10$.

Note that whenever we add a triangle, V remains constant, whereby E and F increase by 1. Hence, χ remains the same.

Triangles are now removed until one remains; e.g., ABC, for which

$$V = 3, \quad E = 3, \quad F = 1$$

Hence,

$$V - E + F = 1$$

for a triangle and

$$V - E + F = 1 + 1 = 2$$

for a cube.

The value of $V - E + F$ remains invariant within any topological equivalence class.

Describe the method of triangulation applied to prove Euler's formula

$$V - E + F = 2$$

for polyhedra.

SOLUTION:

From a polyhedron with V vertices, E edges, and F faces we remove one face. Removal of one face decreases $V - E + F$ by one.

The new surface is deformed until all vertices and edges lie in a plane. We obtain something that is topologically equivalent to a disc. Now we perform the triangulation. Each face is divided into a triangle in such a way that no new vertices are created (Figure 1).

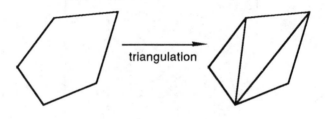

triangulation

FIGURE 1

Note that $V - E + F$ remains unchanged.

We shall remove the triangles in such a way that the triangle removed has at least one edge on the boundary. Two situations are possible (see Figure 2).

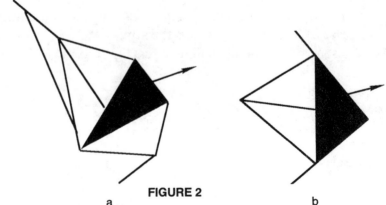

FIGURE 2

a   b

For a) V remains the same, E and F are decreased by one; hence, $V - E + F$ is invariant.

Similarly, for b) $V - E + F$ is invariant because V and F decrease by one and E decreases by two. Finally we are left with one triangle for which

$$V - E + F = 3 - 3 + 1 = 1.$$

Therefore, for a polyhedron, $\chi = 2$.

● **PROBLEM 9-38**

Verify the expression

$$V - E + F = \chi \qquad (1)$$

for the drawings depicted in Figure 1 and Figure 2.

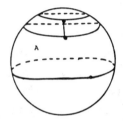

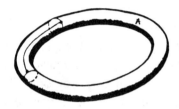

FIGURE 1   FIGURE 2

Explain any discrepancies.

SOLUTION:

For a sphere

$$V - E + F = 2. \qquad (2)$$

From Figure 1

$$V = 3, \quad E = 4, \quad F = 4. \qquad (3)$$

Hence,

$$V - E + F = 3 - 4 + 4 = 3 \neq 2. \qquad (4)$$

All regions of a map should be simply connected. Region A is not simply connected. Thus rule (1) cannot be applied.

For a torus $\chi = 0$. From Figure 2

$$V = 2, \quad E = 3, \quad F = 2. \qquad (5)$$

and

$$V - E + F = 1 \neq 0 \qquad (6)$$

because A is not simply connected.

● **PROBLEM 9–39**

Each surface is characterized by two numbers

1. the genus, g

2. the Euler characteristic, χ.

What is the relationship, if any, between these numbers?

SOLUTION:

The genus of the surface is the greatest number of distinct continuous non-self-intersecting closed curves which may be drawn on a surface without separating it into distinct regions. Each division of a surface to become a map, has to include some edges so that all regions are simply connected. That is related to the genus of the surface.

	Genus	Euler characteristic
Sphere	0	2
Torus	1	0
Two-fold Torus	2	-2
n-fold Torus	n	$2 - 2n$

The relation between the genus and the Euler characteristic of any given surface is

$$V - E + F = \chi = 2 - 2g.$$

● **PROBLEM 9–40**

Suppose you have a "hairy" sphere and a "hairy" torus. What happens when you comb down the hair on a sphere? Is it possible to comb a torus nicely?

SOLUTION:

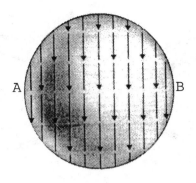

FIGURE 1

Figure 1 depicts a sphere with its hair combed. This "hairdo" has two points of discontinuity, *A* and *B*, as illustrated in Figure 2.

FIGURE 2

Another way of brushing a sphere is shown in Figure 3, which has two points of discontinuity *C* and *D*.

FIGURE 3

It is possible to brush the hair of a torus in a nice way; that is, in such a way that continuity is preserved everywhere (see Figure 4).

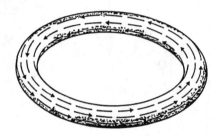 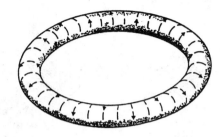

FIGURE 4

Note that all this "brushing" is equivalent to assigning a direction to each point of a surface.

This problem belongs to differential topology.

● PROBLEM 9-41

Instead of combing hair, we can visualize a flow of fluid on the surface. The points of discontinuity are called singular points. Make a sketch of a few singular points.

SOLUTION:

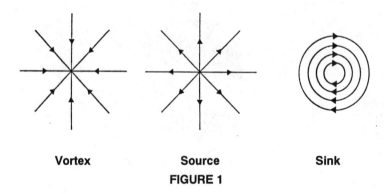

| Vortex | Source | Sink |

FIGURE 1

Figure 1 depicts some often encountered singular points. Some other singular points are illustrated in Figure 2.

In most cases, the name of a singular point explains its nature.

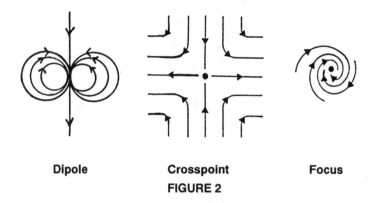

Dipole

Crosspoint

Focus

FIGURE 2

● **PROBLEM 9–42**

For each singular point described in Problem 9-41, find its index.

SOLUTION:

To each singular point, we assign an integer called its index, which is obtained by travelling around this point along a circle in a counter-clockwise direction and counting the number of counter-clockwise revolutions made by an arrow with its base on the path and pointing in the direction of the flow on the surface.

Let us compute the index of a source.

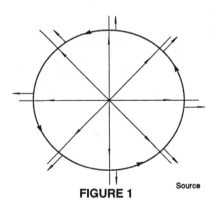

Source

FIGURE 1

The little arrow makes one counter-clockwise revolution, hence the index of a source is one. It is easy to see that for a sink, a vortex and a focus, the index is one.

For a dipole (see Figure 2), the index is 2.

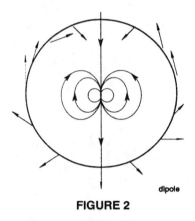

dipole

FIGURE 2

For the crosspoint (see Problem 9-41) the index is − 1. It is easy to see that the little arrow makes one clockwise revolution.

● PROBLEM 9–43

Brush a hairy sphere so that it has only one singular point. Make a sketch and determine the index of this singularity.

SOLUTION:

It is possible to brush a sphere in such a way, that it would have only one singularity, namely a dipole.

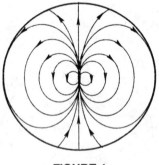

FIGURE 1

The index of a dipole is two.

● PROBLEM 9–44

THEOREM

The Euler characteristic of the surface is equal to the sum of the indices

296

of the singular points of this surface. ∎

Verify this theorem for the surfaces shown in Problems 9-40 and 9-43.

SOLUTION:

Consider the surface of a sphere depicted in Figure 1 of Problem 9-40. It has two vortices. The index of a vortex is one. Hence,

$$1 + 1 = 2$$

which is the Euler characteristic of a sphere.

Similarly, Figure 3 of a sphere of Problem 9-40 has two singular points: a sink with index one and a source with index one. Again,

$$1 + 1 = 2.$$

Both toruses of Figure 4 of Problem 9-40 have no singular points. Hence, the sum of indices is 0 which is the Euler characteristic of a torus.

The sphere depicted in Problem 9-43 has one dipole, whose index is 2.

● **PROBLEM 9-45**

There is some fluid flow on the sphere which contains a crosspoint. Complete the description of this flow. What other singularities contain a flow with a vortex on a torus?

SOLUTION:

The sum of the indices of the singularities on a sphere is two. The index of a crosspoint is −1. Thus,

$$-1 + \Sigma \text{ indices of other singular points} = 2.$$

Here are some possible combinations:

sink and two sources

source and two sinks

dipole and a sink.

The Euler characteristic of a torus is 0. So, if there is a vortex with index 1, there must be a crosspoint with index − 1.

We obtain

$$1 + (-1) = 0.$$

297

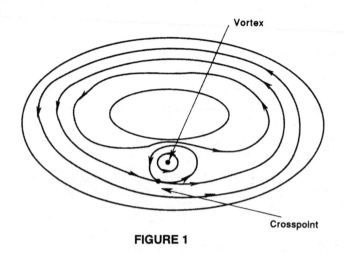

FIGURE 1

Show that the sum of the indices of singular points of a surface is equal to its Euler characteristic.

SOLUTION:

For any map on a given surface the expression

$$V - E + F$$

is invariant and equal to its Euler characteristic. Thus

$$V - E + F = \chi.$$

Suppose there is a map drawn on some surface. This map can be replaced by a flow as follows:

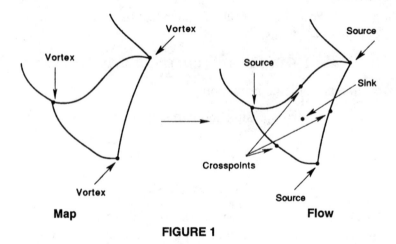

FIGURE 1

1. Replace vertices by sources.

298

2. Put crosspoints at the center of each arc.

3. Put a sink at the center of each region.

We obtain:

$$V \text{ sources of index } 1$$

$$E \text{ crosspoints of index} - 1$$

$$F \text{ sinks of index } 1$$

The sum of indices is

$$V - E + F$$

■

Can you draw these figures without lifting the pencil and passing once and only once through each line. If not, explain why.

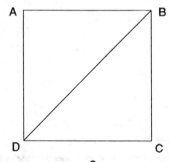 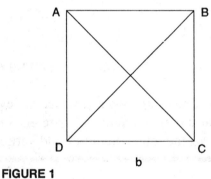

a   b

FIGURE 1

SOLUTION:

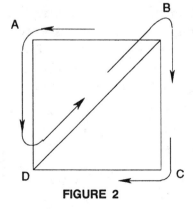

FIGURE 2

Figure 1a can be easily drawn according to the requirements as shown in Figure 2.

We can start from *B* and then move to *A, D, B, C, D*.

It is not possible to draw Figure 2b. Let us return to Figure 2. There are two points *B* and *D* where three lines merge. Then, to make a drawing we have to start with one such point and end with the other. In case of Figure 1b, we have four points, *A, B, C, D* where three lines merge. One can be used to start a drawing, the second to end it; but there is no way we can "get rid" of the remaining two points.

Determine the order of each vertex.

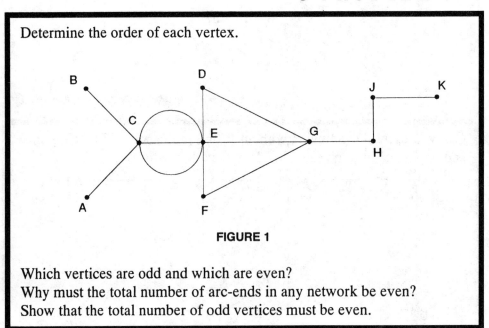

FIGURE 1

Which vertices are odd and which are even?
Why must the total number of arc-ends in any network be even?
Show that the total number of odd vertices must be even.

SOLUTION:

A network consists of a finite number of vertices linked by arcs. The arcs must be non-intersecting. Such a network is sometimes called a linear graph. The number of arc-ends meeting at the vertex is called an order.

A vertex whose order is an even number is called an even vertex. A vertex with an odd order is called an odd vertex.

From the drawing we have

A	1	odd
B	1	odd
C	5	odd
D	2	even

E	6	even
F	2	odd
G	4	even
H	2	even
J	2	even
K	1	odd

The number of arc-ends is twice the number of arcs; hence, it is even.

The number of arc-ends is equal to the sum of the orders of all the vertices. Hence, the total number of odd vertices must be even.

● **PROBLEM 9–49**

Show that the network

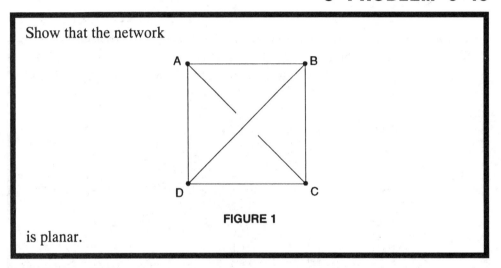

FIGURE 1

is planar.

SOLUTION:

A network that can be mapped onto a simply connected surface in such a way that the arcs don't intersect is called planar.

The network of Figure 1 can be mapped on the plane (see Figure 2).

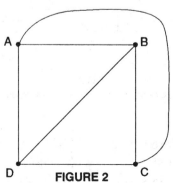

FIGURE 2

This network can also be mapped onto a sphere as shown in Figure 3.

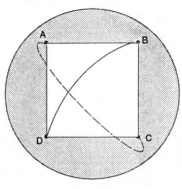

FIGURE 3

● **PROBLEM 9-50**

Suppose there are three houses, *A, B,* and *C.* Each house has to be connected with water, electricity and gas in such a way that the pipes do not intersect. Can it be done?

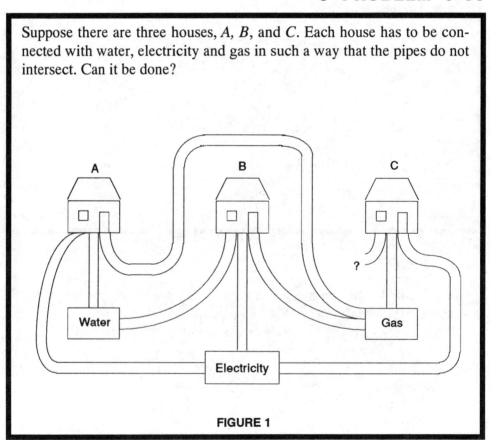

FIGURE 1

SOLUTION:

It cannot be done. Here is why.

Let us connect the first house *A* with water, electricity, and gas, and then proceed through these points to the second house *B,* see Figure 2.

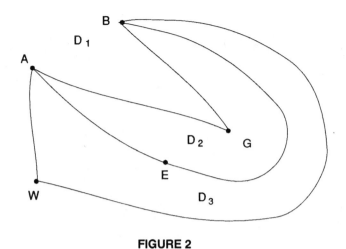

FIGURE 2

We have three lines from A to B which do not cross. These lines divide the whole plane into three areas: D_1, D_2, D_3. The third house C lies in one of the areas.

If it lies in D_1, then E cannot be connected; if in D_2, then W cannot be connected; and if in D_3, then G cannot be connected.

● **PROBLEM 9-51**

Give an example of a non-planar network.

SOLUTION:

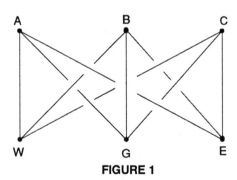

FIGURE 1

Note that the network depicted in Figure 1 illustrates Problem 9-50.

Another example is the complete network on five vertices shown in Figure 2.

It can be proven that every non-planar network must contain either a network of Figure 1 or network of Figure 2 as a sub-network.

303

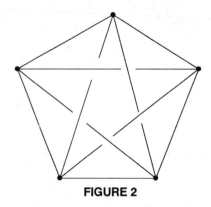

FIGURE 2

A network is complete when all vertices are directly linked to each other by the minimum number of arcs.

● **PROBLEM 9-52**

Give an example of a network that cannot be traversed by a single path.

SOLUTION:

A sequence of arcs which can be followed continuously without any arc being passed more than once is called a path. Figure 1 depicts a path, *ABCD*.

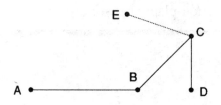

FIGURE 1

A path traverses a network if every arc of the network is included in the path.

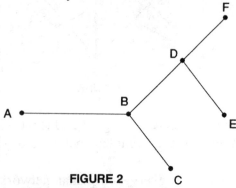

FIGURE 2

No path traverses a network shown in Figure 2.

A closed path (that is such, which starts and finishes at the same vertex) is called a circuit.

A path that is not closed is called open.

● **PROBLEM 9-53**

Show that a network which has more than two odd vertices cannot be traversed by a single path.

SOLUTION:

Figure 1 depicts a network, which

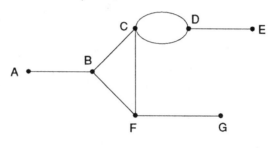

FIGURE 1

cannot be traversed by a single path.

This network has three vertices of the order of one, three vertices of the order of three and one vertex of the order of four. If a path traversing a network is closed, then an even vertex may be the starting and terminating point of the "journey." If a path traversing a network is open, then it starts and terminates at odd vertices. Hence, we conclude that if a network has more than two odd vertices, it cannot be traversed by a single path.

The network in Figure 1 has six odd vertices and cannot be traversed by a single path.

● **PROBLEM 9-54**

Which of the following networks can be traversed by a single path?

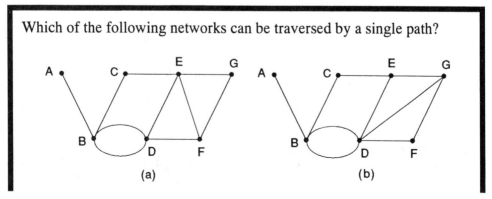

305

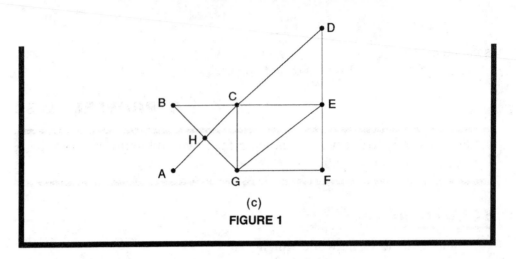

(c)

FIGURE 1

SOLUTION:

A connected network can be traversed by a single open path, if and only if it has exactly two odd vertices.

Network (a) of Figure 1 has two odd vertices A and F. It can be traversed $ABDBCEDFGEF$. Network (b) of Figure 1 has four odd vertices A, E, G, D. It cannot be traversed by a single path.

Network (c) of Figure 1 has two odd vertices A and C, and it can be traversed by a single path $AHCDEFGHBCGEC$.

● **PROBLEM 9-55**

Which of the networks is connected and which is disconnected?

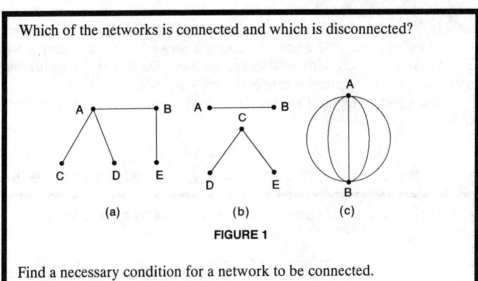

(a) (b) (c)

FIGURE 1

Find a necessary condition for a network to be connected.

SOLUTION:

A network is said to be connected if every pair of vertices belongs to some path; otherwise the network is disconnected.

Networks (a) and (c) are connected.

Network (b) is disconnected, vertices A and C don't belong to any path. Let a be the number of arcs and n the number of vertices. Then a necessary condition for a network to be connected is

$$n - 1 \leq a.$$

This condition is necessary but not sufficient.

● **PROBLEM 9-56**

Reduce a network depicted in Figure 1 to a tree.

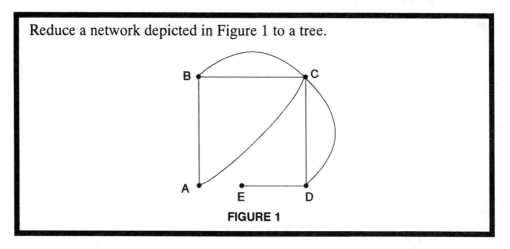

FIGURE 1

SOLUTION:

Any connected network, such that the number of its vertices is one more than the number of its arcs is called a tree. Any connected network can be reduced to a tree by the removal of appropriate arcs. We can remove BC, AC, and CD to obtain a network (see Figure 2) which is a tree.

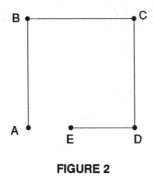

FIGURE 2

● **PROBLEM 9-57**

A planar network is mapped onto a simply connected surface, so that no arcs intersect.

The number of arcs and vertices is given. Determine the number of bounded regions into which the surface is separated.

SOLUTION:

Let v be the number of vertices and a the number of arcs. Then the number of bounded regions is

$$a - n + 1.$$

Indeed, for a triangle

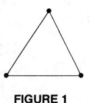

FIGURE 1

we have $3 - 3 + 1 = 1$. For an arc with two vertices $1 - 2 + 1 = 0$.

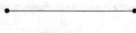

FIGURE 2

According to the network depicted in Figure 2, by adding new arcs and vertices, we can obtain any network.

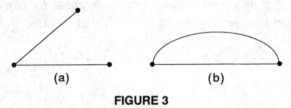

(a)   (b)

FIGURE 3

In Figure 3(a) we added one arc and one vertex, hence the number $a - n + 1$ remains the same. In Figure 3(b) one arc was added and the number of vertices is constant, hence $a - n + 1$ increases by 1.

In such a way, we can obtain any network. That completes the proof.

● **PROBLEM 9-58**

List the tie-sets, which specify the network in Figure 1.

308

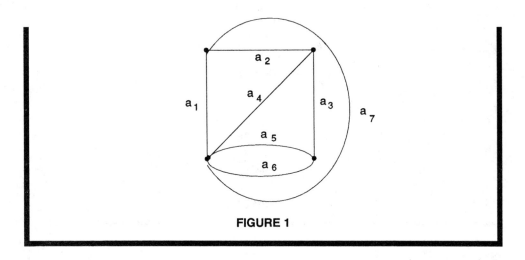

FIGURE 1

SOLUTION:

Any planar network can be completely determined in terms of its independent bounded regions or in terms of paths between its independent pairs of vertices. There are $\frac{V(V-1)}{2}$ pairs but only $v-1$ independent pairs (v is the number of vertices). A network is defined by tie-sets when its structure is determined by paths between its independent pairs of vertices. A tie-set is a single closed path, such that exactly two arcs meet at each vertex. Figure 2 depicts some tie-sets.

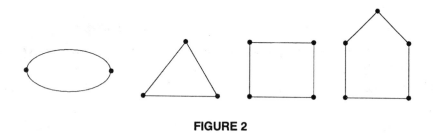

FIGURE 2

Here is one of of possible lists of tie-sets of the network of Figure 1

$$a_5\, a_6,\ a_1\, a_2\, a_4,\ a_3\, a_5\, a_4,\ a_2\, a_3\, a_6\, a_7.$$

Another possibility is

$$a_1\, a_2\, a_3\, a_6,\ a_1\, a_4\, a_3\, a_6\, a_7,\ a_3\, a_4\, a_6,\ a_1\, a_2\, a_3\, a_5$$

or any other combination of four independent closed paths.

● **PROBLEM 9–59**

Is it possible to traverse a network depicted in Figure 1 by a single path?

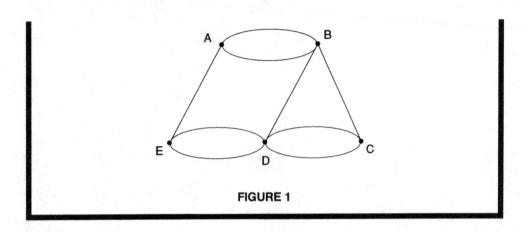

FIGURE 1

SOLUTION:

To be traversed by a single path, a network has to be connected (it is a necessary condition but not sufficient).

If a network has two odd vertices, then these vertices have to be the initial and final vertices of the path. Now suppose a network has an even number $2n > 2$ of odd vertices. Then at least n paths are necessary to traverse it.

The network of Figure 1 has four odd vertices, hence two paths are needed to traverse this network. We can choose for example

$$ABAEDBCDC \quad \text{and} \quad ED.$$

● **PROBLEM 9-60**

Here is the famous problem of the Königsberg bridges solved by Euler. Is it possible to cross them all in turn, passing no more than once over each?

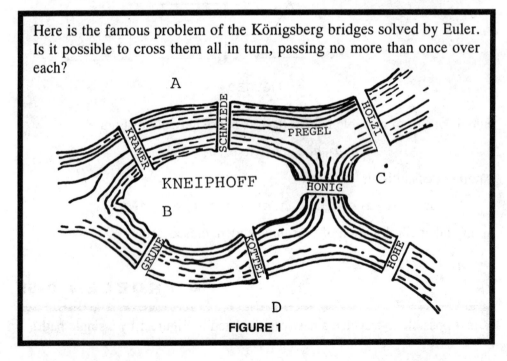

FIGURE 1

310

SOLUTION:

Let us denote the regions as shown in Figure 1.

Then we have to traverse a network shown in Figure 2 with one single path.

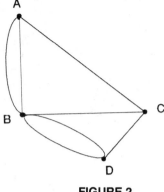

A

C

B

D

FIGURE 2

But this is impossible, because *A, B, C,* and *D* are odd vertices.

● PROBLEM 9-61

Consider again the Königsberg bridges. Where would the eighth bridge be built in order to make the walk possible? Again, we have to cross all bridges by passing only once over each.

SOLUTION:

From Figure 2 of Problem 9-60, we see that there are four odd vertices.

Hence the eighth bridge has to eliminate two odd vertices, for example, *C* and *D* (see Figure 1)

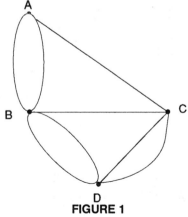

A

B

C

D

FIGURE 1

We have to build a bridge between *C* and *D* as shown in Figure 2.

311

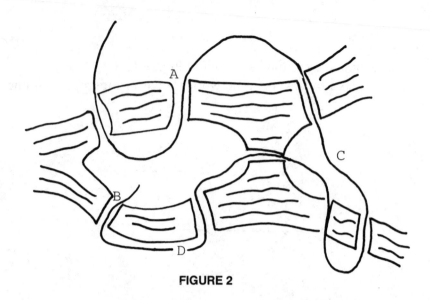

FIGURE 2

The only two odd vertices are now A and B. Hence a walk has to start at A (or B) and end at B (or A).

Is it possible to draw the figure shown in Figure 1 without lifting the pencil and passing once and only once through each point?

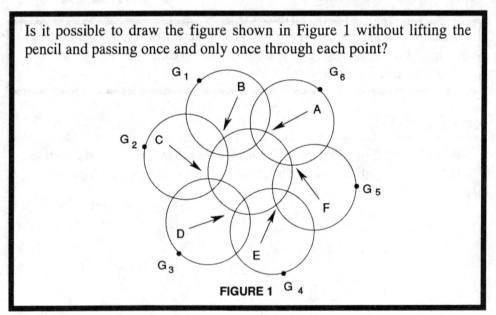

FIGURE 1 G_4

SOLUTION:

Yes. Note that all the vertices are of the order of four.

We can start from A and draw a circle $A\,G_1\,A$, then move to B and draw a circle $B\,G_2\,B$, then move to C and so on, until we draw the last surrounding circle $F\,G_6\,F$. To complete the drawing we sketch FA, which closes the inner circle.

312

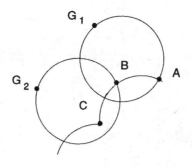

FIGURE 2

It is also possible to draw the Olympic symbol (see Figure 3) without lifting the pencil and passing only once through each point.

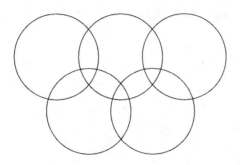

FIGURE 3

● **PROBLEM 9–63**

Here is an odd puzzle. It's quite possible that you already know the solution. If you don't, try to find it.

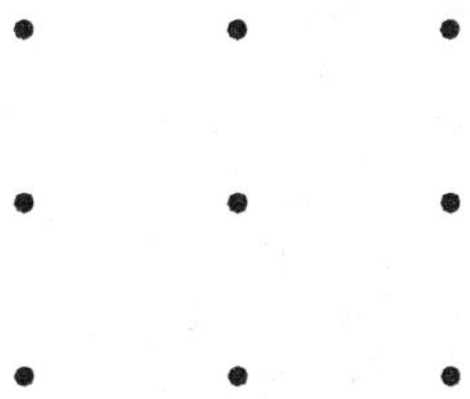

FIGURE 1

The network consists of 9 dots (see Figure 1). Draw 4 lines without lifting the pencil from the paper so as to cross out every dot.

SOLUTION:

To solve the problem, we have to leave the frame of the dots.

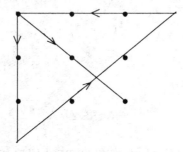

FIGURE 2

Another solution is shown in Figure 3. Here the line crosses each dot once and only once.

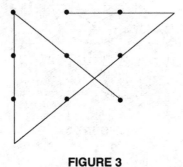

FIGURE 3

● PROBLEM 9-64

Suppose you own a mansion with the floor-plan depicted in Figure 1.

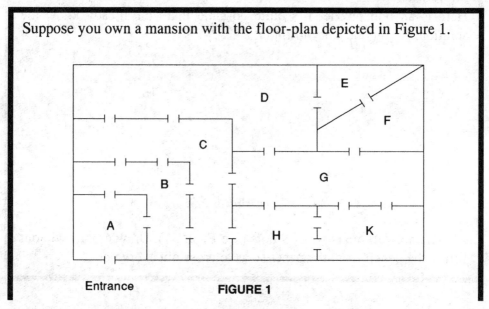

Entrance   **FIGURE 1**

314

To support yourself, you have to guide tourists through each room. Is is possible to traverse the mansion starting at the entrance and passing each door once and only once?

SOLUTION:

Let us denote the rooms A, B, C, D, E, F, G, H, and K.

Each room represents a vertex of a network. The number of doors is the order of the vertex, hence A has order 3, $B - 4$, $C - 4$, $D - 4$, $E - 2$, $F - 2$, $G - 6$, $H - 4$, $k - 4$. There is only one odd vertex. It is possible to traverse the network by a single path. One possible route is shown in Figure 2.

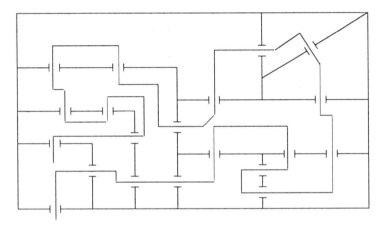

FIGURE 2

● **PROBLEM 9-65**

Prove the following theorem:

THEOREM

Every network with all even vertices, or with, at the most, two odd vertices can be traversed by a single path.

SOLUTION:

Consider first, the case when all vertices are even.

We can start at an arbitrary vertex A. Since A is even, the path must end at A. If parts of the network not already visited remain and the order of A is higher than 2, we can draw another closed path.

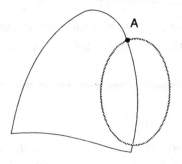

FIGURE 1

Two closed paths with a common point A can be considered as one closed path.

If the task is not yet accomplished, the new path must be linked at B with the rest of the network.

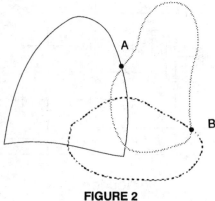

FIGURE 2

We can repeat this procedure until we reach the solution, that is when the path traverses the whole network.

The reasoning is almost the same for the case of a network with two odd vertices A and B.

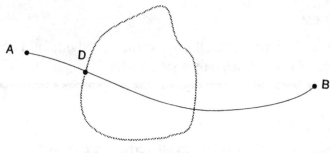

FIGURE 3

We start at A (or B) and end at B (or A). The rest of the network contains only even vertices previously discussed and is linked to AB at some point D.

What regular polygons with all diagonals can you draw with one stroke, that is, without lifting a pen and passing once and only once through each one?

SOLUTION:

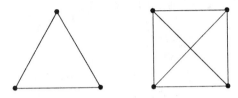

FIGURE 1

It is possible to draw a triangle but not a square, which has four odd vertices.

For a corresponding network to have only even vertices, the polygon must have an odd number of sides. Figure 2 depicts 5-gon and 7-gon.

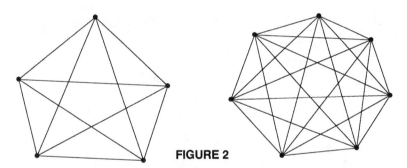

FIGURE 2

If you have enough patience you can draw polygons of higher order 19-, 21-, 23-, 25-gons etc. Figure 3 illustrates 23-gon with all of its diagonals.

FIGURE 3

317

Suppose you have three chains with clips. Arrange them in such a way that they interlock as a whole but individual pairs do not interlock. That is, if you open any one chain, they all become separated.

SOLUTION:

Figure 1 does not offer a solution.

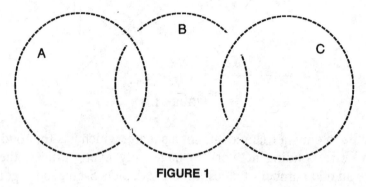

FIGURE 1

Opening *B* separates all of the chains. But opening *A or* C does not. Figure 2 illustrates the solution.

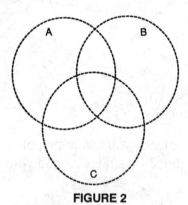

FIGURE 2

Opening any chain separates the remaining two.

The knowledge of knots is part of topology. It has been proven that it is impossible to tie two knots at two ends of a rope in such a way that when brought together they may cancel each other.

● **PROBLEM 9-68**

Prove that five (or more) countries are never neighbors of each other.

SOLUTION:

It is easy to show that four countries can be neighbors of each other, see Figure 1.

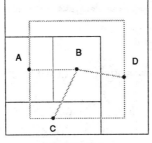

FIGURE 1

Let A, B, C, and D be the capitals or respective countries. Their capitals can be connected by roads that pass through only two countries (Figure 1). Connecting the capitals A, B, and C we get the triangle shown in Figure 2.

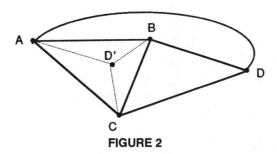

FIGURE 2

D can be interior or exterior of this triangle. In both cases, we obtain a large triangle composed of three small adjacent triangles. E is the capital of the fifth country. It lies in one of the small triangles or outside the large triangle, Figure 3.

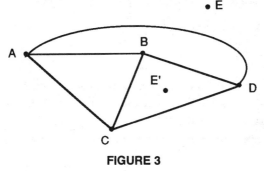

FIGURE 3

In either case, E is separated from one capital by a triangle of roads. To get from E to B (in our case) one has to cross at least one existing road.

Hence, countries with capitals B and E are not neighbors. That completes the proof.

319

Show that the chromatic number of a plane surface is larger than three.

SOLUTION:

Here we deal with the famous four-color problem. In 1976 it was proven that:

every map drawn on a sheet of paper can be colored with only four colors in such a way that countries having a common border receive different colors.

This problem was formulated by Francis Guthrie in 1852.

The least number of colors required to color a map on any given surface is called the chromatic number of that surface. A chromatic number is a topological property of the surface.

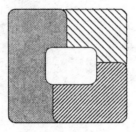

FIGURE 1

The drawing shows that three colors are insufficient to color a map on the plane surface. We will prove later that five colors are sufficient to color any map. So far we know that:

$$\text{chromatic number of a plane surface} \geq 4.$$

Derive the regular map with the least number of additional vertices and arcs from the non-regular map depicted in Figure 1.

SOLUTION:

A map is a network consisting of a number of vertices and non-intersecting arcs linked together in a such a way that an area is separated into simply connected regions. Figures 1 depicts a map.

A map is called regular, if

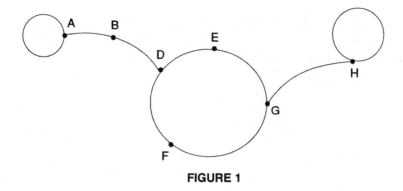

FIGURE 1

1. its vertices are of order three or more.

2. each arc separates two distinct regions.

3. each arc joins two distinct vertices.

The map of Figure 1 is not regular. We shall add some vertices and arcs to make it regular.

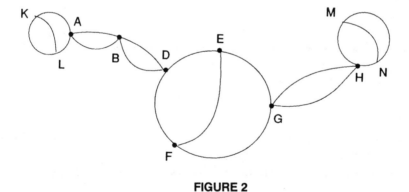

FIGURE 2

All vertices of the map of Figure 2 are of an order higher than or equal to three. The new map is regular.

● PROBLEM 9–71

In problem 9-68, we proved that it is not possible for five countries to be located in such a way that each of them is adjacent to each of the other four. This result led Augustus DeMorgan (who was the first one to prove it) to believe that he solved the four-color problem. Why was he wrong?

SOLUTION:

The fact that five mutually adjacent countries do not exist on a map does not constitute a proof of the four-color conjecture. Quite a few mathematicians have made this mistake. Consider a map of six countries.

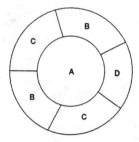

FIGURE 1

Among the six countries, there is no collection of four in which each member is adjacent to the other three. Nevertheless it takes four colors to color this map. From the figure we see that the number of colors required for a map (here 4) is not the same as the maximum number of mutually adjacent countries (here 3).

● **PROBLEM 9–72**

Prove the following:

THEOREM

Any regular map on the plane surface or the surface of a sphere must have at least one region bounded by fewer than six arcs.

SOLUTION:

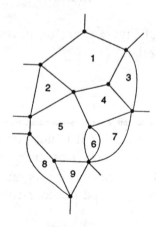

FIGURE 1

322

Let χ be the Euler characteristic, then

$$V - E + F = \chi \tag{1}$$

where V is the number of vertices, E, arcs, and F regions. Hence, for any surface of the Euler characteristic $\chi > 0$, we have

$$V - E + F > 0 \quad \text{or} \quad 6V - 6E + 6F > 0. \tag{2}$$

Since the map is regular, (see Problem 9-70) all vertices are of the order of 3 or higher.

$$2E \geq 3V \tag{3}$$

From (2) and (3), we obtain

$$6F - 2E > 0 \tag{4}$$

Let F_n denote the number of regions, each bounded by exactly n arcs. In Figure 1, regions 2, 3, 7, 8, and 9 are bounded by 3 arcs, hence $F_3 = 5$. While $F_2 = 1$, $F_4 = 1$, $F_5 = 1$, $F_6 = 1$. The map is regular, therefore

$$F_1 = 0 \tag{5}$$

The total number of regions is

$$F = \sum_{n=2,3,4\ldots} F_n \tag{6}$$

From (4), (5), and (6), we find

$$6 \sum_{n=2,3,\ldots} F_n - \sum_n nF_n = \sum_{n=2,3,\ldots} (6-n)F_n > 0 \tag{7}$$

Then

$$4F_2 + 3F_3 + 2F_4 + F_5 - F_7 - \ldots > 0 \tag{8}$$

Hence some $n < 6$ must exist for (8) to hold. Thus, any regular map on a surface of the Euler characteristic > 0 must have at least one region bounded by less than six arcs.

● PROBLEM 9–73

Using the results of Problem 9-72, prove the six color theorem.

SIX COLOR THEOREM

Any regular map on a surface of the Euler characteristic > 0 requires six colors at the most, if no neighboring regions are to be colored the same.

SOLUTION:

Obviously the theorem is true for $F \leq 6$. We shall prove that, if the theorem holds for F', then it holds for $F' + 1$. We proved in Problem 9-72 that at least one region exists which is bounded by less than six arcs, that is by two, three, four, or five arcs. Suppose this region shrinks to a point. The four possibilities are illustrated below.

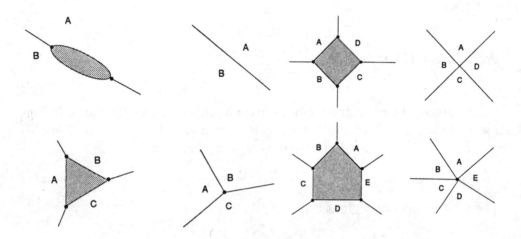

For all cases with a contracted map, which contains F' regions the theorem holds by assumption. That is each of the contracted maps requires six colors. From the drawing, we see that when the region is restored there is a color available for it without the total of six colors being exceeded. Hence, if the theorem is true for F, then it is true for $F +1$. Since it is true for $F \leq 6$, the proof is completed.

● PROBLEM 9–74

Any regular map on a surface of the Euler characteristic χ can be colored by at most β colors, where

$$\beta F > 6(F - \chi) \qquad (1)$$

Show that at least one region must have a boundary consisting of less than β arcs.

SOLUTION:

For any regular map

$$V - E + F = \chi \qquad (2)$$

$$2E \geq 3V$$

Thus

$$6(E - V) = 6(F - \chi) \geq 2E \qquad (3)$$

Combining (1) and (3) we find

$$\beta F > 2E \qquad (4)$$

Since the map is regular

$$\beta \sum_n F_n > \sum_n nF_n \qquad (5)$$

$$\sum_n F_n (\beta - n) > 0 \,, \; n \geq 2 \qquad (6)$$

From (6) we conclude that at least one region must have a boundary consisting of less than β arcs.

● **PROBLEM 9-75**

Determine the smallest integer β_0 satisfying

$$F\beta > 6(F - \chi) \qquad (1)$$

when $F > \beta$. Consider the following values of χ : 0, 1, 2, –1, –2, ..., –10.

SOLUTION:

For $\chi = 0$, we have

$$\beta > 6. \qquad (2)$$

The smallest integer β_0 satisfying (2) is $\beta_0 = 7$.
 Let us write (1) in the form

$$\beta > 6(1 - \tfrac{\chi}{F}) \qquad (3)$$

Then for $\chi = 1$, we obtain $\beta_0 = 6$. Similarly for $\chi = 2$, we find

$$\beta > 6 - \tfrac{12}{F} \qquad (4)$$

The smallest integer β_0 satisfying (4) is $\beta_0 = 6$.
 Note that the same values of β_0 were obtained from the six-color theorem.
 We find:

$$\text{for } \chi = -1, \; \beta > 6 + {}^6/_F, \; \beta_0 = 7 \qquad (5)$$

$$\text{for } \chi = -2, \; \beta > 6 + {}^{12}/_F, \; \beta_0 = 8 \qquad (6)$$

for $\chi = -3$, $\beta > 6 + {}^{18}/_F$, $\beta_0 = 9$

for $\chi = -4$, $\beta > 6 + {}^{24}/_F$, $\beta_0 = 9$

for $\chi = -5$ $\beta > 6 + {}^{30}/_F$, $\beta_0 = 10$

for $\chi = -6$ $\beta > 6 + {}^{36}/_F$, $\beta_0 = 10$

for $\chi = -7$ $\beta > 6 + {}^{42}/_F$, $\beta_0 = 10$

for $\chi = -8$ $\beta > 6 + {}^{48}/_F$, $\beta_0 = 11$

for $\chi = -9$ we get $\beta_0 = 11$

for $\chi = -10$ we get $\beta_0 = 12$ (6)

We can summarize the results:

χ	β_0
2	6
1	6
0	7
-1	7
-2	8
-3	9
-4	9
-5	10
-6	10
-7	10
-8	11
-9	11
-10	12

● **PROBLEM 9-76**

From the inequality

$$\beta F > 6(F - \chi) \text{ for all } F > \beta \tag{1}$$

derive the formula for β_0 for negative values of χ (see Problem 9-75).

SOLUTION:

From (1) we have

$$\beta > 6(1 - \chi/_F) \tag{2}$$

Since $\chi < 0$, we can substitute for F the smallest admissible $\beta + 1$. Thus

$$\beta > 6(1 - \frac{x}{\beta+1}) \tag{3}$$

and

$$(\beta + 1)\beta > 6\beta + 6 - 6\chi \tag{4}$$

or

$$(\beta - \tfrac{5}{2})^2 > \tfrac{49}{4} - 6x \tag{5}$$

$$\beta > \sqrt{\tfrac{49}{4} - 6x} + \tfrac{5}{2} \tag{6}$$

The smallest integer β_0 satisfying (6) is

$$\beta_0 = \text{int}\left[\sqrt{\tfrac{49}{4} - 6x} + \tfrac{5}{2}\right] + 1 \tag{7}$$

where int $x = a$ means a is an integer and $x \geq a$.

For negative values of χ, we obtain:

χ	β_0
-1	7
-2	8
-3	9
-4	9
-5	10
-6	10
-7	10
-8	11
-9	11
-10	12
-11	12
-12	12
-13	13
-14	13
-15	13

327

Show that seven colors are necessary to paint a torus.

SOLUTION:

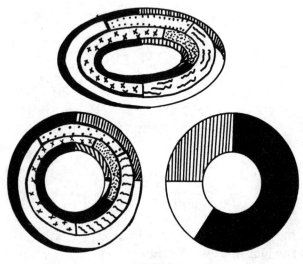

FIGURE 1

Figure 1 shows a torus divided into seven regions. Each country is a neighbor to the remaining six countries. We see that to paint a map on the torus, one needs at least seven colors. Hence the chromatic number β of the torus is

$$\beta \geq 7. \qquad (1)$$

The Euler characteristic χ of the torus is χ = 0. From Problem 9-75, we obtain

$$\beta \leq 7. \qquad (2)$$

Hence, the chromatic number of the torus is 7.

● **PROBLEM 9-78**

Discuss the results of Problems 9-69, 9-71, 9-73, 9-75, 9-77.

SOLUTION:

A chromatic number, defined as the least number of colors required to color a map on any given surface, is a topological property of the surface. We tried to determine the chromatic number as a function of the Euler characteristic of the surface. Here are the results:

from Problem 9-69	for $\chi = 2$	$\beta \geq 4$
from Problem 9-73	for $\chi > 0$	$\beta \leq 6$
from Problem 9-75	for $\chi = 2$ or 1	$\beta \leq 6$
	for $\chi = 0$	$\beta \leq 7$
	for $\chi = -1$	$\beta \leq 7$
	for $\chi = -2$	$\beta \leq 8$
	for $\chi = -3$	$\beta \leq 9$
	for $\chi = -4$	$\beta \leq 9$
	for $\chi = -5$	$\beta \leq 10$
	for $\chi = -6$	$\beta \leq 10$
	for $\chi = -7$	$\beta \leq 10$
	for $\chi = -8$	$\beta \leq 11$
from Problem 9-77	for $\chi = 0$	$\beta \geq 7$

Note that the table in Problem 9-75 gives only the sufficient numbers of colors. Another problem is to determine the necessary number of colors.

It has been proven that for the surfaces of the Euler characteristic $\chi = 1$ or $\chi = 0$ or $\chi =$ an even negative integer, the values obtained in Problem 9-75 are necessary as well as sufficient. For those surfaces, the calculated values of β are the chromatic numbers. Hence

χ	Chromatic Number
1	6
0	7
-2	8
-4	9
-6	10
-8	11
-10	12

● **PROBLEM 9-79**

Suppose the world has the shape of a two-fold torus. What is the least number of colors needed to depict on the model of the world political situation?

329

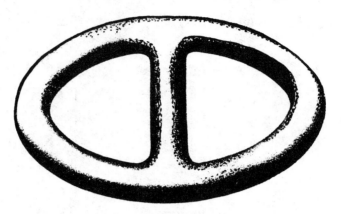

FIGURE 1

SOLUTION:

The Euler characteristic of a two-fold torus is $\chi = -2$. From the table (Problem 9-78), we find that chromatic number of a two-fold torus is $\beta = 8$. Hence, eight colors are needed to paint any map on the surface of a two-fold torus.

● **PROBLEM 9-80**

It has been proven that the maximum number of colors required for regular maps on a sphere or on a plane is six. Using Heawood's argument, show that this number can be reduced to five.

SOLUTION:

Any regular map on a sphere contains a region bounded by less than six arcs.

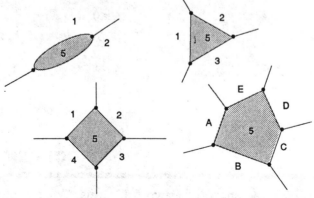

FIGURE 1

330

This region can always be painted with the color number five. It's obvious for the regions bounded by two, three, or four arcs.

For the region having five arcs as a boundary, some pair of the regions A, B, C, D, E must have no common boundary. Indeed, suppose A and C have a common boundary, then B has no common boundary with D (or E). Thus we need four colors to paint A, B, C, D, and E and the fifth one is available.

Suppose one arc is removed from the region bounded by two, three, or four arcs.

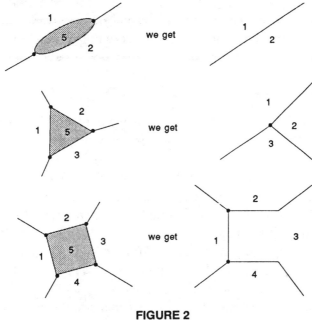

FIGURE 2

The number of regions decreases by one. If the maps in Figure 2 to the right can be colored with five colors, so can the original maps to the left.

If, for instance, A and C are of the same color, then by removing two arcs (Figure 3) we decrease the number of regions by two.

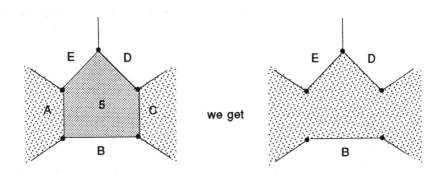

FIGURE 3

331

Again, if the map to the right (Figure 3) can be colored with five colors, then so can the original map. Gradually we reduce the number of regions until it is less than six. Since that can obviously be colored with less than six colors, the original map requires five colors or less. This result is known as Heawood's five color theorem.

● **PROBLEM 9–81**

How many colors are required to color every map on the plane surface?

SOLUTION:

Four! In 1976 the four-color problem was solved. For over a hundred years many mathematicians have tried in vain to prove (or disprove) this simple statement:
Four colors are required to color any map on the plane surface.
Attempting to find the solution, mathematicians like Heawood, Kempe and Birkhoff developed the new branch-graph theory, now used in arranging the airline routes and wiring diagrams.

In 1975 a group of scientists at the University of Illinois, using an IBM 360 computer, proved the four-color conjecture. The method used involved the reducibility of over 1,500 configurations.

It is amazing that the complete proof of this simple statement consists of hundreds of pages, miles of calculations and is impossible to carry out without the use of the computer.

Chances are that one day some bright teenager will find a short and elegant proof. It is also conceivable that no such proof is possible.

● **PROBLEM 9–82**

1. Which of the points A, B, C, and D lie inside the curve and which lie outside the curve?

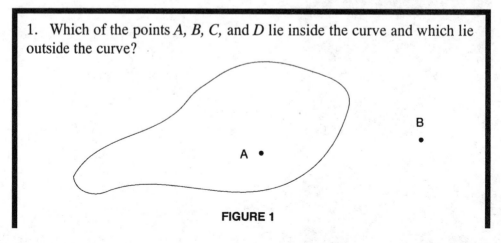

FIGURE 1

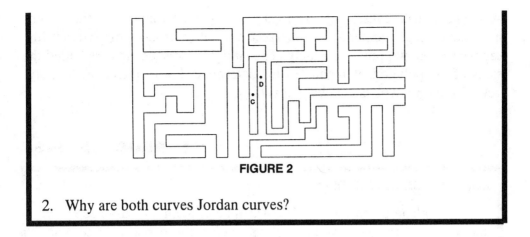

FIGURE 2

2. Why are both curves Jordan curves?

SOLUTION:

1. It is obvious that a triangle (or circle) separates a plane surface into an area inside and outside its perimeter. The same property is assumed for any continuous non-self-intersecting closed curve. For the curve of Figure 1, it is easy to see that A lies inside its perimeter and B outside its perimeter.

For more complicated curves, such as in Figure 2, it takes a while to determine what is inside and what is outside. In this case, C is outside and D is inside.

Even though the terms "inside" and "outside" are intuitively simple, they require a rigorous mathematical treatment.

2. By definition, any curve homeomorphic to a circle is called a Jordan curve.

● **PROBLEM 9–83**

Define and give an example of a polygonal path.

SOLUTION:

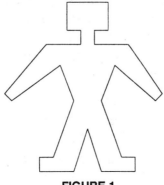

FIGURE 1

A polygonal path consists of n straight line segments joining n distinct points in a plane. Line segments can intersect only at their end points. Each line segment joins two points uniquely. The straight line segments are called the sides of the polygon. Polygonal paths determine the perimeter of the polygon. The figure depicts a polygonal path.

● PROBLEM 9–84

JORDAN CURVE THEOREM

On a plane or on the surface of a sphere, a Jordan curve separates the surface into two disjoint regions having the curve as a common boundary.

■

It is surprising that this theorem is obvious to almost everybody but it requires a fairly complex proof.

Prove the Jordan curve theorem for a polygonal path.

SOLUTION:

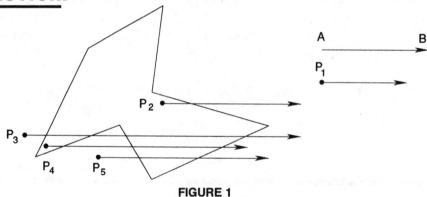

FIGURE 1

For a given polygonal path, it is always possible to find a line segment AB which is not parallel to any of the sides. This separates the plane (with the exception of the polygonal path) into two disjoint sets:

α - the set of points such that a ray from the point in the direction parallel to AB intersects the polygonal path an even number of times;

β - an odd number of times.

$$P_1 \in \alpha, \quad P_3 \in \alpha, \quad P_5 \in \alpha,$$
$$P_2 \in \beta, \quad P_2 \in \beta$$

from Figure 1.

Suppose a point moves along a line segment not intersecting a polygonal path and not parallel to AB, (Figure 2).

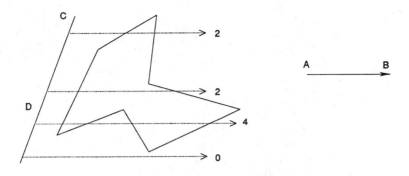

FIGURE 2

The number of intersections of the ray with the polygonal path always changes by a multiple of two. Thus, if any point belonging to α is joined to any point of β by a polygonal path, this new path must cross the old one an odd number of times. See Figure 3.

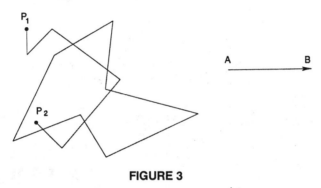

FIGURE 3

If both points P_1 and P_2 belong to the same subset (α or β), they can be joined by a polygonal path not intersecting the original polygonal path.

Thus it is possible to identify α as the set of points outside the polygonal path and β as the set of points inside.

● **PROBLEM 9–85**

Look again at the proof of the Jordan curve theorem for a polygonal path (Problem 9-84). Why does the same method applied to any Jordan curve lead to difficulties?

SOLUTION:

Since a polygonal path contains the finite number of sides of the polygon, it is easy to find a direction not parallel to any side.

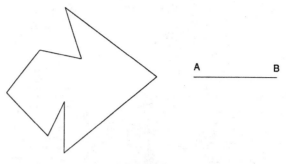

FIGURE 1

AB is not parallel to any side. For a closed curve it may no longer be possible to find a direction which is nowhere tangential to a curve. Obviously for a circle, no straight line in the plane exists, which is not parallel to some tangent to the circle. See Figure 2 where *CD* is parallel to *AB*.

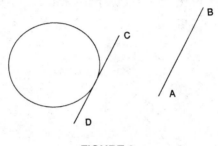

FIGURE 2

● **PROBLEM 9–86**

Another amazing statement in topology is that in a plane there are infinitely many simply connected bounded regions which all have the same boundary. Consider this puzzle for three regions:

FIGURE 1

There is an island. Within the first day you have to dig three canals:

from the sea, from the cold lake, and from the warm lake in such a way that the waters are separated and any point of the island is less than 1 mile from any water by the end of the day.

In the next half day the digging continues according to the same principles and each point has to be less than a $1/2$ mile from each kind of water. The work is carried out in intervals 1 day, $1/2$, $1/4$, $1/8$, $1/16$, ... of the day.

What happens at the end of the second day?

SOLUTION:

The island (or whatever is left of it) forms a closed set A nowhere dense. Each point of A is arbitrarily near to each kind of water.

Note that A is the common boundary of three regions: the sea, the warm lake, and the cold lake.

● **PROBLEM 9–87**

Rotate a disc and an annulus in its own plane about its own center. What point (or points) is always mapped into itself?

SOLUTION:

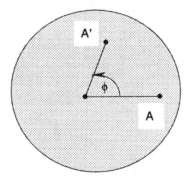

FIGURE 1

Let ϕ be the angle of rotation. The transformation is rigid, one-to-one and continuous. Each point A of the disc is mapped into A'. For ϕ not an integer multiple of 2π, there is one point only, which maps into itself, namely the center of the disc.

For the same rotation of an annulus there is no point which maps into itself.

337

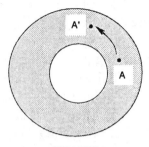

FIGURE 2

● **PROBLEM 9-88**

Explain the result of Problem 9-87 in terms of Brouwer's fixed point theorem.

SOLUTION:

BROUWER'S FIXED POINT THEOREM

For any continuous transformation of a disc to itself, there is at least one point which is mapped to itself.

■

Since rotation is a continuous transformation, at least one point of a disc is mapped into itself.

There is no continuous transformation f of a disc into itself

$$f : D \to D$$

such that, for every $x \in D$, $f(x) \neq X$.

Brouwer's fixed point theorem deals with the topological property, hence it holds for any region homeomorphic to a disc. It can be applied to various situations in life.

Consider for example a pool of oil spilled on the road.

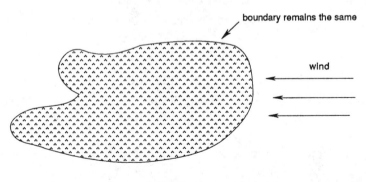

FIGURE 1

338

When the breeze blows, the surface of oil is moved. Assuming the boundary of oil remains the same and the oil is not "broken" in any way, there is at least one point where the oil remains in exactly the same place as it was before the wind began to blow.

A thin sheet of an elastic material is stretched as shown in Figure 1a.

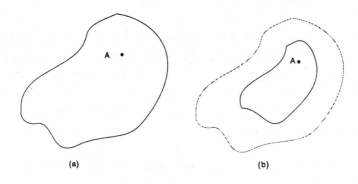

(a)   (b)

FIGURE 1

Show that when the material contracts so as to occupy only a part of its original area (Figure 1b), then there will always be a point A that occupies the same location before and after contraction.

SOLUTION:

Contraction is a continuous transformation, thus we can apply Brouwer's fixed point theorem.

We conclude that at least one point occupies the same place after contraction as before. Note that the same is true if we stretch out an elastic material.

Prove Problem 9-89 without direct reference to Brouwer's fixed point theorem.

SOLUTION:

Suppose the original region is *colored in the form of a chessboard* and point P occupies P' after contraction. All squares can be grouped into three categories:

I. For every point belonging to the square, P' lies nearer the right side than this point.

II. For every point belonging to the square, P' lies nearer the left side than this point.

III. Squares that are neither I nor II.

It is easy to see that the squares of type I and II are never neighbors. Thus, square III goes from top to the bottom and a line can be drawn passing only through the squares of type III, which join the upper edge with the base.

Let us draw an arrow PP' at each point of this line. Since not all P' can lie higher than P not all arrows are directed upward. By the same token not all arrows are directed downward. Hence at least one point Q exists with a horizontal QQ'.

Since Q belongs to square III, there must be point R in this square with a vertical RR'.

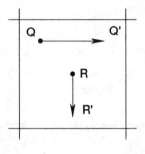

FIGURE 1

Since the square is small, such an abrupt change of direction is impossible (note that transformation is continuous) for a large PP'.

Hence, the arrow PP' for all points of the square must be small. When the number of squares increases, we reach the limit of point P_0, such that

$$P_0 P_0' = 0 \quad \text{that is,} \quad P_0 = P_0'$$

and P_0 is the fixed point of the transformation.

● PROBLEM 9–91

Here is a curious consequence of the Brouwer's fixed point theorem:
At any given time there are two points on the earth which are antipodes and which have the same temperature and the same air pressure.

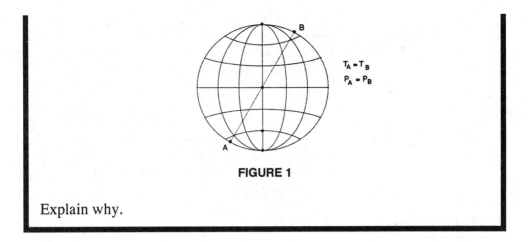

FIGURE 1

Explain why.

SOLUTION:

Suppose you have a sphere made of elastic material, for example, an inflatable ball. If the ball is folded and deflated so as to become flat, then there are two antipodes, A and B, which will lie one upon another.

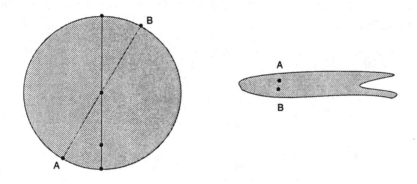

FIGURE 2

It can be proven by using Brouwer's fixed point theorem. Since both temperature and pressure are continuous, the conclusion follows.

● PROBLEM 9-92

Here is a fixed point theorem for one dimension:
  If an interval is continuously transformed to itself, then there is at least one point of the interval which remains fixed. ■

One way to prove it is to divide the interval into small segments. List all complete segments for each division of the interval AB shown below.

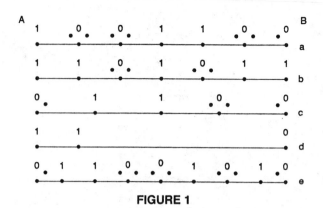

FIGURE 1

SOLUTION:

On the original segment AB we arbitrarily choose any number of points. All these points and the end-points A and B are labeled at random, 0 or 1.

On both sides of every 0, we put a dot. When a zero is at the end-point A or B, it gets one dot.

A small segment (i.e., one that does not include any subsegments) is called a complete segment when it has a 0 at one end and a 1 at the other.

For each division a to e, we list all small segments and circle the complete segments.

a. (10), 00 ,(01), 11 , (10), 00

b.   11 ,(10),(01), (10), (01), 11

c. (01), 11 ,(10), 00

d.   11 ,(10)

e. (01), 11 ,(10), 00 , (01), (10), (01), (10)

Each complete segment contains one dot, other segments contain no dots or two dots.

● PROBLEM 9–93

Show that if the end-points of the original line segment are labeled 0 and 1, then any division of this segment contains at least one complete segment.

342

SOLUTION:

Let n be the number of complete segments. Then the total number of dots is n plus some even positive integer

$$\# \text{ dots} = n + 2k.$$

On the other hand,

$$\# \text{ dots} = \# \text{ of 0's at the end-points} + 2 \cdot l$$

where l is the number of interval 0's.

We conclude that if the end-points of the original segment are labeled 0 and 1, then the number of complete segments must be odd. Thus, an original segment 01 (or 10) contains at least one complete segment.

● **PROBLEM 9-94**

By applying Problems 9-92 and 9-93, prove Brouwer's fixed point theorem for one dimension.

SOLUTION:

Suppose the original line segment, labeled 01, is continuously transformed into itself.

We label the points of the transformed segment 0, if their distance from the end-point 0 has not decreased and 1, if their distance from end-point 1 has not decreased.

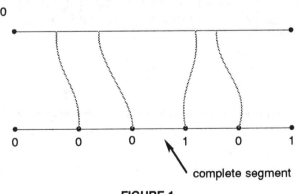

FIGURE 1

There must be at least one complete segment and since the labeling can go on infinitely, this segment can be made arbitrarily small.

At the limit, this segment tends to a single point, which is the fixed point we are seeking.

Prove Brouwer's fixed point theorem for two dimensions.

SOLUTION:

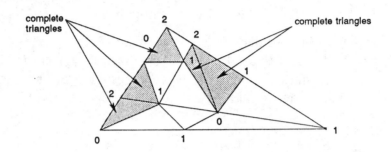

FIGURE 1

We shall use a method similar to the one-dimensional case. Consider a triangle arbitrarily subdivided into smaller triangles. The original triangle is labeled 012. The vertices lying on the side 01 are labeled 0 or 1 only, vertices on 02 are labeled 0 or 2 only and vertices on 12 are labeled 1 or 2 only.

A complete triangle is defined as a small triangle with vertices labeled 012.

This division of an original triangle contains at least one complete triangle. Next we continuously transform and divide the original triangle. Diminishing the complete triangle becomes the fixed point.

The above theorem is sometimes called Sperner's lemma.

In the same manner, we can prove Brouwer's fixed point theorem for any dimension.

Make a necessary cut of the cylinder (Figure 1) and draw the corresponding plane diagram.

SOLUTION:

Curves x and y are the boundaries of the cylinder. By cutting the cylinder along the line AB and then opening it, we obtain a rectangle. To ensure that the rectangle can be folded back, so as to give the cylinder we described its vertices; and use arrows to indicate that there is no twisting.

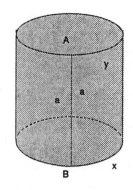

FIGURE 1

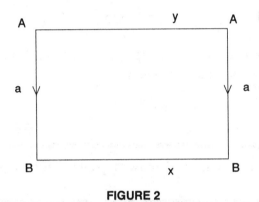

FIGURE 2

The diagram obtained as such is called a plane diagram.

For the torus depicted in Figure 1, make the needed cuts and sketch its corresponding plane diagram.

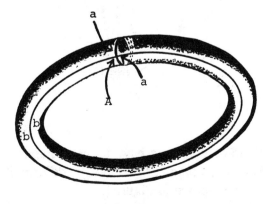

FIGURE 1

SOLUTION:

Two cuts, *a, b* are shown in Figure 1. The first cut, *a*, enables us to straighten the torus so as to obtain a cylinder.

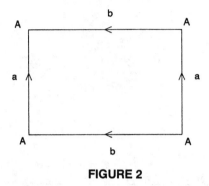

Here again arrows indicate that there is no twisting. Also note that all four vertices are labeled with the same letter.

● PROBLEM 9-98

Make the necessary cuts and draw a plane diagram (rectangle) of a sphere.

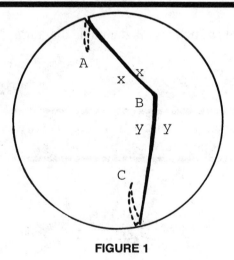

FIGURE 1

SOLUTION:

To obtain a plane diagram, we have to cut the sphere twice as shown in Figure 1. To obtain a rectangle, we have to deform the cuts. The plane diagram of the sphere is shown in Figure 2.

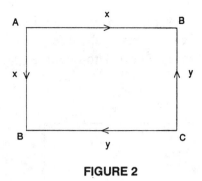

FIGURE 2

Sketch the plane diagrams representing a Möbius band and a Klein bottle.

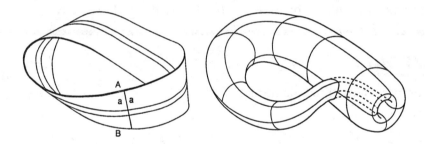

FIGURE 1

SOLUTION:

We cut a Möbius band along the line AB and then twist it (Figure 2). Note that we used two different letters, x and y, since the edges are not joined together.

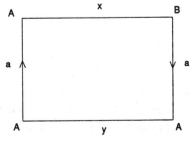

FIGURE 2

Note that from the plane diagram of a Klein bottle it is impossible to obtain a Klein bottle in three-dimensional space.

We have to make two cuts in order to obtain a plane diagram of a Klein bottle (Figure 3).

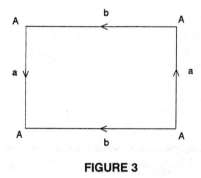

FIGURE 3

Draw the plane diagram of the real projective plane. Compare the plane diagrams of a Möbius band, Klein bottle and the real projective plane.

SOLUTION:

Compare these three diagrams:

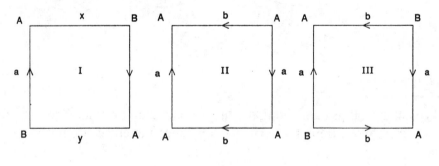

FIGURE 1

I is the plane diagram of a Möbius band. By adding arrows to avoid twisting, we obtain a diagram of a Klein bottle. If the opposite sides of the rectangle are to be joined together, then we should have to place the arrows in opposite directions as in diagram III. III is the plane diagram of the real projective plane, where each pair of opposite sides is joined in the opposite direction (with a twist).

The real projective plane can be represented as follows:

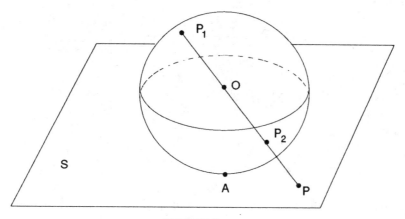

FIGURE 2

There is a sphere and a plane, which is tangential to the sphere at point A. Points P_1 and P_2 on the sphere are mapped to the single point P on the plane. Every great circle of the sphere is mapped to a line, with the exception of the great circle parallel to S. This circle is mapped into a line at infinity. The Euclidean plane, with the line of infinity added, is called the real projective plane.

● **PROBLEM 9–101**

From the plane diagrams, obtain the Euler characteristic χ of

1. a cylinder

2. a one-fold torus

3. a sphere

4. a Möbius band

5. a Klein bottle

6. a real projective plane.

SOLUTION:

We shall apply the expression

$$\chi = V - E + F. \tag{1}$$

For a cylinder

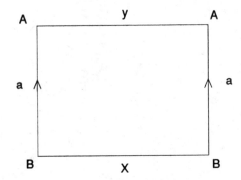

there are two distinct vertices, $V = 2$, three distinct sides, $a, x, y, E = 3$, and one face $F = 1$. Thus

$$\chi = 2 - 3 + 1 = 0.$$

For a one-fold torus

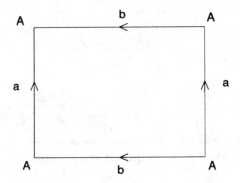

$V = 1, E = 2, F = 1$, hence

$$\chi = 1 - 2 + 1 = 0.$$

For a sphere

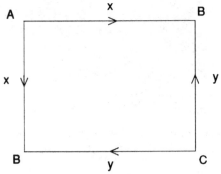

$V = 3, E = 2, F = 1$, hence

$$\chi = 3 - 2 + 1 = 2.$$

For a Möbius band

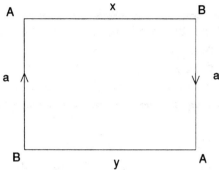

$V = 2, E = 3, F = 1$, hence

$$\chi = 2 - 3 + 1 = 0.$$

For a Klein bottle

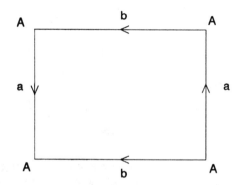

$V = 1, E = 2, F = 1$, hence

$$\chi = 1 - 2 + 1 = 0.$$

For a real projective plane

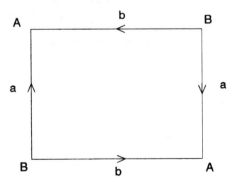

$V = 2, E = 2, F = 1$, hence

$$\chi = 1.$$

351

We already proved that the maximum number of colors required for any map on the surface of a torus is seven.

By using a plane diagram of a torus, show that seven is a necessary number. Thus, prove that the chromatic number of a torus is seven.

SOLUTION:

By remembering that a plane diagram will be folded into a torus, we can color the rectangle as follows:

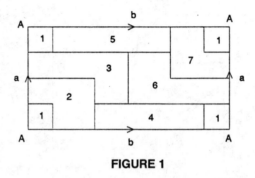

FIGURE 1

Note that when folded, each color borders with the remaining 6. This result is known as the seven color theorem for a one-folded torus.

Find the symbolic representation of a torus and a sphere.

SOLUTION:

The plane diagrams are

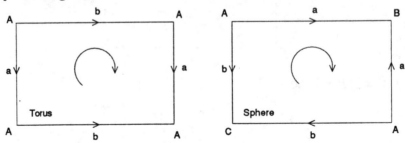

FIGURE 1

352

Let us establish the reference orientation as clockwise (we can choose counter-clockwise as well).

Starting from, for instance, the left upper corner and assigning + to each edge when the arrow moves clockwise or − when counter-clockwise we obtain

$$+ b + a - b - a$$

for a torus. And

$$+ a - a + b - b$$

for a sphere. Note that we can start from any vertex, hence, for a sphere equivalent representations are

$$- a + b - b + a$$

$$+ b - b + a - a$$

$$- b + a - a + b$$

● **PROBLEM 9-104**

What is the Euler characteristic of a sphere with s holes?

SOLUTION:

The Euler characteristic of a sphere is $\chi = 2$.

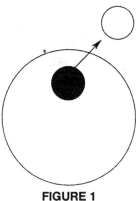

FIGURE 1

Suppose a hole is removed from a sphere (Figure 1). Then the number of the arc increases by one. The Euler characteristic

$$2 = \chi = V - E + F \tag{1}$$

decreases by one. Hence for a sphere with one disc removed

353

$$\chi = 2 - 1 = 1. \tag{2}$$

When two discs are removed

$$\chi = 2 - 2 = 0. \tag{3}$$

The sphere with two holes is homeomorphic to an open cylinder.

In general, a sphere with s holes has the Euler characteristic

$$\chi = 2 - s. \tag{4}$$

● **PROBLEM 9-105**

What is the Euler characteristic of a sphere with n handles and no holes?

SOLUTION:

A sphere with two handles and no holes (see Figure 1) is homeomorphic to a two-fold torus.

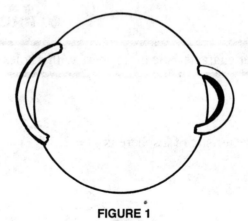

FIGURE 1

The Euler characteristic of a two-fold torus is -2. Hence, the Euler characteristic of a sphere with two handles is -2.

In general, a sphere with n handles and no holes left open is homeomorphic to an n-fold torus. The genus of the surface of an n-fold torus is n. But

$$\chi = 2 - 2g$$

Then

$$\chi = 2 - 2n$$

is the Euler characteristic of a sphere with n handles.

The Euler characteristic of a sphere with no handles is

$$\chi = 2 - 2 \cdot 0 = 2$$

with one handle

$$\chi = 2 - 2 = 0$$

with two handles

$$\chi = 2 - 4 = -2$$

etc.

● **PROBLEM 9-106** P.356

Suppose a sphere with no holes and n handles is given with a map drawn on it. Since every region of the map must be simply connected, at least one arc must be drawn along every handle (Figure 1).

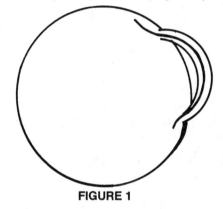

FIGURE 1

By using the procedure described below, determine the Euler characteristic of a sphere with n handles from

$$\chi = V - E + F.$$

SOLUTION:

We shall detach one end of each handle in order to obtain a sphere with n holes and n pipes, called cuffs.

I. **First Stage**

We have a map on a sphere with n handles and no holes, such that $\chi = V - E + F$.

II. Additional vertices are added to the map, one at every intersection of an arc and a boundary where a handle is to be disengaged. Such a boundary is now treated as an arc. Hence

355

$$V \text{ becomes } V + k$$
$$E \text{ becomes } E + 2k$$
$$F \text{ becomes } F + k$$

where k is the number of additional vertices to be added.

III. Detachment

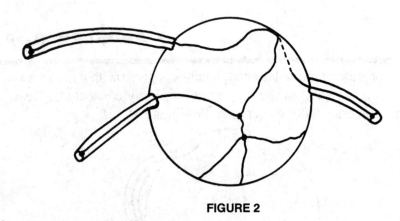

FIGURE 2

Figure 2 shows a sphere with 3 handles detached. In order to make a surface closed, we have to cover n holes and n open cuffs with $2n$ discs. When the handles are detached, k arcs and k vertices are added.

Now we have

$$V = 2k \text{ vertices, } E + 3k \text{ arcs, and } F + K + 2n \text{ regions.}$$

The thereby obtained surface is homeomorphic to a sphere. Hence,

$$\chi = 2 = (V + 2k) - (E + 3k) + (F + k + 2n) \tag{2}$$

and

$$V - E + F = 2 - 2n \tag{3}$$

$$\chi = 2 - 2n. \tag{4}$$

P355 Cf. Problem 9-106.

● PROBLEM 9–107

The Euler characteristic of a cylinder is zero and of a disc is one. By attaching and removing various surfaces, find the Euler characteristic of a sphere with 7 handles and 4 holes.

SOLUTION:

It is easy to calculate the Euler characteristic from the formula

$$\chi = 2 - 2n - k \tag{1}$$

where n is the number of handles and k is the number of holes

$$\chi = 2 - 2 \cdot 7 - 4 = -16. \tag{2}$$

The other way is to start with a brand new sphere $\chi = 2$. Then make $2 \cdot 7 = 14$ holes for handles $\chi = 2 - 2 \cdot 7$, then 4 holes $\chi = 2 - 2 \cdot 7 - 4$, then put 7 cylinders to make handles, $\chi = 2 - 2 \cdot 7 - 4 + 7 \cdot 0$. Thus,

$$\chi = \chi\,[\text{sphere}] - \chi\,[\text{disc}] \cdot (2n + k) + \chi\,[\text{cylinder}] \cdot n =$$

$$= 2 - 14 - 4 = -16. \tag{3}$$

● **PROBLEM 9–108**

Generally, if n open surfaces $P_1, P_2, ..., P_n$ are joined together along the boundaries, then the Euler characteristic χ of the resulting surface is

$$\chi = \chi[P_1] + \chi[P_2] + ... + \chi[P_n] . \tag{1}$$

By applying (1), obtain the Euler characteristic of a sphere, putting together

1. n cylinders and two discs.

2. a sphere with k holds and k discs.

SOLUTION:

1. By putting n open cylinders, end to end, we obtain a single cylinder. By adding two discs, we obtain a surface homeomorphic to the surface of a sphere.

$$\chi\,[\text{sphere}] = \chi\,[\text{cylinder with two discs}] =$$

$$= n\chi\,[\text{cylinder}] + 2\chi\,[\text{disc}] = n \cdot 0 + 2 \cdot 1 = 2. \tag{2}$$

2. The Euler characteristic of a sphere with k holes is

$$\chi = 2 - k.$$

By covering k holes with k discs, we obtain a sphere. Hence

$$\chi \text{ [sphere]} = \chi \text{ [sphere with } k \text{ holes]} +$$

$$+ k\chi[\text{disc}] = 2 - k + k = 2. \tag{3}$$

Determine the Euler characteristic of a cross-cap.

SOLUTION:

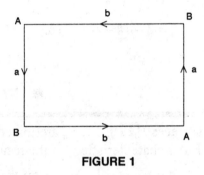

FIGURE 1

The plane diagram (Figure 1) of the real projective plane (see Problem 9-100) can be topologically deformed into a sphere with the appropriate hole shown in Figure 2.

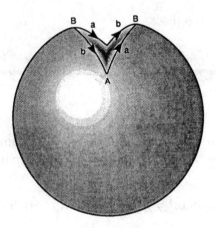

FIGURE 2

By closing the hole in such a way that the sides are joined according to the arrows, we obtain a closed surface intersecting itself (see Figure 3). The surface is one-sided. Points A and B are single; all other points of the segment AB are double points.

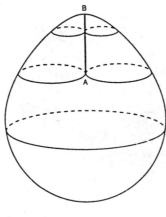

FIGURE 3

By removing the lower hemisphere which is homeomorphic, to a disc, we obtain a surface called a cross-cap (Figure 4).

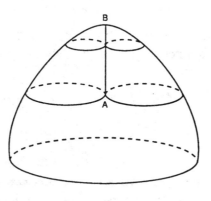

FIGURE 4

Hence,

$$\chi \text{ [cross-cap]} = \chi \text{ [real projective plane]} - \chi[\text{disc}] =$$

$$= 1 - 1 = 0.$$

The Euler characteristic of a cross-cap and a Möbius band are the same. Indeed, a Möbius band can be topologically deformed into a cross-cap.

● **PROBLEM 9–110**

Derive the formula for the Euler characteristic of a sphere with n handles, k holes and l cross-caps.

SOLUTION:

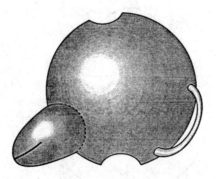

FIGURE 1

A sphere with one handle, one cross-cap and two holes is shown in the figure.

Note that each handle requires two holes and each cross-cap requires one hole. Thus, we have to make

$$2n + l + k$$

holes. The Euler characteristic of a cylinder is zero and of a cross-cap is also zero. Thus, the Euler characteristic of a sphere with n handles, l cross-caps and k holes is

$$\chi = 2 - (2n + l + k) = 2 - 2n - l - k.$$

● **PROBLEM 9–111**

Present in standard model form disc, sphere, cylinder, real projective plane, Möbius band, torus, two-fold torus and n-fold torus.

SOLUTION:

A surface described as a sphere with n handles, l cross-caps and k holes is said to be represented in standard model form.

	n-handles	l cross-caps	k holes	χ
sphere	0	0	0	2
disc	0	0	1	1
cylinder	0	0	2	0
real projective plane	0	1	0	1
Möbius band	0	1	1	0

torus	1	0	0	0
two-fold torus	2	0	0	−2
n-fold torus	n	0	0	2 − 2n

Determine the rank and Euler characteristic of a torus with a hole.

SOLUTION:

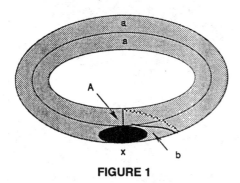

FIGURE 1

The rank of an open surface is defined as requiring the least number of cuts to make the surface homeomorphic to a disc.

From Figure 1, we see that two cuts are required to make a torus with a hole homeomorphic to a disc. The cut surface can be deformed to yield a plane diagram (Figure 2) with the symbolic notation

$$\pm x + a + b - a - b.$$

The torus with a hole has 1 handle, 1 hole and 0 cross-caps. Hence

$$\chi = 2 - (2n + k + l) = 2 - 2 - 1 - 0 = -1.$$

The same value can be obtained from a plane diagram

$$\chi = V - E + F = 1 - 3 + 1 = -1.$$

FIGURE 2

Find the rank and Euler characteristic of a sphere with one cross-cap and one hole.

SOLUTION:

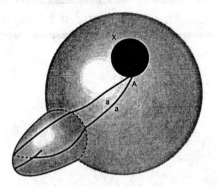

FIGURE 1

This is an open surface. It requires one cut (Figure 1) to become homeomorphic to a disc.

The rank of a sphere with one cross-cap and one hole is one. The surface obtained can be deformed to give a triangle.

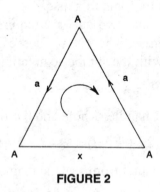

FIGURE 2

Its symbolic representation

$$\pm x - a - a$$

The Euler characteristic

$$\chi = 2 - (2n + k + l) = 0$$

or from the plane diagram

$$\chi = V - E + F = 1 - 2 + 1 = 0.$$

Why is predicting the physical and chemical properties of chemical substances before they are synthesized so important?

SOLUTION:

More than seven million different molecules have been synthesized. At least some of their properties are known with fairly good accuracy. In some sense it forms a basis for future research. The cost of obtaining new molecules can sometimes be enormous. It can happen that the final product turns out to be a disappointment. A new molecule is not exactly what we have been looking for.

Hence, it is important, not only from the scientific point of view, to be able to predict the properties of chemical substances before they are synthesized.

Describe briefly the topological method of making chemical predictions.

SOLUTION:

The heart of this method is the topology of individual molecules. We neglect the nature and lengths of the chemical bonds holding the atoms of a molecule together. We take into account the number of atoms in the molecule and how the atoms are interconnected within the molecule (single straight chain, chain with branches, several chains, a ring, etc.).

As a first step we take a small number of known molecules, for instance, $A_1, A_2, ..., A_k$. Then we establish a procedure which converts the topological structure of each molecule A into a single, characteristic number α called index.

$$A_1 \rightarrow \alpha_1$$

$$A_2 \rightarrow \alpha_2$$

$$\vdots$$

$$A_k \rightarrow \alpha_k$$

Suppose we are interested in a certain chemical property, say the boiling point. For each $A_1, A_2, ..., A_k$ we determine its boiling point $T_1, T_2, ..., T_k$. By using the Cartesian coordinates on one axis, we mark the value of the index and on the other, the value of the chemical (physical) property.

Thus, we obtain a plot which can be used to predict the future.

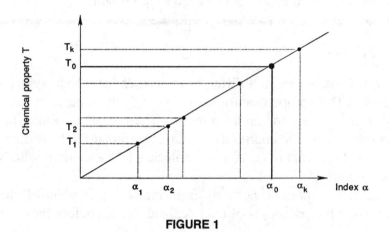

FIGURE 1

Suppose we want to know the boiling point of a molecule whose index is α_0. From the plot, we obtain T_0.

● **PROBLEM 9-116**

What is a chemical graph?

SOLUTION:

A chemical graph is a drawing in which the atoms of the molecule are depicted as points and the bonds linking the atoms are depicted as straight lines. The length of lines and the angles are immaterial.

Thus, graphs represent structures in an abstract manner. Graph theory, a mathematical discipline, studies graphs in detail.

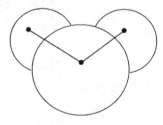

FIGURE 1–Graph of H_2O

Generally, in chemical graphs the hydrogen atoms are omitted.

364

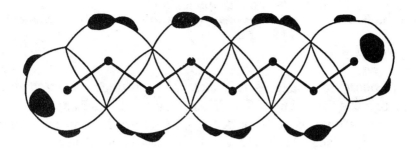

FIGURE 2-Graph of an n-octane.

Large balls depict carbon atoms and small balls depict hydrogen atoms.

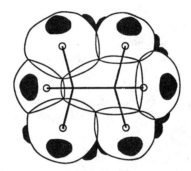

FIGURE 3–The structure of 2, 2, 3, 3,-tetramethylbutane.

The graph is shown inside with the hydrogen atoms omitted.

● **PROBLEM 9–117**

Describe how the carbon number is determined.

SOLUTION:

Once the chemical graph of a molecule has been drawn, it is fairly easy to compute its characteristic number - index. Such a number is known as a graph invariant. One of the most simple graph invariants is the vertex number, which chemists refer to as the carbon number. It is the number of vertexes in the graph. For hydrocarbon molecules (i.e., molecules consisting only of hydrogen and carbon atoms) it is the number of carbon atoms. It should be remembered that the carbon number has low discriminating power because many different molecules can have the same carbon number.

Describe the Wiener index and its applications.

SOLUTION:

delete

The carbon number is a useful index for analyzing straight-chain molecules for branched molecules. The Wiener index is more appropriate. This index uses the notion of distance between two vertices, which is defined as follows: The distance between any two vertices is equal to the number of edges traversed while taking the shortest route within the graph between these vertices.

The Wiener index of a molecule is equal to the sum of the distances between all pairs of atoms in the molecule.

Consider for example isopentane.

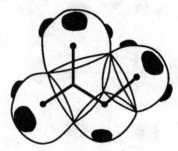

FIGURE 1

Its graph and distances are shown in Figure 2.

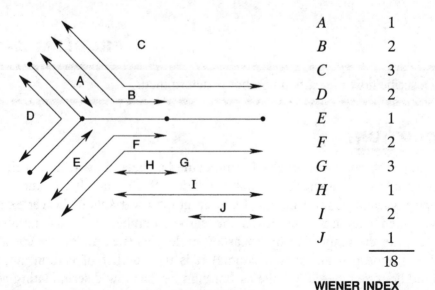

A	1
B	2
C	3
D	2
E	1
F	2
G	3
H	1
I	2
J	1
	18

WIENER INDEX

FIGURE 2

Show how the Wiener index can be applied to determine the chemical or physical property of a molecule that has not yet been synthesized.

SOLUTION:

The Wiener index as described in Problem 9-118 converts the topological structure of a molecule into a single number-index. It has been determined that the Wiener index correlates with many physical properties.

First, we establish the molecular structure of some existing molecules and draw their graphs.

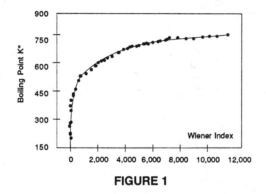

FIGURE 1

Suppose we are interested in the boiling point. For each molecule, we know its Wiener index and its boiling point. Thus, obtained points yield a fairly smooth curve (see Figure 1). The correlation is more evident when we use the logarithms of the Wiener index and boiling point (see Figure 2).

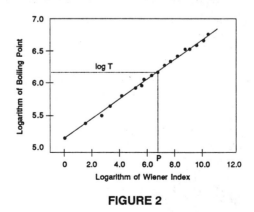

FIGURE 2

Suppose we want to synthesize a molecule whose logarithm of the Wiener index is P. From the plot we find the logarithm of its boiling point.

Compute the molecular-connectivity index of a molecule whose graph is shown below:

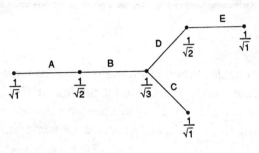

SOLUTION:

The molecular-connectivity index is the most universal index devised so far. It depends on the topological concept of degree. The degree of any vertex is equal to the number of other vertices it is attached to.

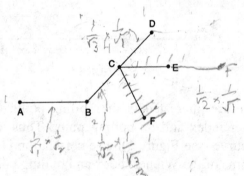

Hence the degree of A is 1, of B is 2, of C is 4, of D is 1, of E is 1 and of F is 1.

Each edge has a value: the product of the reciprocals of the square roots of the degrees of the vertices it joins.

FOR EDGE   Molecular – connectivity Index $= \sum \dfrac{1}{\sqrt{N_i} \times \sqrt{N_j}}$

JOINING VERTICE v_i & v_j

HYPHEN

where the sum is taken over all the edges.

Hence, the molecular-connectivity index for our molecule is

$$\left(\frac{1}{\sqrt{1}} \times \frac{1}{\sqrt{2}}\right) + \left(\frac{1}{\sqrt{2}} \times \frac{1}{\sqrt{3}}\right) + \left(\frac{1}{\sqrt{3}} \times \frac{1}{\sqrt{1}}\right) + \left(\frac{1}{\sqrt{3}} \times \frac{1}{\sqrt{2}}\right) +$$

$$+ \left(\frac{1}{\sqrt{2}} \times \frac{1}{\sqrt{1}}\right) = \frac{2}{\sqrt{2}} + \frac{1}{\sqrt{3}} + \frac{2}{\sqrt{6}} = 2.80806$$

● PROBLEM 9-121

In some cases, better correlations are obtained by considering only the fragments of larger molecules. Compute the molecular-connectivity index for the fragments of molecules indicated by dotted lines.

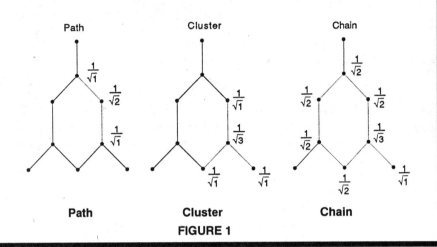

FIGURE 1

SOLUTION:

From the definition of the molecular-connectivity index, we obtain:

Path

$$\left(\frac{1}{\sqrt{1}} \times \frac{1}{\sqrt{2}}\right) + \left(\frac{1}{\sqrt{2}} \times \frac{1}{\sqrt{1}}\right) = 1.4142$$

Cluster

$$\left(\frac{1}{\sqrt{1}} \times \frac{1}{\sqrt{3}}\right) + \left(\frac{1}{\sqrt{1}} \times \frac{1}{\sqrt{3}}\right) + \left(\frac{1}{\sqrt{1}} \times \frac{1}{\sqrt{3}}\right) = 1.73205$$

Chain

$$4 \times \left(\frac{1}{\sqrt{2}} \times \frac{1}{\sqrt{2}}\right) + 2 \times \left(\frac{1}{\sqrt{2}} \times \frac{1}{\sqrt{3}}\right) + \left(\frac{1}{\sqrt{1}} \times \frac{1}{\sqrt{3}}\right) = 3.39384$$

● PROBLEM 9-122

Compute the Balaban centric index for a molecule whose graph is shown below.

369

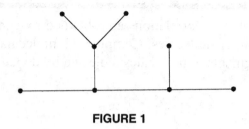

FIGURE 1

SOLUTION:

This index emphasizes the degree of branching in a molecule. All vertices that are connected to only one other vertex are counted and removed from the molecule's graph. The number of vertices removed at each step is squared and added to a running total. The procedure is repeated until all vertices are removed.

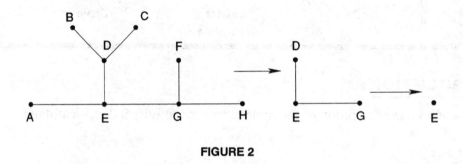

FIGURE 2

As a first step we remove vertices A, B, C, F, and H and get 5^2. Then we remove D and G and obtain 2^2. Finally, we remove E and obtain 1^2.

Thus, the index is $5^2 + 2^2 + 1^2$.

● **PROBLEM 9-123**

List a few applications of molecular-connectivity index.

SOLUTION:

The molecular-connectivity index has numerous and diversified applications. It is used in developing new drugs and in predicting the taste and smell of new substances. This index correlates well with such physical properties as the boiling point, density, heat of vaporization and solubility in water. It is known that many biological responses are launched when an appropriate stimulating molecule docks with a receptor on the surface of a cell. Very often the specific shape of the molecule is less important than its volume and surface area. Molecular-connectivity indices correlate well with volume and surface area. Thus, the molecular-connectivity indices can predict the ability of molecules to act as anesthetics, hallucinogens or narcotics.

It is also possible to predict the smell of molecules as well as whether a molecule will taste bitter or sweet.

● PROBLEM 9-124

Describe the applications of molecular-connectivity indices in environmental studies.

SOLUTION:

It is a very costly and time-consuming process to determine how fast pollutants spread into the environment. Here, molecular-connectivity indices are very helpful.

The indices correlate well with the ability of many pollutants to spread within the air, water or soil. The relationship exists between the indices and the tendencies of substances to concentrate within living organisms. To test these properties without the indices is extremely difficult and expensive. That is why the U.S. Environmental Protection Agency applies the indices to predict the toxic potentiality of unknown or untested pollutants.

● PROBLEM 9-125

What does the octane number of a fuel describe? What methods are used to predict it?

SOLUTION:

In general, the octane number describes the efficiency with which a fuel burns, that is, its tendency not to "knock." The fuel is mixed with oxygen and then ignited by a spark.

A "knock" appears when during compression, oxygen atoms combine with fuel before it has been ignited. Usually the octane number is determined under standardized conditions in a test engine.

Straight-chain molecules tend to "knock" more than branched molecules.

Attempts to establish the correlation between the tendency of the fuel to "knock" and its Wiener index yielded only fair results.

An index introduced by A. Balaban, called the centric index, emphasized the branching of molecules. It correlated very well with the octane number of hydrocarbon molecules.

We defined the centric index in Problem 9-123. 9-122/P.370

● PROBLEM 9-126

An index called hydrogen-deficiency index was devised to predict the amount of soot produced by burning hydrocarbon molecules. Define this index.

SOLUTION:

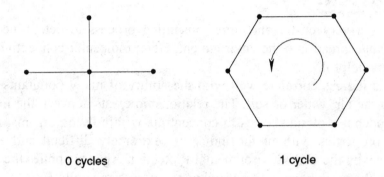

0 cycles   1 cycle

FIGURE 1

The hydrogen-deficiency number is equal to the number of independent cycles, or nips, and double bonds in a molecule.

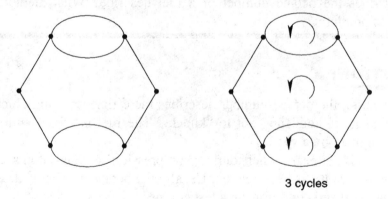

3 cycles

FIGURE 2

The hydrogen-deficiency number combined with the averaged-distance-sum connectivity index yields an index which provides a very good correlation with the soot production for almost 100 hydrocarbon molecules.

Describe how the cancer-causing tendency of a molecule (carcinogen-icity) can be predicted by means of topological methods.

SOLUTION:

The task of finding a correlation between carcinogenic behavior of molecules and topological indices is an extremely difficult one.

First of all, the experimental measurements are not necessarily accurate or sufficient. Then the growth of cancer is a process consisting of many stages.

Thus, the index should take into account not only the original molecules but also the molecules produced during successive stages. So far the best results were achieved by William C. Herndon and László van Szentpály. They introduced the combination of simple indices to predict the carcinogen-icity of polycyclic aromatic hydrocarbons. Since the molecules, below and above certain size limits, are not carcinogenic, the final index includes the carbon number and the square of the carbon number (which depends on the size of the molecules). Another aspect is that certain regions of hydrocarbons are more important (more carcinogenic) than other regions.

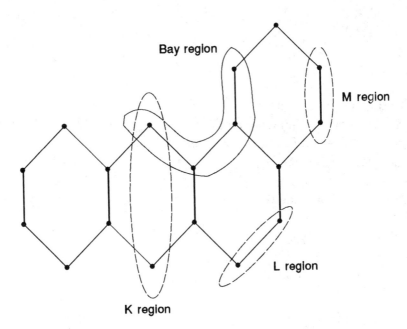

FIGURE 1

The Bay, L and M regions must be fairly active chemically, while the K region must remain fairly inactive.

CHAPTER 10

METRIC SPACES

Let R represent the set of real numbers. Show that one of the possible metrics for R is the absolute value

$$d(x, y) = |x - y| \qquad (1)$$

SOLUTION:

DEFINITION OF A METRIC SPACE

Let X be a set. Function $d(x,y) \in R$

$$d : X \times X \to R \qquad (2)$$

is said to be a metric on X if for all x, y, $z \in X$

1. $d(x, y) \geq 0$.

2. $d(x, y) = 0$ iff $x = y$.

3. $d(x, y) = d(y, x)$.

4. $d(x, z) \leq d(x, y) + d(y, z)$.

The set X with a metric d is called a metric space and is denoted by (X, d). The function $d(x, y)$ is called the distance between x and y.

■

For any real numbers x, y, and z we have

$$|x - y| \geq 0$$

$$|x - y| = 0 \quad \text{iff} \quad x - y = 0 \quad \text{iff} \quad x = y$$

$$|x - y| = |y - x|$$

$$|x - z| \leq |x - y| + |y - z|.$$

Hence (1) defines a metric.

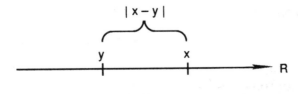

FIGURE 1

If x and y are real numbers on the real axis, then $|x - y|$ is the distance between them.

1. Is it possible to define a metric on any set?

2. Show that the set R^n with the distance defined by

$$d(x, y) = \max \{ |x_i - y_i|, 1 \le i \le n \} \qquad (1)$$

is a metric space.

SOLUTION:

1. Let X be any set. Then, function d defined by

$$d(x,y) = \begin{cases} 1 & \text{if } x \ne y \\ 0 & \text{if } x = y \end{cases} \qquad (2)$$

is a metric. Therefore a metric may be defined for any set.

2. We have

$$d(x, y) \ge 0 \quad \text{because } |x_i - y_i| \ge 0$$

$$0 = d(x, y) = \max \{ |x_i - y_i|, 1 \le i \le n \} \quad \text{iff}$$

$$|x_i - y_i| = 0 \quad \text{for} \quad 1 \le i \le n \quad \text{iff} \quad x_i = y_i \quad \text{for}$$

$$1 \le i \le n \quad \text{iff} \quad x = y.$$

We denote

$$x = (x_1, \ldots, x_n)$$

$$y = (y_1, \ldots, y_n) \qquad (3)$$

Symmetry is obvious

$$d(x, y) = d(y, x) \qquad (4)$$

Triangle inequality.
Suppose for some $1 \le l \le n$

$$d(x, y) = |x_i - y_i|$$

we have

$$d(x, y) = |x_i - y_i| \le |x_l - z_l| + |z_l - y_l| \le$$

$$\le d(x, z) + d(z, y). \tag{5}$$

● **PROBLEM 10-3**

Show that in the set R^n,

$$d_p(x,y) = \sqrt[p]{\sum_1^n |x_i - y_i|^p} \tag{1}$$

is a metric for each $p \ge 1$.

SOLUTION:

We shall prove the triangle inequality.
For $p \ge 1$, the function

$$f(x) = x^p \tag{2}$$

for $x \in R^1$, $x \ge 0$ satisfies

$$f''(x) \ge 0 \tag{3}$$

and hence it is convex. Thus, $f(x)$ satisfies

$$f(\alpha x + [1 - \alpha]y) \le \alpha f(x) + (1 - \alpha) f(y) \tag{4}$$

for $x, y \ge 0$.

For any set of real numbers $a_1, \ldots, a_n, b_1, \ldots, b_n$ let

$$A = \sqrt[p]{\sum_1^n |a_i|^p} \qquad B = \sqrt[p]{\sum_1^n |b_i|^p} \tag{5}$$

Define $x, y,$ and α as; $\forall i \in [1, n]$

$$x_i = \frac{a_i}{A}, \qquad y_i = \frac{b_i}{B}, \qquad \alpha_i = \frac{A}{A + B}. \tag{6}$$

Substituting (6) for each $i = 1, 2, \ldots, n$ into (4) and adding all n inequalities, we obtain

$$\sqrt[p]{\sum_1^n |a_i - b_i|^p} \le \sqrt[p]{\sum_1^n |a_i|^p} + \sqrt[p]{\sum_1^n |b_i|^p} \tag{7}$$

Equation (7) is called Minkowski's inequality.

377

Substituting

$$a_i = x_i - y_i$$

$$b_i = y_i - z_i \tag{8}$$

for $n = 1, 2, \ldots, n$ into (7) we obtain

$$d_p(x, z) \le d_p(x, y) + d_p(y, z). \tag{9}$$

In particular for $n = 2$ and $p = 2$, we obtain

$$d(x,y) = \sqrt{(x_1 - y_1)^2 + (x_2 - y_2)^2} \tag{10}$$

which is a distance on R^2.

Let d be a metric on a set X. Show that

$$d_0(x,y) = \frac{d(x,y)}{1 + d(x,y)} \tag{1}$$

where $x, y \in X$ is also a metric on X.

SOLUTION:

We have to show that d_0 satisfies the triangle inequality. All remaining conditions for a metric are obviously satisfied.

Since d is a metric we have

$$\frac{d(x,y)}{1 + d(x,y) + d(y,z)} \le \frac{d(x,y)}{1 + d(x,y)} = d_0(x,y) \tag{2}$$

$$\frac{d(y,z)}{1 + d(x,y) + d(y,z)} \le \frac{d(y,z)}{1 + d(y,z)} = d_0(y,z) \tag{3}$$

Function d is a metric. Hence $d(x, z) \le d(x, y) + d(y, z)$ and

$$d_0(x,z) = \frac{d(x,z)}{1 + d(x,z)} \le \frac{d(x,y) + d(y,z)}{1 + d(x,y) + d(y,z)} =$$

$$= \frac{d(x,y)}{1 + d(x,y) + d(y,z)} + \frac{d(y,z)}{1 + d(x,y) + d(y,z)} \le d_0(x,y) + d_0(y,z) \tag{4}$$

Thus d_0 is a metric.

Let X represent the set of all functions from the closed interval $[0, 1]$ into itself. Show that for any $f, g \in X$

$$d(f, g) = \text{least upper bound } \{\,|f(x) - g(x)| : x \in [0, 1]\} \qquad (1)$$

is a metric.

SOLUTION:

Any subset of the real numbers which has an upper bound has a least upper bound and

$$0 \le |f(x) - g(x)| \le 1 \quad \text{for all } f, g \in X \text{ and } x \in [0, 1]. \qquad (2)$$

Hence $d(f, g)$ is defined for all $f, g \in X$.

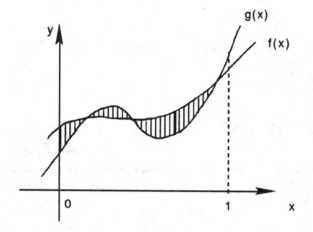

FIGURE 1

1. Since

$$|f(x) - g(x)| \ge 0, x \in [0, 1] \qquad (3)$$

the least upper bound of (3) is non-negative, i.e.,

$$d(f, g) \ge 0 \quad \text{for all } f, g \in X. \qquad (4)$$

2. $\qquad\qquad\qquad d(f, g) = 0 \quad \text{iff} \quad f = g. \qquad (5)$

If $d(f, g) = 0$, then a least upper bound

$$\{\,|f(x) - g(x)| : x \in [0, 1]\,\} = 0 \qquad (6)$$

and

379

$$\{ \, |f(x) - g(x)| : x \in [0, 1] \} = \{ \, 0 \, \}. \tag{7}$$

Hence

$$|f(x) - g(x)| = 0 : x \in [0, 1] \tag{8}$$

and $f(x) = g(x)$ for all $x \in [0, 1]$.

If $f = g$, then $f(x) = g(x)$ for all $x \in [0, 1]$ and

$$\{ \, |f(x) - g(x)| : x \in [0, 1] \} = \{ \, 0 \, \}. \tag{9}$$

Hence

$$d(f, g) = 0. \tag{10}$$

3. $d(f, g) = d(g, f)$ for all $f, g \in X$ because

$$|f(x) - g(x)| = |g(x) - f(x)|. \tag{11}$$

4. $$d(f, h) \le d(f, g) + d(g, h) \tag{12}$$

because for all $f, g, h \in X$

$$|f(x) - h(x)| \le |f(x) - g(x)| + |g(x) - h(x)| \tag{13}$$

for all $x \in [0, 1]$.

● PROBLEM 10-6

1. Let (X_1, d_1) and (X_2, d_2) represent metric spaces. Show that $(X_1 \times X_2, d)$ where

$$d((x_1, x_2), (y_1, y_2)) = d_1(x_1, y_1) + d_2(x_2, y_2) \tag{1}$$

for $x_1, y_1 \in X_1$ and $x_2, y_2 \in X_2$, is a metric space.

2. Applying (1), define a metric on a Cartesian product of n metric spaces $X_1 \times X_2 \times \ldots \times X_n$.

SOLUTION:

1. Since d_1 and d_2 are metrics

$$d_1(x_1, y_1) \ge 0 \text{ and } d_2(x_2, y_2) \ge 0 \tag{2}$$

and

$$d((x_1, x_2), (y_1, y_2)) \ge 0$$

for all $x_1, y_1 \in X_1$ and for all $x_2, y_2 \in X_2$.

Similarly since $d_1(x_1, y_1) = d_1(y_1, x_1)$ and $d_2(x_2, y_2) = d(y_2, x_2)$ we have

$$d((x_1, x_2), (y_1, y_2)) = d((y_1, y_2), (x_1, x_2)) \qquad (4)$$

$$d((x_1, x_2), (y_1, y_2)) = 0 \quad \text{iff} \quad (x_1, x_2) = (y_1, y_2). \qquad (5)$$

Suppose $d((x_1, x_2), (y_1, y_2)) = 0$ then

$$d_1(x_1, y_1) = 0 \quad \text{and} \quad d_2(x_2, y_2) = 0.$$

Thus

$$x_1 = y_1 \quad \text{and} \quad x_2 = y_2 \text{ or } (x_1, x_2) = (y_1, y_2). \qquad (6)$$

If $(x_1, x_2) = (y_1, y_2)$, then

$$d_1(x_1, y_1) = 0 \quad \text{and} \quad d_2(x_2, y_2) = 0. \qquad (7)$$

Thus

$$d((x_1, x_2), (y_1, y_2)) = 0. \qquad (8)$$

Now, we will show that

$$d((x_1, x_2), (z_1, z_2)) \leq d((x_1, x_2), (y_1, y_2)) + d((y_1, y_2), (z_1, z_2)). \qquad (9)$$

Indeed, since d_1 and d_2 are metrics

$$d((x_1, x_2), (z_1, z_2)) = d_1(x_1, z_1) + d_2(x_2, z_2) \leq$$

$$\leq d_1(x_1, y_1) + d_1(y_1, z_1) + d_2(x_2, y_2) + d_2(y_2, z_2) =$$

$$= d((x_1, x_2), (y_1, y_2)) + d((y_1, y_2), (z_1, z_2)). \qquad (10)$$

2. If $(X_1, d_1), (X_2, d_2), \ldots, (X_n, d_n)$ are metric spaces, then

$$d(x, y) = d_1(x_1, y_1) + \ldots + d_n(x_n, y_n) \qquad (11)$$

is a metric on $X_1 \times X_2 \times \ldots \times X_n$.

● **PROBLEM 10-7**

The following metrics are defined on R^2

$$d_1((x_1, y_1), (x_2, y_2)) = \begin{cases} 0 & \text{if } (x_1, y_1) = (x_2, y_2) \\ 1 & \text{if } (x_1, y_1) \neq (x_2, y_2) \end{cases}$$

$$d_2((x_1, y_1), (x_2, y_2)) = \sqrt{(x_1 - x_2)^2 + (y_1 - y_2)^2}$$

$$d_3((x_1, y_1), (x_2, y_2)) = |x_1 - x_2| + |y_1 - y_2|$$
$$d_4((x_1, y_1), (x_2, y_2)) = \max\left(|x_1 - x_2|, |y_1 - y_2|\right).$$

Find the ball of radius 1 and center (0, 0) with respect to metrics d_1, d_2, d_3, and d_4.

SOLUTION:

DEFINITION OF A BALL

The convex set

$$B_d(a, r) = \{x : d(x, a) < r\}$$

is called an open ball of radius r and center a.

■

We have $B_{d_1}((0,0),1) = \{(0,0)\}$.

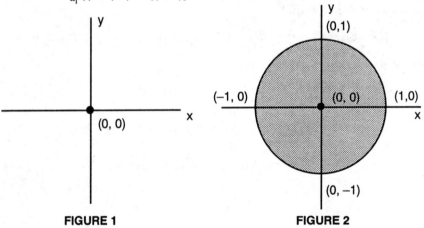

FIGURE 1   FIGURE 2

Figure 2 illustrates the d_2-ball, $B_{d_2}((0,0), 1)$.

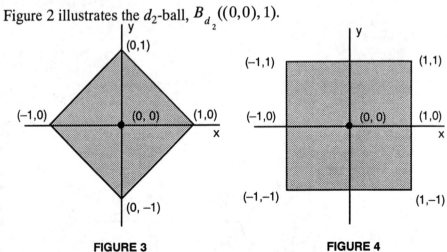

FIGURE 3   FIGURE 4

B_{d_3} is shown in Figure 3.

B_{d_4} is shown in Figure 4.

Sometimes when it is obvious what metric is used we shall write $B(a, r)$ instead of $B_d(a, r)$.

● PROBLEM 10-8

1. Describe explicitly the ball $B(x_0, r)$ in the metric space (R, d) where $d(x, y) = |x - y|$.

2. Let $C[0, 1]$ be the collection of all continuous functions on $[0, 1]$ with the metric defined by

$$d(f, g) = \sup \{ |f(x) - g(x)| : x \in [0, 1] \}. \tag{1}$$

Find the ball $B(f_0, \varepsilon)$ where f_0 is a continuous function and $\varepsilon > 0$.

SOLUTION:

1. The ball $B(x_0, r)$ is shown in Figure 1.

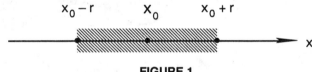

FIGURE 1

The ball consists of all the points x on the real axis such that

$$|x - x_0| < r \quad \text{or} \quad x_0 - r < x < x_0 + r. \tag{2}$$

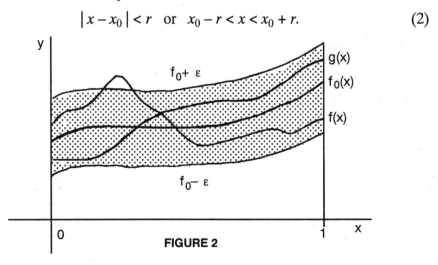

FIGURE 2

383

2. Let f_0 be a continuous function as given in Figure 2. The open ball $B(f_0, \varepsilon)$ consists of all continuous functions which lie in the area bounded by, (and excluding), the lines at $f_0 - \varepsilon$ and $f_0 + \varepsilon$.

● **PROBLEM 10-9**

Prove this theorem:
THEOREM

   Let $B(x_0, r)$ be a ball. For every point $y \in B(x_0, r)$, there exists a ball $B(y, r')$ such that

$$B(y, r') \subset B(x_0, r). \tag{1}$$

■

SOLUTION:

   Point $y \in B(x_0, r)$ hence

$$d(x_0, y) < r. \tag{2}$$

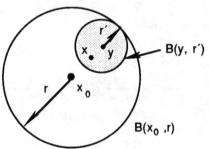

FIGURE 1

   Let us define

$$r' = r - d(x_0, y) > 0. \tag{3}$$

We will show that

$$B(y, r') \subset B(x_0, r). \tag{4}$$

Suppose

$$x \in B(y, r') \tag{5}$$

then

$$d(x, y) < r' \tag{6}$$

384

or

$$d(x, y) < r - d(x_0, y)$$

$$d(x, y) + d(x_0, y) < r. \tag{7}$$

But d is a metric, so

$$d(x, x_0) \le d(x, y) + d(y, x_0) < r. \tag{8}$$

Hence

$$d(x, x_0) < r$$

and

$$x \in B(x_0, r). \tag{9}$$

● **PROBLEM 10–10**

1. Prove that if B_1 and B_2 are balls with the same center, then one of them is a subset of the other.

2. Prove this theorem:
THEOREM

Let B_1 and B_2 be balls and let $x \in B_1 \cap B_2$. Then, a ball B exists such that

$$x \in B \subset B_1 \cap B_2 \tag{1}$$

■

SOLUTION:

1. Let $B_1(x_0, r_1)$ and $B_2(x_0, r_2)$. Since r_1 and r_2 are real numbers, we have either $r_1 \le r_2$ or $r_2 \le r_1$. Suppose $r_1 \le r_2$. Then if $x \in B_1(x_0, r_1)$ we obtain $d(x_0, x) < r_1 \le r_2$.
Thus $d(x_0, x) < r_2$ and $x \in B_2(x^0, r_2)$.
If $r_2 \le r_1$, then $B_2(x_0, r_2) \subset B_1(x_0, r_1)$.

2. Let x be a point, such that

$$x \in B_1 \cap B_2. \tag{2}$$

Then

$$x \in B_1 \quad \text{and} \quad x \in B_2.$$

385

By Problem 10–9, if $x \in B_1(x_0, r_1)$, then a ball $B'_1(x, r'_1)$ exists such that

$$x \in B'_1 \subset B_1. \qquad (3)$$

Similarly, a ball $B'_2(x, r'_2)$ exists such that

$$x \in B'_2 \subset B_2 \qquad (4)$$

as shown in Figure 1.

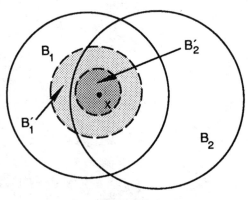

FIGURE 1

The balls B'_1 and B'_2 both have the same center. By part 1 of this problem, one of them is contained in the other. Suppose

$$B'_2 \subset B'_1. \qquad (5)$$

Then

$$x \in B'_2 \subset B'_1 \subset B_1. \qquad (6)$$

Let $B = B'_2$, then

$$x \in B \subset B_1 \cap B_2. \qquad (7)$$

● PROBLEM 10–11

1. Let R be the set of real numbers with the absolute value metric. Show that the open interval $(0, 1)$ is an open set.

2. Show that the ball

$$B(x_0, r) = \{\, x : d(x, x_0) < r \,\} \qquad (1)$$

is an open set.

386

SOLUTION:

DEFINITION OF AN OPEN SET

Let (X, d) be a metric space. A subset $A \subset X$ is said to be open, if for every $x \in A$, a ball $B(x, r)$ exists such that

$$x \in B(x, r) \subset A.$$

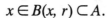

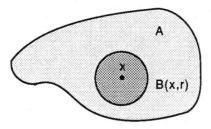

FIGURE 1

1. Let $x \in (0, 1)$, then

$$0 < x < 1. \tag{2}$$

We define

$$r = \min \{ x, 1 - x \}. \tag{3}$$

Then

$$B(x, r) \subset (0, 1) \tag{4}$$

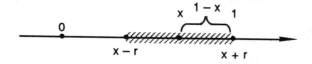

FIGURE 2

Any open interval (a, b) is an open set.

2. Consider ball (1). Let

$$y \in B(x_0, r). \tag{5}$$

Then $d(x_0, y) < r$.

By virtue of Problem 10–9, we conclude that a ball $B'(y, r')$ exists such that

$$y \in B'(y, r') \subset B(x_0, r) \tag{6}$$

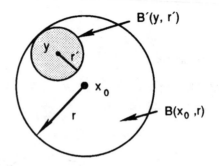

FIGURE 3

Prove the following:
THEOREM

Let (X, d) be a metric space. Then

1. X, ϕ are open sets,

2. the intersection of any two open sets is an open set,

3. the union of any family of open sets is an open set. ■

SOLUTION:

1. If $x \in X$ and $r > 0$, then

$$B(x, r) \subset X. \qquad (1)$$

Hence X is an open set.
   For each $x \in \phi$ and any $r > 0$,

$$B(x, r) \subset \phi. \qquad (2)$$

Hence ϕ is an open set.

2. Suppose A and B are open subsets of (X, d) and $x \in A \cap B$. Since both A and B are open, balls exist such that

$$B_1(x, r_1) \subset A$$

$$B_2(x, r_2) \subset B. \qquad (3)$$

Setting

$$r = \min(r_1, r_2) \tag{4}$$

we obtain

$$B(x, r) \subset A \cap B. \tag{5}$$

Hence $A \cap B$ is an open set.

3. Let $\bigcup_\alpha A_\alpha$ be a union of a family of open sets. Suppose

$$x \in \bigcup_\alpha A_\alpha \tag{6}$$

Then an open set A_α exists such that is a member of this family and

$$x \in A_\alpha \tag{7}$$

and since A_α is open

$$x \in B(x, r) \subset A_\alpha \subset \bigcup_\alpha A_a \tag{8}$$

● PROBLEM 10–13

Let (X, d) be a metric space. Show that a subset $A \subset X$ is open if and only if A is the union of a family of balls.

SOLUTION:

We shall prove

$$(A \text{ is open}) \Leftrightarrow (A = \bigcup_{i \in I} B_i) \tag{1}$$

⇐ In Problem 10–12, we proved that the union of any family of open sets is an open set. In Problem 10–11, we proved that a ball is an open set. Hence $A = \bigcup_{i \in I} B_i$ is an open set.

⇒ Suppose A is an open set. Then

$$\forall x \in A \quad \exists B(x, r) \subset A. \tag{2}$$

Therefore

$$\bigcup_{x \in A} B(x, r) \subset A \tag{3}$$

But the ball $B(x, r) \subset A$ exists for every $x \in A$. Hence

$$A = \bigcup_{x \in A} B(x, r). \tag{4}$$

Show that the closed interval $[0, 1]$ is a closed subset of R with the absolute value metric.

SOLUTION:

DEFINITION OF A CLOSED SET

Let (X, d) be a metric space. A subset B of X is said to be closed in (X, d) when its complement $X - B$ is an open set in (X, d).

■

We shall show that the set $R - [0, 1]$ is open in R.
  Suppose

$$x \in R - [0, 1] \tag{1}$$

Then, either $x > 1$ or $x < 0$. We can assume that $x > 1$, (the proof for $x < 0$ is similar).

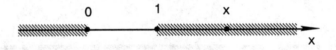

FIGURE 1

We have

$$0 < \delta = x - 1 \tag{2}$$

and

$$B(x, \,{}^{\delta}/_{2}) \subset R - [0, 1]. \tag{3}$$

Therefore $R - [0, 1]$ is an open set and $[0, 1]$ is a closed set.
  Similarly we can show that a closed ball in the metric space (X, d)

$$B(x, r) = \{\, y : d(x, y) \leq r \,\} \tag{4}$$

is a closed set.

● **PROBLEM 10-15**

Prove the following:
THEOREM

  Let (X, d) be a metric space. Then

1. X and ϕ are closed subsets of X,

2. the union of two closed sets, hence of a finite number of such sets, is a closed set,

3. the intersection of any family of closed sets is a closed set.

■

SOLUTION:

1. ϕ is an open set, therefore $X - \phi = X$ is closed, because it is a complement of an open set. X is an open set, therefore ϕ is a closed set because $\phi = X - X$.

2. Suppose B_1 and B_2 are closed sets. Then $X - B_1$ and $X - B_2$ are open sets. We have

$$X - (B_1 \cup B_2) = (X - B_1) \cap (X - B_2) \tag{1}$$

Since $X - B_1$ and $X - B_2$ are open sets, $(X - B_1) \cap (X - B_2)$ is an open set. Hence $X - (B_1 \cup B_2)$ is a closed set.

3. We shall show that

$$\bigcap_{\alpha} B_{\alpha} \tag{2}$$

is a closed set when all B_{α}'s are closed sets.

$$X - \left(\bigcap_{\alpha} B_{\alpha}\right) = \bigcup_{\alpha} (X - B_{\alpha}) \tag{3}$$

Suppose all sets B_{α} are closed. Then all sets $X - B_{\alpha}$ are open sets and their union $\bigcup_{\alpha} (X - B_{\alpha})$ is an open set.

Therefore $X - \left(\bigcap_{\alpha} B_{\alpha}\right)$ is an open set and $\bigcap_{\alpha} B_{\alpha}$ is a closed set.

● PROBLEM 10–16

Which of the following subsets of (R^2, d) where d is the Pythagorean metric are closed?

1. $\{(x, y) : x = 2\}$.

2. $\{(x, y) : x, y \text{ are integers}\}$.

3. $\{(x, y) : y = x^2\}$.

391

4. $\{(x, y) : x^2 + y^2 \le 1 \text{ and } x < 1\}$.

5. $\{(x, y) : x^2 + y^2 > 2\}$.

SOLUTION:

1.

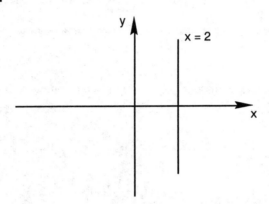

FIGURE 1

This is a closed set because $R^2 - \{(x, y) : x = 2\}$ is open.

2.

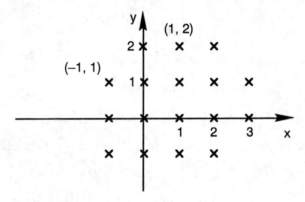

FIGURE 2

This set is closed.

3.

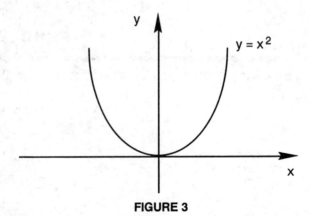

FIGURE 3

This set is closed.

392

4.

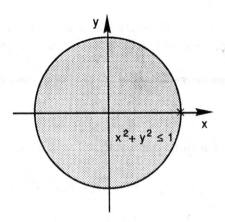

FIGURE 4

This set is neither closed nor open.

5.

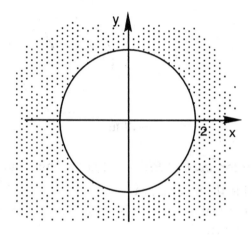

FIGURE 5

Since the circle $x^2 + y^2 = 2$ is not in the set, then the set is open..

● **PROBLEM 10–17**

DEFINITION OF CONVERGENCE

A sequence of points (x_n) of a metric space (X, d) is convergent to the point x of this space, if the sequence of real numbers $d(x_n, x)$ is convergent to zero. The point $x \in X$ is called the limit of the sequence (x_n). We write

$$x = \lim_{n \to \infty} x_n \qquad (1)$$

■

Write down this definition using the symbolism of logic.

Given an example to illustrate that a given sequence can be convergent with respect to one metric and not convergent with respect to another metric.

SOLUTION:

The definition of a limit point may be written in symbolism of logic as follows:

$$\left(\lim_{n \to \infty} x_n = x \right) \equiv \left(\lim_{n \to \infty} d(x_n, x) = 0 \right) \equiv$$

$$\equiv \forall \varepsilon > 0 \quad \exists k \in N \quad \forall n \in N$$

$$(n > k) \Rightarrow (d(x_n, x) < \varepsilon) \tag{2}$$

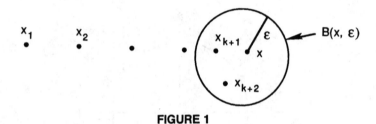

FIGURE 1

If $\lim_{n \to \infty} x_n = x$, then for any ball $B(x, \varepsilon)$ all but a finite number of elements of (x_n) are located inside the ball $B(x, \varepsilon)$.

Suppose the set of real numbers R is equipped with the metric $d_1(x, y) = |x - y|$ and with the metric

$$d_2(x, y) = \begin{cases} 0 & \text{when } x = y \\ 1 & \text{when } x \neq y \end{cases}$$

The sequence $(^1/_n)$ is given. In the space (R, d_1), this sequence is convergent. In the space (R, d_2), this sequence is not convergent.

● **PROBLEM 10–18**

Is it possible for a convergent sequence to have two different limits?

SOLUTION:

Let (X, d) be a metric space. Suppose sequence (X_n) has two different limits x_1 and x_2, $x_1 \neq x_2$. Then

394

$$\lim_{n \to \infty} x_n = x_1 \tag{1}$$

$$\lim_{n \to \infty} x_n = x_2 \tag{2}$$

Since $x_1 \ne x_2$ we have

$$d(x_1, x_2) = \alpha > 0. \tag{3}$$

From the triangle inequality

$$d(x_1, x_2) \le d(x_n, x_1) + d(x_n, x_2). \tag{4}$$

From (1) and (2) we obtain

$$\lim_{n \to \infty} d(x_n, x_1) = \lim_{n \to \infty} d(x_n, x_2) = 0. \tag{5}$$

This is a contradiction with (3). Hence $x_1 = x_2$. The sequence can have only one limit.

The same result can be reached "graphically." Since $x_1 \ne x_2$

$$d(x_1, x_2) = \alpha > 0. \tag{6}$$

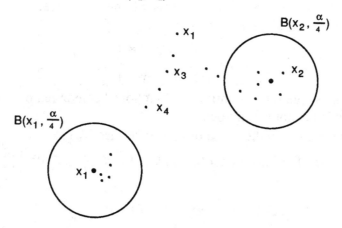

FIGURE 1

Balls $B(x_1, \alpha/u)$ and $B(x_2, \alpha/u)$ are disjoint and each contains all but a finite number of elements of (x_n). Contradiction.

● PROBLEM 10–19

Let X be the set of all functions from $[0,1]$ into itself with the metric d defined as follows

$$d(f, g) = \text{least upper bound} \ \{ \, |f(x) - g(x)| : x \in [0, 1] \, \}. \tag{1}$$

Check the convergence of the sequence (f_n) where $f_n(x) = x^n$.

SOLUTION:

The elements of the sequence are

$$x, x^2, x^3, x^4, \ldots, x^n, \ldots \qquad (2)$$

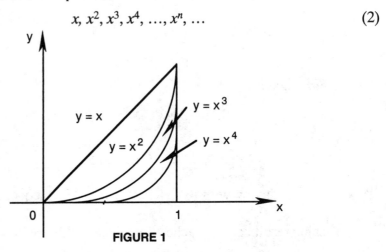

FIGURE 1

Figure 1 shows the first few elements of the sequence. We see that the sequence "converges" to the function

$$f(x) = \begin{cases} 0 & \text{for } x \neq 1 \\ 1 & \text{for } x = 1 \end{cases} \qquad (3)$$

Here, the sequence of continuous functions "converges pointwise" to the function which is not continuous.

It converges to f in the sense of metric defined by (1).

$$d(f_n, f) = \text{least upper bound } \{ \, |x^n - f(x)| : x \in [0, 1] \, \} \qquad (4)$$

and

$$\lim_{n \to \infty} d(f_n, f) = 0. \qquad (5)$$

● **PROBLEM 10–20**

Prove the following:

THEOREM

A sequence of points $z_n = (x_n, y_n)$ of the space $Z = X \times Y$ is convergent to the point $z = (x, y)$, if and only if

$$x = \lim_{n \to \infty} x_n \qquad (1)$$

$$y = \lim_{n \to \infty} y_n \qquad (2)$$

■

396

SOLUTION:

Suppose $x = \lim\limits_{n \to \infty} x_n$ and $y = \lim\limits_{n \to \infty} y_n$.

Let $\varepsilon > 0$, then $k \in N$ exists such that for $n > k$

$$d_1(x_n, x) < \varepsilon \quad \text{and} \quad d_2(y_n, y) < \varepsilon. \tag{3}$$

Let us define

$$d(z_n, z) = \sqrt{d_1^2(x_n, x) + d_2^2(y_n, y)}. \tag{4}$$

Then

$$d(z_n, z) < \varepsilon\sqrt{2} \tag{5}$$

and

$$\lim\limits_{n \to \infty} z_n = z. \tag{6}$$

Then k exists such that for $n > k$

$$d(z_n, z) < \varepsilon \tag{7}$$

and

$$d(z_n, z) = \sqrt{d_1^2(x_n, x) + d_2^2(y_n, y)} < \varepsilon. \tag{8}$$

Thus

$$d_1(x_n, x) < \varepsilon \quad \text{for} \quad n > k \tag{9}$$

i.e.,

$$\lim\limits_{n \to \infty} x_n = x \tag{10}$$

Similarly, we obtain

$$d_2(y_n, y) < \varepsilon \quad \text{for} \quad n > k \tag{11}$$

i.e.,

$$\lim\limits_{n \to \infty} y_n = y. \tag{12}$$

● **PROBLEM 10–21**

Show that the function $f(x) = 2x + 1$ is continuous,

$$f : R \to R \tag{1}$$

SOLUTION:

DEFINITION OF A CONTINUOUS FUNCTION

A function $f : X \rightarrow Y$, where (X, d) and (Y, d_1) are metric spaces, is said to be continuous, if for every $f(a) \in Y$ and any positive number $\varepsilon > 0$, there is a positive number $\delta > 0$ such that if $x \in B(a, \delta)$, then $f(x) \in B(f(a), \varepsilon)$.

■

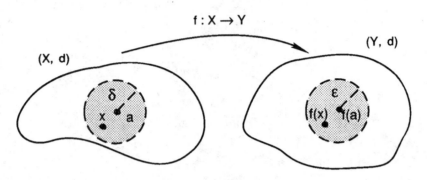

FIGURE 1

$$(x \in B(a, \delta)) \Rightarrow (f(x) \in B(f(a), \varepsilon))$$

In practical applications, this definition is rather difficult to deal with. Some other methods of establishing the continuity of functions will be shown later. Let us rewrite the definition using symbolic notation.

$f : (X, d) \rightarrow (Y, d_1)$ is continuous if

$$\forall \, f(a) \in Y \quad \forall \, \varepsilon > 0 \quad \exists \, \delta > 0 \quad \text{such that}$$

$$(d(x, a) < \delta) \Rightarrow (d_1(f(a), f(x)) < \varepsilon)$$

Let $f(a) = 2a + 1$ and $\varepsilon > 0$. We can choose $\delta = {}^{\varepsilon}/_2$, if $|x - a| < \delta = {}^{\varepsilon}/_2$, then

$$|f(x) - f(a)| = |2x + 1 - 2a - 1| =$$

$$= |2x - 2a| = 2|x - a| < 2 \cdot {}^{\varepsilon}/_2 = \varepsilon.$$

Hence, the function is continuous.

● **PROBLEM 10-22**

To determine whether or not a function is continuous, this theorem can be used:

THEOREM

$f: (X, d) \rightarrow (Y, d_1)$

    (f is continuous) $\Leftrightarrow$ ($\forall\, B \subset Y$, B is open $f^{-1}(B)$ is open)   (1)

                                                                   ■

Prove it.

SOLUTION:

$\Rightarrow$ f is continuous and A is an open subset of Y. We need to show that

$$f^{-1}(A) = \{\, x \in X : f(x) \in A \,\} \tag{2}$$

is an open subset of X.

    Let $x \in f^{-1}(A)$, then $f(x) \in A$. Since A is open there exists an $\varepsilon > 0$ such that

$$B(f(x), \varepsilon) \subset A. \tag{3}$$

Since f is continuous, then there exists a $\delta > 0$ such that

$$f[B(x, \delta)] \subset B(f(x), \varepsilon) \subset A \tag{4}$$

as shown in Figure 1.

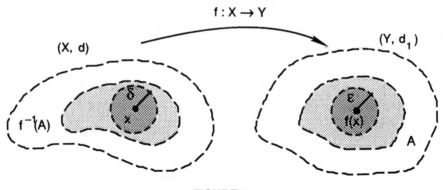

$f : X \rightarrow Y$

FIGURE 1

Therefore

$$B(x, \delta) \subset f^{-1}(A) . \tag{5}$$

For $x \in f^{-1}(A)$, a $\delta > 0$ exists such that (5) is true; hence $f^{-1}(A)$ is an open subset of X.

$\Leftarrow$ Suppose that if A is an open subset of Y, then $f^{-1}(A)$ is an open subset of X. If $f(a) \in Y$ and there exists an $\varepsilon > 0$, then

$$B(f(a) , \varepsilon) \tag{6}$$

399

is an open subset of Y.

Thus,

$$f^{-1}[B(f(a), \varepsilon)] \tag{7}$$

is an open subset of X. There exists a $\delta > 0$ such that

$$B(a, \delta) \subset f^{-1}[B(f(a), \varepsilon)] \tag{8}$$

or

$$f(B(a, \delta)) \subset B(f(a), \varepsilon). \tag{9}$$

Therefore f is continuous.

● **PROBLEM 10–23**

Prove the following useful:
THEOREM

Let $f: (X, d) \rightarrow (Y, d_1)$ then

$$(f \text{ is continuous}) \Leftrightarrow$$

(for any $B(y, r)$ in Y, $f^{-1}(B)$ is an open set of X) $\qquad$ (1) ∎

SOLUTION:

$\Rightarrow$ Suppose f is continuous and B is any ball in Y. Then B is an open subset of Y. By theorem of Problem 10–22, $f^{-1}(A)$ is an open subset of X.

$\Leftarrow$ Suppose for any ball $B \subset Y$, $f^{-1}(A)$ is an open subset of X.

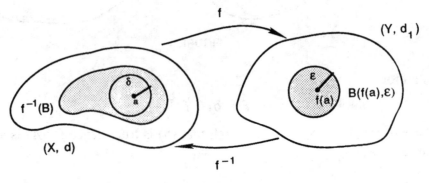

FIGURE 1

Let $f(a)$ be any point in Y and ε be any positive number. Then $B(f(a), \varepsilon)$ is a ball in Y.

The set $f^{-1}[B(f(a), \varepsilon)]$ is an open subset of X. We have

$$a \in f^{-1}[B(f(a), \varepsilon)]. \tag{2}$$

Set $f^{-1}(A)$ is an open subset of X (not necessarily a ball). Hence a ball $B(a, \delta)$ exists such that

$$B(a, \delta) \subset f^{-1}[B(f(a), \varepsilon)]. \tag{3}$$

Thus

$$(x \in B(a, \delta)) \Rightarrow (f(x) \in B(f(a), \varepsilon)).$$

Function f is continuous.

● **PROBLEM 10-24**

Consider the function

$$f(x, y) = x \tag{1}$$

$$f : (R^2, d_1) \rightarrow (R, d_2) \tag{2}$$

where

$$d_1((x_1, y_1), (x_2, y_2)) = \sqrt{(x_1 - x_2)^2 + (y_1 - y_2)^2} \tag{3}$$

$$d_2(x, y) = |x - y| \tag{4}$$

Show that f is continuous.

SOLUTION:

We shall use the theorem proved in Problem 10–23.

Let $B(a, r)$ be any ball in (R, d_2). Then $B(a, r)$ is an open interval $(a - r, a + r)$ in R.

The set $f^{-1}[B(a, r)]$ is

$$f^{-1}[(a - r, a + r)] = \{(x, y) ; a - r < x < a + r \}.$$

This set is shown in Figure 1; it is an open subset of R^2.

Hence, the function defined by (1) and (2) is continuous.

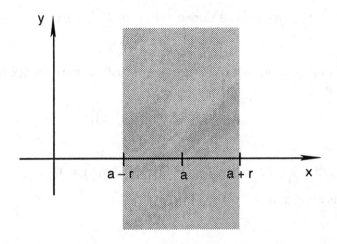

FIGURE 1

● **PROBLEM 10–25**

Find the limit of the sequence

$$\sin (1 + 1/n)^n \qquad (1)$$

SOLUTION:

Consider the sequence $(1 + 1/n)^n$. Its limit is

$$\lim_{n \to \infty} (1 + \tfrac{1}{n})^n = e. \qquad (2)$$

Now we shall apply the following:

THEOREM

Let $f : (X, d) \to (Y, d_1)$. Function f is continuous, if and only if given any sequence (x_n), such that

$$\lim_{n \to \infty} x_n = x, \quad \lim_{n \to \infty} f(x_n) = f(x). \qquad \blacksquare$$

A continuous function preserves the convergence of a sequence. Function $f(x) = \sin x$ is continuous. Therefore, since

$$\lim_{n \to \infty} (1 + \tfrac{1}{n})^n \to e \qquad (3)$$

we have

$$\lim_{n \to \infty} \{\sin(1 + \tfrac{1}{n})^n\} = e. \qquad (4)$$

Sometimes we shall use

$$x_n \underset{n \to \infty}{\to} x \qquad (5)$$

402

to indicate

$$\lim_{n \to \infty} x_n = x.$$

● **PROBLEM 10-26**

1. Which of the following sets are bounded?

    closed interval $[0, 1]$ in R

    ball $B(0, 1)$ in R^2

    half-line $x < 0$ in R

2. Which of the following functions are bounded?

    $f(x) = x^2$

    $f(x) = \sin x$

SOLUTION:

1. Let A be a subset of (X, d).
DEFINITION

Consider the set A. The diameter of A, denoted by *delta* (A) is the least upper bound of the distances $d(x, y)$ between all pairs of points $x, y \in A$. Sets with finite diameter are said to be bounded. For simplicity, we write "least upper bound" as "lub." ∎

Interval $[0, 1]$ is bounded in (R, d) where $d(x, y) = |x - y|$.
The ball $B(0, 1)$ is bounded.
Half-line, $x < 0$ is not bounded.

2. **DEFINITION**
A mapping $f : X \to Y$ where Y is a metric space, is called bounded if the set $f(x)$ is bounded. ∎

Function

$$f : R \to R$$

403

$$f(x) = x^2$$

maps R onto $[0, \infty)$. Hence

$$f(R) = [0, \infty).$$

The set $[0, \infty)$ is not bounded, hence the function is not bounded.

Function $f(x) = \sin x$ maps R onto $[-1, 1]$ hence the function is bounded.

● **PROBLEM 10–27**

Prove the following:

THEOREM 1

The set $\Phi(X, Y)$ of all bounded mappings $f : X \to Y$, where X is an arbitrary set and (Y, d) a metric space, is a metric space with metric defined by

$$d_1(f, g) = \sup d(f(x), g(x)). \tag{1}$$

■

SOLUTION:

First let us prove:

THEOREM 2

If f and g are bounded mappings to the set X into the metric space (Y, d), then the metric

$$d_1(f, g) = \sup d(f(x), g(x)) \tag{2}$$

is finite.

■

Let a be a given element of X. Then

$$d(f(x), g(x)) \le d(f(x), f(a)) + d(f(a), g(x)) \le$$

$$\le d(f(x), f(a)) + d(f(a), g(a)) + d(g(a), g(x)). \tag{3}$$

Hence

$$d_1(f, g) \le \delta[f(X)] + d(f(a), g(a)) + \delta[g(X)]. \tag{4}$$

■

We have

$$d_1(f, g) \geq 0 \quad \text{for all } f, g \tag{5}$$

$$d_1(f, g) = 0 \quad \text{iff } f = g \tag{6}$$

Indeed, if sup $d(f(x), g(x)) = 0$ then $f(x) = g(x)$ for all $x \in X$.

$$d_1(f, g) = d_1(g, f) \text{ for all } f, g \tag{7}$$

$$d_1(f, g) = \sup d(f(x), g(x)) \leq$$

$$\leq \sup [d(f(x), h(x)) + d(f(x), g(x))] \leq$$

$$\leq \sup d(f(x), h(x)) + \sup d(h(x), g(x)) =$$

$$= d_1(f, h) + d_1(h, g). \tag{8}$$

Hence d_1 defined in (1) is a metric.

---

1. Show that if $A \neq \phi$ and $B \neq \phi$ and $A \subset B$, then

    $$\delta(A) \leq \delta(B). \tag{1}$$

2. Show that if $A \cap B \neq \phi$, then

    $$\delta(A \cup B) \leq \delta(A) + \delta(B). \tag{2}$$

SOLUTION:

1. We have

    $$\delta(A) = \text{lub } d(x, y) \tag{3}$$

    where $x, y \in A$.
    Note that since $A \subset B$

    $$d(x, y) \leq d(x, y) \tag{4}$$

    for $x, y \in A$.
    Then,

    $$\begin{array}{ccc} \text{lub } d(x, y) & \leq & \text{lub } d(x, y) \\ \text{where } x, y \in A & & \text{where } x, y \in B \end{array} \tag{5}$$

and

$$\delta(A) \leq \delta(B). \tag{6}$$

2.

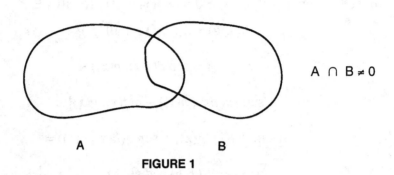

$A \cap B \neq 0$

A   B

FIGURE 1

Let us define the distance between sets A and B

$$\delta(A, B) = \text{greatest lower bound } \{d(a, b) : a \in A, b \in B\}. \tag{7}$$

Then, for any subsets A and B of (X, d)

$$\delta(A \cup B) \leq \delta(A) + \delta(B) + \delta(A, B). \tag{8}$$

But $A \cap B \neq 0$ therefore

$$\delta(A, B) = 0. \tag{9}$$

Hence

$$\delta(A \cup B) \leq \delta(A) + \delta(B). \tag{10}$$

● PROBLEM 10-29

Let (X, d) be a metric space. Show that a subset A of X is closed, if and only if for any $x \in X - A$,

$$\delta(x, A) \neq 0 \tag{1}$$

where

$$\delta(x, A) =: \text{greatest lower bound or (glb) } \{d(x, a) : a \in A\}. \tag{2}$$

SOLUTION:

Suppose A is a closed set, then $X - A$ is open. Hence for any $x \in X - A$, a ball $B(x, r)$ exists such that

$$B(x, r) \subset X - A \tag{3}$$

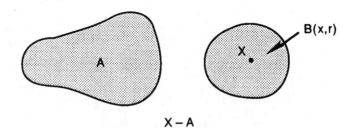

X – A

FIGURE 1

Then

$$\delta(x, A) \geq r \tag{4}$$

and

$$\delta(x, A) \neq 0. \tag{5}$$

Now suppose, for any $x \in X - A$, $\delta(x, A) \neq 0$.
Let

$$r = \delta(x, A) \tag{6}$$

then

$$B(x, r) \subset X - A. \tag{7}$$

Hence $X - A$ is open and A is closed

$$A = X - (X - A). \tag{8}$$

● **PROBLEM 10–30**

Let (X, d) be a metric space. Show that if A and A' are closed subsets of X, such that $A \cap A' = \phi$, then open sets B and B' exist such that

$$A \subset B, \quad A' \subset B', \quad B \cap B' = \phi. \tag{1}$$

SOLUTION:

Let us define for each $x \in A$

$$r_x = {}^1/_2 \, \delta(x, A') \tag{2}$$

and for each $x' \in A'$

$$r'_{x'} = {}^1/_2 \, \delta(x', A) \tag{3}$$

407

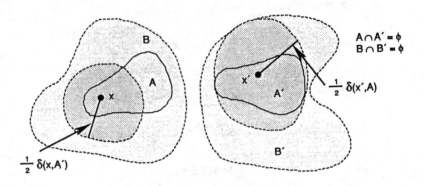

FIGURE 1

We define

$$B = \bigcup_A B(x, r_x)$$

$$B' = \bigcup_A B(x', r'_{x'}) \tag{4}$$

Both sets B and B' are open because they are the union of a family of open sets (balls).

Since the union has taken over all the elements of A

$$A \subset B \tag{5}$$

similarly

$$A' \subset B'. \tag{6}$$

We will show that $B \cap B' = \phi$.

Suppose on the contrary

$$x \in B \cap B'. \tag{7}$$

Then for some $y \in A$

$$\delta(y, x) < r_y \tag{8}$$

and for some $y' \in A'$

$$\delta(y', x) < r'_{y'} \tag{9}$$

Suppose $r_y \geq r'_{y'}$, then

$$\delta(y, y') \leq \delta(y, x) + \delta(y', x) < r_y + r'_{y'} \leq$$

$$\leq 2r_y = \delta(y, A') \tag{10}$$

where $\delta(y, y') = d(y, y')$. But

$$\delta(y, y') \geq d(y, A'). \tag{11}$$

408

Hence a contradiction,

$$B \cap B' = \phi. \tag{12}$$

Two closed disjoint sets are well separable, in the sense that it is possible to put them into two open sets, which again are disjoint.

Consider the metric space (R^2, d) where d is the Pythagorean metric. Let A be the hyperbola $y = 1/x$ and A' the x -axis. Find the open sets B and B' as described in Problem 10–30.

SOLUTION:

Both sets

$$A = \{ (x, y) : y = 1/x, x \neq 0 \} \tag{1}$$

and

$$A' = \{ (x, y) : y = 0 \} \tag{2}$$

are closed and

$$A \cap A' = \phi. \tag{3}$$

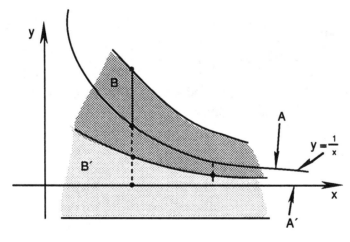

FIGURE 1

At first glance, it seems impossible to find the separating open sets B and B' because

$$\delta(A, A') = 0 \tag{4}$$

409

i.e., the distance between A and A' is zero. Function $y = 1/x$ has the x-axis for an asymptote. Such sets B and B', however, exist.

$$B = \{ (x, y) : 1/2x < y < 1/x + 1 \} \tag{5}$$

$$B' = \{ (x, y) : -1 < y < 1/2x \}. \tag{6}$$

Both B and B' are open. We have

$$A \subset B \text{ and } A' \subset B' \tag{7}$$

$B \cap B' = \phi$.

● **PROBLEM 10-32**

1. Let (R, d) be the space of real numbers with the absolute value metric and let

$$A = \{1/n : n \in N\}. \tag{1}$$

Find the closure of A.

2. Show that for any subset A of X, the closure of A is a closed subset of X.

SOLUTION:

DEFINITION OF THE CLOSURE OF A

Let (X, d) be the metric space and $A \subset X$. The closure of A, denoted by $\overline{A}$ (sometimes by $C(A)$), is defined by

$$\overline{A} =: \{x \in X : \delta(x, A) = 0\}. \tag{2}$$

∎

1. Since $1/n \rightarrow 0$, we have

$$\delta(0, A) = 0. \tag{3}$$

For each $x \in A$,

$$\delta(x, A) = 0 \tag{4}$$

Therefore

$$A \subset \overline{A}. \tag{5}$$

410

From (3) and (5) we obtain

$$A \cup \{0\} \subset \overline{A}. \tag{6}$$

Suppose $x \in \overline{A}$ and $x \notin A \cup \{0\}$. Then

$$\delta(x, A) > 0. \tag{7}$$

We conclude that

$$A \cup \{0\} = \overline{A}. \tag{8}$$

2. Suppose $\overline{A}$ is not closed. Then $X - \overline{A}$ is not open and there is an element $x \in X - \overline{A}$ such that for every $\varepsilon > 0$

$$B(x, \varepsilon) \cap A \neq \phi. \tag{9}$$

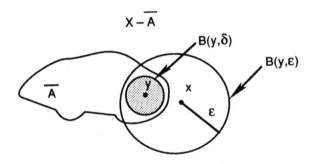

$X - \overline{A}$

$B(y,\delta)$

$B(y,\varepsilon)$

$\overline{A}$

y

x

ε

FIGURE 1

Choose $$y \in \overline{A} \cap B(x, \varepsilon). \tag{10}$$

Since $B(x, \varepsilon)$ is open, $\delta > 0$ exists such that

$$b(y, \delta) \text{ A } B(x, \varepsilon). \tag{11}$$

On the other hand $y \in \overline{A}$, then

$$\delta(y, A) = 0. \tag{12}$$

Therefore, a exists such that

$$a \in A \cap B(y, \delta) \subset B(x, \varepsilon). \tag{13}$$

Hence, for any $\varepsilon > 0$, there is $a \in A$ such that $\delta(x, a) < \varepsilon$

$$\text{greatest lower bound } \{ \delta(x, a) : a \in A \} = \delta(x, A) = 0. \tag{14}$$

Therefore $x \in \overline{A}$, but we assumed that $x \in X - \overline{A}$. Contradiction.

Prove:

THEOREM

The limit of a uniformly convergent sequence of bounded mappings is bounded.

■

SOLUTION:

DEFINITION OF UNIFORM CONVERGENCE

Let $f_n : X \to Y$, $n \in N$, where X is an arbitrary set and (Y, d) is a metric space. The sequence (f_n) is said to converge uniformly to f of

$$\forall \, \varepsilon > 0 \quad \exists \, k \in N \quad \forall \, x \in X \quad \forall \, n \geq k \quad d(f_n(x), f(x)) < \varepsilon. \tag{1}$$

■

Consider the space $\Phi(X, Y)$ of all bounded mappings $f : X \to Y$ with the metric defined by

$$d(f, g) =: \sup \{d(f(x), g(x)) : x \in X\}. \tag{2}$$

We have

$$\left(\lim_{n \to \infty} f_n = f \right) \equiv \left(\lim_{n \to \infty} d(f_n, f) = 0 \right) \equiv$$

$$\equiv \forall \varepsilon > 0 \, \exists k \in N \, \forall n \geq k \, \sup\{d(f_n(x), f(x))\} < \varepsilon \equiv$$

$$\equiv \forall \varepsilon > 0 \, \exists k \in N \, \forall n \geq k \forall x \in X \, d(f_n(x), f(x)) < \varepsilon. \tag{3}$$

Thus, in the space $\Phi(X, Y)$ the condition $\lim f_n = f$ means that the sequence (f_n) converges uniformly to f.

Let $\varepsilon > 0$ and let $f_n \to f$, (f_n) be a uniformly convergent sequence of bounded mappings. We choose n such that

$$d(f_n, f) < \varepsilon. \tag{4}$$

We have

$$d(f(x_1), f(x_2)) \leq d(f(x_1), f_n(x_1)) + d(f_n(x_2), f(x_2)) +$$

$$+ d(f_n(x_1), f_n(x_2)). \tag{5}$$

Thus

$$\delta[f(X)] \leq \delta[f_n(X)] + 2\varepsilon \tag{6}$$

and f is bounded.

CHAPTER 11

TOPOLOGICAL SPACES

1. Let X denote any set and $P(X)$ a family of all subsets of X. Show that $P(X)$ is a topology.

2. Find a topology on X, which contains as few as possible subsets of X.

SOLUTION:

DEFINITION OF TOPOLOGY

T is a topology on X, if and only if

1. $T \subset P(X)$

2. $\phi \in T$ and $X \in T$

3. If $A_1, A_2 \in T$, then $A_1 \cap A_2 \in T$, i.e., each finite intersection of members of T is also a member of T.

4. Each union of members of T is also a member of T. ∎

DEFINITION OF TOPOLOGICAL SPACE

A couple (X, T) consisting of a set X and a topology T on X is called a topological space. ∎

1. Since $P(X)$ is a family of all subsets of X, $P(X)$ is a topology. This topology is called the discrete topology on X. It contains the maximum possible number of sets.

2. Let X represent any set. The collection $T = \{\phi, X\}$ is a topology on X.
  This topology is called the indiscrete topology (or trivial topology) on X. It contains the fewest possible number of sets.
  Elements of topological spaces are called points. The members of T are called the open sets of the topological space (X, T).

Let $X = \{a, b, c, d\}$ and $T_0 = \{\{a\}, \{a, b\}, \{a, c, d\}\}$. Find a few possible topologies T on X, such that

$$T_0 \subset T. \tag{1}$$

SOLUTION:

Each union of members of T is also a number of T (see Problem 11–1). Hence

$$\{a, b\} \cup \{a, c, d\} = \{a, b, c, d\} = X. \tag{2}$$

One of the possible topologies on X is

$$T_1 = \{\phi, X, \{a\}, \{a, b\}, \{a, c, d\}\}$$

$$T_0 \subset T_1. \tag{3}$$

Suppose $\{b\} \in T_2$, then

$$T_2 = \{\phi, X, \{a\}, \{b\}, \{a, b\}, \{a, c, d\}\}. \tag{4}$$

T_2 is a topology and $T_0 \subset T_2$.
Suppose $\{c\} \in T_3$, then

$$T_3 = \{\phi, X, \{a\}, \{a, b\}, \{a, c, d\}, \{c\}, \{a, c\}, \{a, b, c\}\}. \tag{5}$$

Suppose $\{d\} \in T_4$ then

$$T_4 = \{\phi, X, \{a\}, \{a, b\}, \{a, c, d\}, \{d\}, \{a, d\}, \{a, b, d\}\}. \tag{6}$$

Suppose $\{b\} \in T_5$ and $\{c\} \in T_5$ then

$$T_5 = \{\phi, X, \{a\}, \{a, b\}, \{a, c, d\}, \{b\}, \{c\}, \{a, c\},$$

$$\{a, b, c\}, \{b\,c\}\}. \tag{7}$$

Suppose $\{b\} \in T_6$ and $\{d\} \in T_6$ then

$$T_6 = \{\phi, X, \{a\}, \{a, b\}, \{a, c, d\}, \{b\}, \{d\}, \{a, d\},$$

$$\{b, d\}, \{a, b, d\}\}. \tag{8}$$

Suppose $\{c\} \in T_7$ and $\{d\} \in T_7$ then

$$T_7 = \{\phi, X, \{a\}, \{a, b\}, \{a, c, d\}, \{c\}, \{d\}, \{a, c\},$$

$$\{a, d\}, \{a, b, c\}, \{a, b, d\}, \{c, d\}\}. \tag{9}$$

Now let $T_8 = P(X)$

$$T_8 = \{\phi, X, \{a\}, \{b\}, \{c\}, \{d\}, \{a, b\}, \{a, c\}, \{a, d\}, \{b, c\},$$

$\{b, d\}, \{c, d\}, \{a, b, c\}, \{a, b, d\}, \{a, c, d\}, \{b, c, d\}\}.$ (10)

Suppose $\{a, c\} \in T_9$ then

$$T_9 = \{\phi, X, \{a\}, \{a, b\}, \{a, c\}, \{a, c, d\}, \{a, b, c\}\}. \qquad (11)$$

● PROBLEM 11-3

Let R represent the set of real numbers. Subset $A \subset R$ is called open if for each $x \in A$, there is an $r > 0$, such that

$$B(x, r) =: \{\, y : |\, y - x \,| < r \,\} \subset A \qquad (1)$$

(R, d), where $d(x, y) = |\, x - y \,|$ is a metric space. Let T denote the family of all open sets (in the sense defined above). Verify that T is a topology on R.

SOLUTION:

Obviously ϕ and R are open sets.

$$\phi \in T, \quad R \in T. \qquad (2)$$

Suppose $A_1, ..., A_n$ are open sets, then

$$\bigcap_1^n A_k \in T.$$

Indeed

$$\left(x \in \bigcap_1^n A_k \right) \Rightarrow (\forall k, x \in A_k) \Rightarrow$$

$$\Rightarrow (\forall k \exists r_k > 0 : B(x, r_k) \subset A_k) \Rightarrow$$

$$\Rightarrow \left(\begin{array}{c} B(x, r) \subset \bigcap_1^n A_k \\ \text{where } r = \min\{r_1, r_2, ..., r_n\} \end{array} \right) \qquad (3)$$

Suppose each member of $\{A_\omega : \omega \in \Omega\}$ is open, then $\bigcup_\omega A_\omega$ is also open.

$$\left(x \in \bigcup_\omega A_\omega \right) \Rightarrow (\exists w : x \in A_\omega) \Rightarrow$$

$$\Rightarrow (\exists r > 0 : B(x, r) \subset A_w \subset \bigcup_\omega A_\omega) \qquad (4)$$

Topology defined above is called the Euclidean topology on R'. The topological space (R', T) is called the Euclidean 1-space.

Let (X, d) denote a metric space. Consider the family T of all d-open subsets of X, i.e., subsets which are open in the sense of metric d.
Show that T is a topology on (X, d).

SOLUTION:

Let us recall the definition of an open set A. A is open iff

$$\forall x \in A \quad \exists B(x, r): B(x, r) \subset A. \tag{1}$$

An empty set and the whole space are open sets

$$\phi \in T, \quad X \in T \tag{2}$$

Suppose $A_1, ..., A_n$ are open sets. Then $\overset{n}{\underset{1}{\bigcap}} A_k$ is an open set.

$$\left(x \in \overset{n}{\underset{1}{\bigcap}} A_k\right) \Rightarrow (\forall k : x \in A_k) \Rightarrow$$

$$\Rightarrow (\forall k \exists r_k : B(x, r_k) \subset A_k) \Rightarrow$$

$$\Rightarrow \left(\forall k : B(x, r) \subset A_k \text{ where } r = \min\{r_1, ..., r_n\}\right) \Rightarrow$$

$$\Rightarrow \left(B(x, r) \subset \overset{n}{\underset{1}{\bigcap}} A_k\right). \tag{3}$$

Now, let A_α denote a family of open sets, then $\underset{\alpha}{\bigcup} A_\alpha$ is an open set.

$$\left(x \in \underset{\alpha}{\bigcup} A_\alpha\right) \Rightarrow (\exists \alpha: x \in A_\alpha) \Rightarrow$$

$$\Rightarrow \left(\exists B(x, r): B(x, r) \subset A_\alpha \subset \underset{\alpha}{\bigcup} A_\alpha\right). \tag{4}$$

Topology T described here for the metric space (X, d) is called the metric topology induced on X by d. Each metric space (X, d) can become an "automatically" topological space (X, T). Different metrics on X will induce in general different topologies on X.

? NOT DONE (ON NEXT PAGE)

1. Define the topological space Z and the Sierpinski space.

2. Suppose T is a topology on X, consisting of

$$T = \{\phi, X, A, B\} \tag{1}$$

> where A and B are non-empty distinct proper subsets of X. Determine conditions on A and B.

SOLUTION:

1. Let $X = \{0, 1\}$. Then the discrete topology on X, $T = P(X)$, is denoted by Q.

$$T = \{\phi, \{0, 1\}, \{0\}, \{1\}\} \tag{2}$$

The set $\{0, 1\}$ with topology

$$T = \{\phi, \{0, 1\}, \{0\}\} \tag{3}$$

is called the Sierpinski space.

2. Since T defined by (1) is a topology

$$A \cap B \in T = \{\phi, X, A, B\}. \tag{4}$$

We have either

$$A \cap B = \phi \tag{5}$$

or

$$A \cap B = A \tag{6}$$

or

$$A \cap B = B. \tag{7}$$

Suppose $A \cap B = \phi$. Then

$$(5) \Rightarrow A \cup B \neq A \text{ and } A \cup B \neq B \text{ and } A \cup B \neq \phi. \quad \text{BUT } A \cup B \in T$$

Hence

$$A \cup B = X \tag{8}$$

$\{A, B\}$ is a partition of X.

Suppose now $A \cap B = A$ then

$$(6) \Rightarrow \phi \subset A \subset B \subset X. \tag{9}$$

If $A \cap B = B$ then

$$(7) \Rightarrow \phi \subset B \subset A \subset X. \tag{10}$$

Let $f: X \to Y$ denote a function of $X \neq \phi$ onto a topological space (Y, T_0). Show that the family T of subsets of X, defined by

$$T = \{f^{-1}(B) : B \in T_0\} \qquad (1)$$

is a topology on X.

SOLUTION:

T_0 is a topology on Y. Therefore

$$\phi, Y \in T_0 \qquad (2)$$

Since f is onto

$$f^{-1}(Y) = X. \qquad (3)$$

Also

$$f^{-1}(\phi) = \phi. \qquad (4)$$

Thus

$$\phi, X \in T. \qquad (5)$$

Let $\{A_\omega : \omega \in \Omega\}$ represent a family of sets in T. Then for each ω

$$f^{-1}(B_\omega) = A_\omega \quad B_\omega \in T_0 \qquad (6)$$

We have

$$\bigcup_\omega A_\omega = \bigcup_\omega f^{-1}(B_\omega) = f^{-1}\left(\bigcup_\omega B_\omega\right) \qquad (7) \quad ✳$$

Since T_0 is a topology

$$\bigcup_\omega B_\omega \in T_0 \qquad (8)$$

therefore $\bigcup_\omega A_\omega \in T$.

Let $A_1, A_2 \in T$. Then sets $B_1, B_2, \in T_0$ exist, such that

$$f^{-1}(B_1) = A_1 \quad \text{and} \quad f^{-1}(B_2) = A_2 \qquad (9)$$

Then

$$A_1 \cap A_2 = f^{-1}(B_1) \cap f^{-1}(B_2) = f^{-1}(B_1 \cap B_2). \qquad (10)$$

Since T_0 is a topology and $B_1, B_2 \in T_0$, the intersection $B_1 \cap B_2 \in T_0$. Therefore

$$f^{-1}(B_1 \cap B_2) \in T \quad \text{and} \quad A_1 \cap A_2 \in T. \qquad (11)$$

Thus (X, T) is a metric space.

419

Consider set N with the family T of its subsets consisting of ϕ and all sets of the form

$$A_k = \{k, k+1, k+2, ...\} \; k = 1, 2, 3, ... \; . \tag{1}$$

Show that T is a topology on N.

SOLUTION:

$$A_1 = \{1, 2, 3, ...\} \tag{2}$$

Thus

$$\phi \in T \quad \text{and} \quad N \in T \tag{3}$$

Suppose $A_k, A_l \in T$. Then either $k > l$ or $k < l$ or $k = l$. If $k > l$, then

$$A_k \cap A_l = A_k \in T \tag{4}$$

if $k < l$, then $\qquad A_k \cap A_l = A_l \in T. \tag{5}$

Also if $k = l$, then $A_k \cap A_k = A_k \in T$. Thus $A_k \cap A_l = T$.
Let $\{A_i\}$ represent a family of sets of the form (1). Then

$$p = \min \{i_1, i_2, ...\} \tag{6}$$

and

$$\bigcup_i A_i = A_p = \{p, p+1, p+2, ...\}. \tag{7}$$

Thus T is a topology on N.

● **PROBLEM 11-8**

P.430

1. Show that intersection of any family of topologies on X is also a topology on X.

2. Show that the union of topologies does not have to be a topology.

SOLUTION:

1. Let $\{T_\omega : \omega \in \Omega\}$ be a family of topologies on X.

$$\forall \, \omega \in \Omega : \phi, X \in T_\omega$$

Hence $$\phi, x \in \bigcap_{\omega} T_\omega .$$ (1)

Suppose $A, B \in \bigcap_{\omega} T_\omega$.

$$(A, B \in \bigcap_{\omega} T_\omega) \Rightarrow (\forall \omega : A, B \in T_\omega) \Rightarrow$$

$$\Rightarrow (\forall \omega : A \cap B \in T_\omega) \Rightarrow (A \cap B \in \bigcap_{\omega} T_\omega).$$ (2)

Suppose $\{A_i\}$ is a family of sets such that
$$(\forall i : A_i \in \bigcap_{\omega} T_\omega) \Rightarrow (\forall i \, \forall \omega \, A_i \in T_\omega) \Rightarrow$$

$$\Rightarrow (\forall \omega : \bigcup_i A_i \in T_\omega) \Rightarrow (\bigcup_i A_i \in \bigcap_{\omega} T_\omega).$$ (3)

Hence if $\{T_\omega : \omega \in \Omega\}$ is a family of topologies on X then $\bigcap_{\omega} T_\omega$ is a topology on X.

2. Consider the set

$$X = \{a, b, c\}$$ (4)

and two topologies on X

$$T_1 = \{\phi, X, \{a\}\}$$ (5)

and

$$T_2 = \{\phi, X, \{b\}\}$$ (6)

Then

$$T_1 \cup T_2 = \{\phi, X, \{a\}, \{b\}\}$$ (7)

is not a topology because

$$\{a\} \cup \{b\} = \{a, b\} \notin T_1 \cup T_2.$$ (8)

● PROBLEM 11-9

How many distinct topologies can a set consisting of three elements have? What is their partial ordering?

SOLUTION:

Let

$$X = \{a, b, c\}.$$ (1)

The possible topologies on X are

$$T_1 = \{\phi, X\}$$

$$T_2 = \{\phi, X, \{a\}\}, \qquad T_3 = \{\phi, X, \{b\}\},$$

$$T_4 = \{\phi, X, \{c\}\}, \qquad T_5 = \{\phi, X, \{a, b\}\}$$

$$T_6 = \ldots \tag{2}$$

and so on.

Consider the set $P(X)$

$$P(X) = \{\phi, X, \{a\}, \{b\}, \{c\}, \{a, b\}, \{a, c\}, \{b, c\}\} \tag{3}$$

$| = 1$ topology $\{\phi, X\}$

$\binom{3}{1} = 3$ topologies of the type $\{\phi, X, \{\cdot\}\}$

where $\{\cdot\}$ indicates the set consisting of one element

$\binom{3}{2} = 3$ topologies of the type $\{\phi, X, \{\cdot, \cdot\}\}$

$\binom{3}{1} \times \binom{3}{2}' = 9$ topologies of the type $\{\phi, X, \{\cdot\}, \{\cdot, \cdot\}\}$

$9 = \binom{3}{1} \times \binom{3}{2} \neq 3$ topologies of the type $\{\phi, X, \{\cdot\}, \{\cdot, \cdot\}, \{\cdot, \cdot\}\}$

$9 = \binom{3}{1} \times \binom{3}{1} \neq 3$ topologies of the type $\{\phi, X, \{\cdot\}, \{\cdot\}, \{\cdot, \cdot\}\}$

$9 = \binom{3}{2} + \binom{3}{2} \neq 6$ topologies of the type $\{\phi, X, \{\cdot\}, \{\cdot\}, \{\cdot, \cdot\}, \{\cdot, \cdot\}\}$

$| = 1$ topology $P(X)$

$29 =$ The numbers must be added up in order to obtain the total number of topologies on $X = \{a, b, c\}$.

We can introduce the relation of inclusion among topologies. For example,

$$\{\phi, X, \{a\}\} \subset \{\phi, X, \{a\}, \{b\}, \{a, b\}\}. \tag{4}$$

This relation defines a partial order. Indeed, Let T_1, T_2, T_3 represent topologies on X, then

$$(T_1 \subset T_2) \wedge (T_2 \subset T_3) \Rightarrow (T_1 \subset T_3)$$

$$\forall T : T \subset T$$

$$(T_1 \subset T_2) \wedge (T_2 \subset T_1) \Rightarrow (T_1 = T_2). \tag{5}$$

● **PROBLEM 11–10**
SET

Prove that T is the discrete topology on X if and only if every point is an open set.

422

SOLUTION:

Let $T = P(X)$, then T is called the discrete topology on X. Every set is an open set. We shall prove that

$$(T \text{ is the discrete topology}) \Leftrightarrow (\text{every point is an open set}). \qquad (1)$$

(set)

$(\Rightarrow)$ Since T consists of all subsets of X

$$\forall\, x \in X : \{x\} \in T. \qquad (2)$$

$(\Leftarrow)$ Suppose T is a topology on X, such that

$$\forall\, x \in X : \{x\} \in T. \qquad (3)$$

Let A denote any subset of X. Then USING A TO INDEX A,

$$A = \bigcup_{a \in A} \{a\}. \qquad (4)$$

But all $\{a\}$ are open sets, $\{a\} \in T$ and since T is a topology, the union of open sets is an open set.

Hence, $A \in T$, we conclude, that $T = P(X)$.

● **PROBLEM 11–11**

Let X represent any set and S a family of its subsets, such that

1. $X, \phi \in S$.

2. The union of any two members of S is a member of S.

3. The intersection of any family of members of S is a member of S.

Let T denote a family of subsets of X, such that

$$A \in T \quad \text{iff} \quad X - A \in S. \qquad (1)$$

Show that T is a topology on X.

SOLUTION:

Since $X, \phi \in S$ and $X - X = \phi$, $X - \phi = X$ $X, \phi \in T$.
Suppose $A_1, A_2 \in T$, then $X - A_1 \in S$ and $X - A_2 \in S$

423

$$(X - A_1) \cup (X - A_2) = X - (A_1 \cap A_2) \in S. \tag{2}$$

Therefore $A_1 \cap A_2 \in T$.

Suppose $\{A_\alpha\}$ is a family of subsets of T. Then $\{X - A_\alpha\}$ is a family of subsets of S and

$$\bigcap_\alpha (X - A_\alpha) \in S. \tag{3}$$

But

$$\bigcap_\alpha (X - A_\alpha) = X - \bigcup_\alpha A_\alpha. \tag{4}$$

Hence $X - \bigcup_\alpha A_\alpha \in S$ and $\bigcup_\alpha A_\alpha \in T$.

Thus T is a topology on X. Elements of T are called open sets.

DEFINITION OF A CLOSED SET

Subset B of (X, T) is closed, if $X - B$ is an open set, that is, if $X - B \in T$.

why Here Repeated on P. 438

● **PROBLEM 11-12**

√?

> Give an example of a basis for the Euclidean topological space (R^n, T).

SOLUTION:

DEFINITION OF A BASIS

Let (X, T) denote a topological space. A family $B \subset T$ is called a basis for T if each open set (i.e., an element of T) is the union of members of B.

■

Note that if T is a topology on X, then T is a basis for T. In (R^n, T), the topology is the family of sets open in the sense of the Euclidean metric. We shall show that in R^n

$$B = \{B(x, r) : x \in R^n, r > 0\} \tag{1}$$

is a basis for the Euclidean topology.

Let $A \in T$ denote an open subset of R^n. Then

$$A = \bigcup_{x \in A} B(x, r). \tag{2}$$

Each open set can be represented as a union of balls. Note that (1) is not the only possible basis for the Euclidean topology in R^n.

$\mathcal{B}$

> Let $B \subset T$, where T is a topology on X. Show that the following properties of B are equivalent:
>
> 1. B is a basis for T.
>
> $\mathcal{B}$
>
> 2. For each $A \in T$ and each $x \in A$, there is $B_\omega \in B$, such that
>
> A is open in X $x \in B_\omega \subset A.$

$x \in A \in T \Rightarrow B_\omega \in \mathcal{B} \left[x \in B_\omega \subset A \right]$

SOLUTION:

$(1) \Rightarrow (2)$ Let $A \in T$. Since B is a basis for T

$\mathcal{B}$
$$A = \bigcup_\omega B_\omega \tag{1}$$

where $B = \{B_\omega : \omega \in \Omega\}$. $x \in A$, therefore at least one B_ω exists, such that

$$x \in B_\omega \subset A. \tag{2}$$

$(1) \Leftarrow (2)$ Suppose $A \in T$. Since for each $x \in A$, set B_ω exists, such that
$$B_\omega \in B \tag{3}$$

and

$$x \in B_\omega \subset A \tag{4}$$

we have

$\mathcal{B}$
$$A = \bigcup_{x \in A} B_\omega, \text{ where } x \in B_\omega. \tag{5}$$

Thus B is a basis for T.

The following theorem is useful in describing open sets.

THEOREM

Let $B \subset T$ represent a basis for T. Then set A is open, i.e., $A \in T$, if and only if, for each $x \in A$, there is a $B_\omega \in B$, such that $x \in B_\omega \subset A$.

> Show that R^n with the Euclidean topology has a countable basis.

SOLUTION:

Consider the family of all open balls $B_\omega(x, r)$, such that r is a rational number and all coordinates of $x = (x_1, x_2, \ldots, x_n)$ are rational numbers. This family forms a countable set.

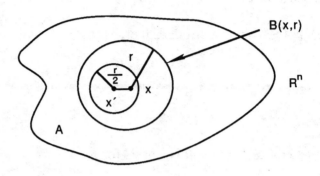

FIGURE 1

Let A denote any open subset of R^n and let $x \in A$. There is a ball $B(x, r)$, such that

$$x \in B(x, r). \tag{1}$$

We can assume that $r > 0$ is a rational number. We can always find a point x', such that

$$x' = (x'_1, x'_2, \ldots, x'_n)$$

where all coordinates of x' are rational and

$$d(x, x') < {}^r/_3. \tag{2}$$

Then

$$x \in B(x', {}^r/_2) \subset B(x, r). \tag{3}$$

Thus

$$B = \{B_\omega(x, r) : \text{all coordinates of } x \text{ are rational,}$$

$$\text{and } r > 0 \text{ is rational}\} \tag{4}$$

is a countable basis of R^n.

Each open set in R^n is the union of, at the most, countably many balls. If T is the Euclidean topology of R^n, then

$$\text{card } T = 2^{\text{card } N}. \tag{5}$$

$(see\ (1-44/P.\ 453)$

426

Let B represent a basis for topology T and let B' represent a family of open sets, such that

$$B \subset B' \subset T. \qquad (1)$$

Show that B' is also a basis for T.

SOLUTION:

Suppose $A \in T$ is an open subset of X.
Since B is a base for T

$$A = \bigcup_\omega B_\omega \qquad (2)$$

where $B_\omega \in B$. Since $B \subset B'$, each $B_\omega \in B$ also belongs to B', $B_\omega \in B'$.
Therefore

$$A = \bigcup_\omega B_\omega \text{ where } B_\omega \in B'. \qquad (3)$$

Hence B' is a base for T.

A set X and a family of all open subsets of X — so called topology T — are given. This pair (X, T) is the topological space. Sometimes we are given the basis B for topology T, then taking all unions of elements of B, we find T.

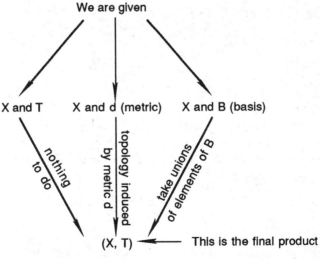

FIGURE 1

Consider this situation. A set X and a family B of its subsets are given. The question is: is this family a basis for any topology T on X? Prove the following theorem which answers this question:

THEOREM

Let B denote a family of subsets of X, such that

1. X is the union of members of B.

2. The intersection of any two members of B is the union of members of B.

Then B is a basis for topology T on X defined by

$$T = \{A \subset X : A \text{ is the union of members of } B\}. \tag{1}$$

SOLUTION:

1. $X \in T$ because X is the union of members of B. Also $\phi \in T$, as the union of the empty subfamily of B.

2. Let $\{A_\alpha\}$ denote a family of members of T, i.e., for each α, $A_\alpha \in T$. But each A_α is the union of members of B

$$A_\alpha = \bigcup_{\beta(\alpha)} B_{\beta(\alpha)} \tag{2}$$

where $B_{\beta(\alpha)} \in B$. Then

$$\bigcup_\alpha A_\alpha = \bigcup_\alpha \bigcup_{\beta(\alpha)} B_{\beta(\alpha)} \in T. \tag{3}$$

3. Suppose A_1 and $A_2 \in T$. Then

$$A_1 = \bigcup_\alpha B_\alpha \tag{4}$$

$$A_2 = \bigcup_\beta B_\beta \tag{5}$$

where $B_\alpha, B_\beta \in B$ and

$$A_1 \cap A_2 = \bigcup_{\alpha,\beta} (B_\alpha \cap B_\beta). \tag{6}$$

The intersection of any two members of B is the union of members of B. Hence, for each α and β

$$B_\alpha \cap B_\beta = \bigcup_\gamma B_\gamma \tag{7}$$

and

$$A_1 \cap A_2 = \bigcup_{\alpha,\beta} \bigcup_\gamma B_{\gamma(\alpha,\beta)} \in T \tag{8}$$

where each γ depends on α and β. Therefore, T defined by (1) is a topology and B described in the theorem is a basis for T.

Let R represent the set of real numbers.

1. Show that the collection of open intervals is the basis for a topology T on R.

2. (R, d) where $d(x, y) = |x - y|$ is a metric space and hence a topological space (R, T') with topology induced by the metric.

Show that both topologies T and T' are the same.

SOLUTION:

1. R is the union of open intervals. Any intersection of two open intervals is either empty or again an open interval. By virtue of theorem of Problem 11–16, the collection of open intervals in R forms a basis for a topology T on R.

2. Let

$$T = \{A : A \text{ is the union of open intervals}\} \tag{1}$$

$$T' = \{B : B \text{ is open in sense of the absolute value metric}\} \tag{2}$$

We will show that

$$(A \in T) \Rightarrow (A \in T'). \tag{3}$$

If $A \in T$, then A is the union of open interals. Each open interval is a set which is open in the sense of the absolute value metric. A union of open sets is again an open set. Hence, $A \in T'$

$$(A \in T') \Rightarrow (A \in T). \tag{4}$$

Suppose A is open in the sense of the absolute value metric. Then

$$\forall x \in A \quad \exists B(x, r) : B(x, r) \subset A \tag{5}$$

But

$$B(x, r) = (x - r, x + r). \tag{6}$$

Hence

$$A = \bigcup_{x \in A} B(x, r) = \bigcup_{x \in A} (x - r, x + r). \tag{7}$$

A is the union of open intervals, $A \in T$. We conclude

$$T = T'. \tag{8}$$

Let A represent a family of subsets of X, $A \subset P(X)$. In general, there are many topologies T on X which contain A, for example $A \subset T = P(X)$. We will show that among these topologies a unique, smallest topology $T(A) \supset A$ exists.

THEOREM

Let $A = \{A_\alpha\}$ be a family of subsets of X. Then $T(A)$ is a unique, smallest topology containing A, when

$$T(A) = \{\phi, X, \text{ all finite intersections of } A_\alpha,$$

$$\text{all unions of finite intersections of } A_\alpha\} \qquad (1)$$

$T(A)$ is said to be generated by A, and A is a subbasis for $T(A)$.

■

Prove this theorem.

SOLUTION:

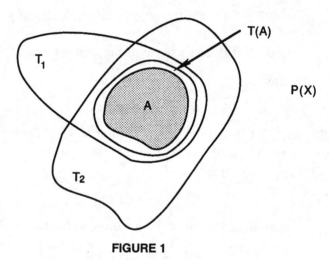

FIGURE 1

Let $T(A)$ be the intersection of all topologies containing A ($P(X)$ is one of such topologies). By Problem 11–8 $T(A)$ is a topology. Since it is the intersection of all topologies containing A, $T(A)$ is the smallest topology, $A \subset T(A)$. By definition $T(A)$ is unique.

We will prove (1). Since $A \subset T(A)$ and $T(A)$ is a topology, $T(A)$ must contain all the sets listed in (1).

On the other hand, the sets listed in (1) form a topology containing A and, therefore, containing $T(A)$.

430

● PROBLEM 11–19

Let R denote the set of all real numbers and let A denote all the sets of the form

$$\{x : x > a\} \quad (1)$$

$$\{x : x < b\} \quad (2)$$

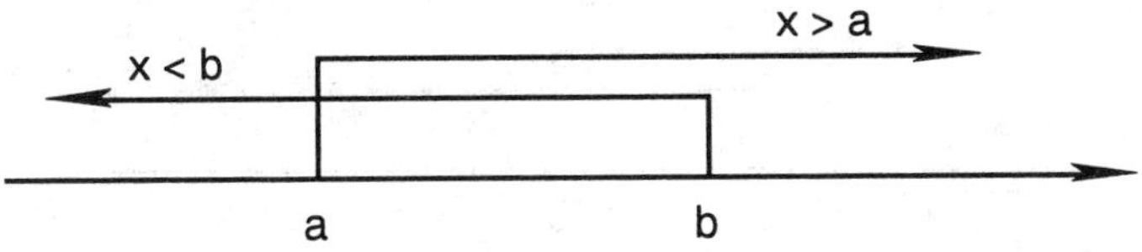

FIGURE 1

Find $T(A)$.

SOLUTION:

By the theorem of Problem 11–18, $T(A)$ consists of

$$T(A) = \{\phi, R, \text{all finite intersections of members of } A,$$
$$\text{all arbitrary unions of finite intersections of members of } A\} \quad (3)$$

All finite intersections of members of A are open intervals (a, b). Hence, we conclude that the

$$\text{set of all open intervals} \subset T(A). \quad (4)$$

The family of all open intervals forms a basis for $T(A)$.

By Problem 11–17 the family of all open intervals forms a basis for the Euclidean topology.

Therefore, $T(A)$ is the Euclidean topology.

● PROBLEM 11–20

R is the set of all real numbers and A consists of all sets of the form

$$\{x : x > a\}$$

$$\{x : x \leq b\} \quad (1)$$

Find $T(A)$ (called the upper limit topology).

SOLUTION:

To find $T(A)$, we must first obtain all finite intersections of elements of A (see Problem 11–18). The sets $(a, b]$

$$(a, b] = \{x : a < x \le b\} \tag{2}$$

form a basis for topology $T(A)$, which is not Euclidean because $(a, b]$ does not belong to the Euclidean topology.

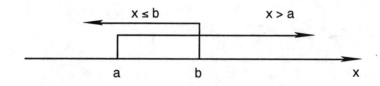

FIGURE 1

Note that we were dealing with the situation, when a family of sets A was given and we had to find topology $T(A)$, such that $A \subset T(A)$. Conversely, for a given topology T, a family of sets $A \subset T$ is called a subbasis for T, if

$$T = T(A).$$

● **PROBLEM 11-21**

Let $X = \{a, b, c, d, e\}$ and let $A \subset P(X)$ where

$$A = \{\{a, b, c\}, \{a, c, e\}, \{c, d\}\}. \tag{1}$$

Find the topology $T(A)$ on X generated by A.

SOLUTION:

By taking all finite intersections of sets in A and the sets ϕ and X, we find the basis $B(A)$

$\checkmark \{a\} = \{abc\} \cap \{ac,e\}$

$$B(A) = \{\phi, X, \{a, b, c\}, \{a, c, e\}, \{c, d\}, \{a, c\}, \{c\}\}. \tag{2}$$

Note that $X \in B$ and $\phi \in B$.

X can be considered the empty intersection of members of A. In order to obtain the topology generated by A, we take all unions of members of $B(A)$.

$$T(A) = \{\phi, X, \{a, b, c\}, \{a, c, e\}, \{c, d\}, \{a, c\}, \{c\}, \{a, b, c, e\},$$

$$\{a, b, c, d\}, \{a, c, e, d\}, \{a, c, d\}\}. \tag{3}$$

$\{a\}$

432

Show that a basis determines one and only one topology.

SOLUTION:

Let X represent any set and B a collection of subsets of X. Suppose B is a basis for two topologies T and T', which are different.

Hence, at least one element $A \subset X$ exists, such that $A \in T$ but $A \notin T'$ (or vice versa).

B is a basis for topology T and $A \in T$, therefore

$$A = \bigcup_{\alpha} B_\alpha \in T \qquad (1)$$

where $B_\alpha \in B$ for each α.

Since B is a basis for topology T', any union of elements of B must be an element of T'. Thus,

$$A = \bigcup_{\alpha} B_\alpha \in T', \qquad (2)$$

which is a contradiction. We conclude that

$$T = T'.$$

A basis as well as a subbasis determines a unique topology.

The following theorem offers a simple method of determining if a given family of sets is a basis.

THEOREM

Let $B = \{B_\omega : \omega \in \Omega\}$ denote a family of subsets of X, such that for each $x \in B_\alpha \cap B_\beta$ and each $\alpha, \beta \in \Omega$ $B_\gamma \in B$ exists, such that

$$x \in B_\gamma \subset B_\alpha \cap B_\beta \qquad (1)$$

Then $B \cup \{\phi\} \cup \{X\}$ is a basis for some topology $T(B)$ on X. $T(B)$ is unique and is the smallest topology containing B.

■

Prove this theorem.

SOLUTION:

By using B as a subbasis, we obtain a topology $T(B)$. We will show that

B is a basis for topology $T(B)$.

$$T(B) = \{\phi, X, \text{all unions of members of } B\} \tag{2}$$

It is enough to prove that each finite intersection of members of B is a union of members of B. We will prove that for each $x \in B_1 \cap \ldots \cap B_n$, there is a $B_0 \in B$, such that

$$x \in B_0 \subset B_1 \cap \ldots \cap B_n, \tag{3}$$

compare Problem 11–13.

For $n = 2$, (3) is true by the hypothesis. By induction, suppose (3) is true for $n - 1$. Then for n

$$x \in B_1 \cap B_2 \cap \ldots \cap B_{n-1} \cap B_n \tag{4}$$

B_0 exists, such that

$$x \in B_0 \cap B_n \tag{5}$$

$$B_0 \subset B_1 \cap \ldots \cap B_{n-1}. \tag{6}$$

Hence

$$x \in B'_0 \subset B_0 \cap B_n \subset B_1 \cap \ldots B_n \tag{7}$$

for some $B'_0 \in B$.

Thus B is a basis for topology $T(B)$ defined by (2).

● **PROBLEM 11–24**

Let $C([0, 1])$ represent the set of all continuous functions on $[0, 1]$. For each $f \in C$ and $r > 0$, we define

$$(f, r) = \{g \in C : \int_0^1 |f - g| < r\}. \tag{1}$$

Show that the family

$$\{K(f, r) : f \in C, r > 0\} \tag{2}$$

forms a basis for some topology T on C.

SOLUTION:

We shall apply the theorem of Problem 11–23. Suppose

$$h \in K(f_1, r_1) \cap K(f_2, r_2). \tag{3}$$

Let

$$t_1 = \int_0^1 |f_1 - h|, t_2 = \int_0^1 |f_2 - h| \tag{4}$$

and let

$$R = \min[r_1 - t_1, r_2 - t_2]. \tag{5}$$

Then $R > 0$ and

$$K(h, R) \, A \, K(f_1, r_1) \cap K(f_2, r_2). \tag{6}$$

Indeed, suppose $F \in K(h, R)$ then

$$\int_0^1 |F - h| < R \tag{7}$$

and

$$\int_0^1 |f_1 - F| \le \int_0^1 |f_1 - h| + \int_0^1 |h - F| < t_1 + (r_1 - t_1) = r_1. \tag{8}$$

Therefore

$$F \in K(f_1, r_1) \quad \text{and} \quad F \in K(f_2, r_2) \tag{9}$$

because

$$\int_0^1 |f_2 - F| \le \int_0^1 |f_2 - h| + \int_0^1 |h - F| < t_2 + (r_2 - t_2) = r_2. \tag{10}$$

● **PROBLEM 11–25**

Each basis in X yields a unique topology. But distinct bases may yield the same topology.

DEFINITION OF EQUIVALENT BASES

Two bases B and B' in X are equivalent if

$$T(B) = T(B') \tag{1}$$

■

Prove this:

THEOREM

Two bases B and B' in X are equivalent if and only if

1. for each $B_\alpha \in B$ and each $x \in B_\alpha$, there is a $B'_\beta \in B'$, such that

$$x \in B'_\beta \subset B_\alpha$$

and

2. for each $B'_\beta \in B'$ and each $x \in B'_\beta$, there is a $B_\alpha \in B$, such that

$$x \in B_\alpha \subset B'_\beta$$

SOLUTION:

Suppose $\qquad\qquad\downarrow\qquad\qquad T(B) = T(B').$ (2)

Then for each $B_\alpha \in B \subset T(B)$ and each $x \in B_\alpha$, since B' is a basis for $T(B)$, by Problem 11–13, there is $B'_\beta \in B'$, such that

$$x \in B'_\beta \subset B_\alpha.$$ (3)

Similarly we show that condition 2. holds.

Now suppose condition 1 holds. Since each $A \in T(B)$ is a union of $\{B_\alpha\}$ belonging to B, *then* it follows that $A \in T(B')$. yielding $T(B) \subset T(B')$. Similarly we have $T(B') \subset T(B)$ and therefore $T(B) = T(B')$.

● **PROBLEM 11-26**

Let S represent a subbase for a topology T on X and let A denote any subset of X. Show that the family

$$S_A = \{A \cap S_\alpha : S_\alpha \in S\}$$ (1)

is a subbase for the relative topology T_A on A.

SOLUTION:

Let D denote an open subset of A with respect to topology T_A. Then

$$D = A \cap E$$ (2)

where E is a T-open subset of X. Family S is a subbase for T, therefore

$$E = \bigcup (S_{n_1} \cap S_{n_2} \cap ... \cap S_{n_k})$$ (3)

where $S_{n_1}, S_{n_2}, ..., S_{n_k} \in S$.

Therefore

$$D = A \cap E = A \cap [\bigcup (S_{n_1} \cap ... \cap S_{n_k})] =$$

$$= \bigcup [(A \cap S_{n_1}) \cap ... \cap (A \cap S_{n_k})].$$ (4)

Thus D is the union of finite intersections of elements of S_A, and S_A is a subbase for T_A.

1. Let $A =]0, 1]$ and $B = \{1/_n : n \in N\}$ be subsets of R. Find the closure of A and B.

2. Find the closure of 0 and 1 in Sierpinski space, where $X = \{0, 1\}$ and $T = \{\phi, X, 0\}$.

SOLUTION:

DEFINITION OF A NEIGHBORHOOD

Let (X, T) denote a topological space. A neighborhood of an $x \in X$ is any open set N_0 containing x, $x \in N_0$. A neighborhood of x is denoted by $N_0(x)$.

■

DEFINITION OF ADHERENT POINT   OR CLOSURE PT

Let $A \subset X$. A point $x \in X$ is adherent to A, if for each $N_0(x)$

$$N_0(x) \cap A \neq \phi. \tag{1}$$

■

DEFINITION OF THE CLOSURE OF THE SET

The closure of A, denoted by $\overline{A}$, is the set of all points in X adherent to A.

$$\overline{A} = \{x \in X : \forall N_0(x) : N_0(x) \cap A \neq \phi\}. \tag{2}$$

■

1. The closure of $A =]0, 1]$ is $\overline{A} = [0, 1]$. The closure of $B = \{1/_n : n \in N\}$ is

$$\overline{B} = \{0, 1/_n\} = \{0\} \cup B \tag{3}$$

2. The open sets are ϕ, X and 0. We have

$$\overline{0} = \{0, 1\} = X$$

$$\overline{1} = 1 \tag{4}$$

correct
But ?
why?
not 0?

$\{1\}$ is the smallest
closed set
that contains 1

$G = \{X \phi \{1\}\}$

$\overline{A} \subset A_{x \in \overline{A}}^{x \in A}$
contrapositive:
$x \notin A \Rightarrow x \notin \overline{A}$

● **PROBLEM 11-28**

P.446

Show that:

1. for every set A, $A \subset \overline{A}$.

2. A is closed iff $A = \overline{A}$.

SOLUTION:

DEFINITION

A set $A \subset X$ is called closed if $X - A$ is an open set.

■

1. Obviously, if $x \in A$, then

$$x \in N_0(x) \cap A \neq \phi. \qquad (1)$$

NBD
P.437

Hence $\qquad x \in \overline{A}, \quad A \subset \overline{A}. \qquad (2)$

from 2 (P.437)

2. $(A \text{ closed}) \Rightarrow (A = \overline{A})$. $\qquad x \in X-A \quad e.g. \not{X \subseteq A}$

If A is closed, then $X - A$ is open. Each $x \notin A$ has a neighborhood $N_0(x)$, such that

$$N_0(x) \cap A \neq \phi. \qquad (3)$$

HUH?

Hence $\qquad \boxed{x \notin \overline{A}} \text{ and } \overline{A} \subset A. \qquad (4)$

From (4) and (2), we obtain $A = \overline{A}$.

$(A = \overline{A}) \Rightarrow (A \text{ closed})$.

Since $A = \overline{A}$, each $x \notin A$ has a $N_0(x)$, such that

$$N_0(x) \cap A \neq \phi. \qquad (5)$$

Hence, $X - A$ is open and A is closed.

● **PROBLEM 11-29**

Show that this is an alternative definition of a closed set:

$\overline{A}$ is the smallest closed set containing A; i.e.,

$$\overline{A} = \cap \{D : (D \text{ is closed}) \wedge (A \subset D)\}. \qquad (1)$$

438

SOLUTION:

?

$$\text{ASSUME} \qquad \overline{A} \subset \bigcap D. \qquad (2)$$

Suppose $x \in \overline{A}$ and $x \notin \bigcap D$. Since each set D is closed, $\bigcap D$ is a closed set

$$(x \notin \bigcap D) \Rightarrow x \in X - \bigcap D, \text{ where } X - \bigcap D \text{ is an open set} \Rightarrow$$

$$\Rightarrow x \text{ has a nbd } G, \text{ such that } G \cap \bigcap D = \phi \Rightarrow$$

$$\Rightarrow G \cap A = \phi \text{ (because } A \subset \bigcap D) \Rightarrow x \notin \overline{A}. \quad \because x \in \bigcap D \quad (3)$$
contradiction

Now we will prove that $\bigcap D \subset \overline{A}$. Suppose $x \notin \overline{A}$, then there is a neighborhood of x, $N_0(x)$, such that

$$N_0(x) \cap A \neq \phi. \qquad (4)$$

Therefore, $X - N_0(x)$ is closed and contains A. Thus $X - N_0(x)$ is one of the sets (denoted by D) in (1).

Since

$$x \notin X - N_0(x) \quad \text{also} \quad x \notin \bigcap D. \qquad (5)$$

See Figure 1.

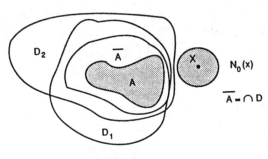

FIGURE 1

● **PROBLEM 11-30** P.446

Prove that:

1. $\overline{\overline{A}} = \overline{A}$.

2. $A \subset B \Rightarrow \overline{A} \subset \overline{B}$.

3. $\overline{A \cup B} = \overline{A} \cup \overline{B}$.

4. $\overline{\phi} = \phi$.

SOLUTION:

1. $\overline{A} = \bigcap D$ (where D is closed). Since the intersection of any family of closed sets is a closed set, $\overline{A}$ is a closed set. A is closed if and only if $A = \overline{A}$. Hence

$$\overline{\overline{A}} = \overline{A}. \tag{1}$$

2. $\overline{A} = \bigcap D$ and $\overline{B} = \bigcap D'$, where each D contains A and each D' contains B. Since $A \subset B$ each D' contains B and contains A. Hence,

$$\overline{A} \subset \overline{B}. \tag{2}$$

3. Set $\overline{A \cup B}$ is closed. Since for each A, $A \subset \overline{A}$, we obtain

$$\overline{A} \cup \overline{B} \subset \overline{A \cup B}. \tag{3}$$

On the other hand, set $\overline{A} \cup \overline{B}$ is closed and

$$A \cup B \subset \overline{A} \cup \overline{B}. \tag{4}$$

Hence

$$\overline{A \cup B} \subset \overline{A} \cup \overline{B}. \tag{5}$$

From (3) and (5) we get

$$\overline{A \cup B} = \overline{A} \cup \overline{B}. \tag{6}$$

4. Set ϕ is a closed set, hence

$$\overline{\phi} = \phi. \tag{7}$$

● PROBLEM 11-31

1. Find the derived set of

   $A = \{^1/_n : n \in N\}$

   $B =]0, 1]$

   $C = \{x : x \in (0,1), x \text{ is a rational number}\}.$

2. Let X denote an indiscrete topological space. Find the derived set A' of any subset $A \subset X$.

440

SOLUTION:

DEFINITION OF CLUSTER POINT

Let $A \subset X$. A point $x \in X$ is called a cluster point of A if each neighborhood of x, N_0 of x contains at least one point of A distinct from x. ■

It is easy to see that 0 is the only cluster point of $\{1/n\}$.

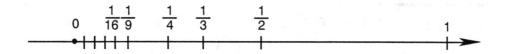

FIGURE 1

Indeed each neighborhood of 0 contains at least one (as a matter of fact, infinitely many) point of $A = \{1/n\}$ distinct from 0.

DEFINITION OF A DERIVED SET

The set

$$A' = \{x \in X : \forall 0(x) : 0(x) \cap (A - x) \neq \phi\} \quad (1) \quad \times$$

of all cluster points of A is called the derived set of A. ■

1. We obtain

$$A' = \{0\}$$

$$B' = [0, 1]$$

$$C' = [0, 1].$$

NO PARENS $\times$

2. We have $T = (X, \phi)$. For any point $x \in X$, X is the only open set containing x. We have $\times$

$$A' = \begin{cases} \phi & \text{if } A = \phi \\ X & \text{if } A \text{ contains two or more points} \\ X - \{a\} & \text{if } A = \{a\} \end{cases}$$

441

Show that

$$\bar{A} = A \cup A'. \qquad (1)$$

Set A is closed if and only if $A' \subset A$.

SOLUTION:

Suppose $x \in A'$ then $x \in \bar{A}$. Hence $A' \subset \bar{A}$. Also $A \subset \bar{A}$, therefore

$$A' \cup A \subset \bar{A}. \qquad (2)$$

To prove the converse inclusion suppose $x \in A$. If $x \in \bar{A}$, then the proof is finished.

If $x \notin A$, then each neighborhood of x intersects A at a point distinct from x, hence $x \in A'$. Thus

$$\bar{A} \subset A \cup A'. \qquad (3)$$

We conclude that

$$\bar{A} = A \cup A'. \qquad (4)$$

Set A is closed if and only if $A = \bar{A}$.
  Thus, A is closed if and only if

$$A' \subset A \qquad (5)$$

i.e., if and only if A contains all of its cluster points.

Prove that

$$(A \cup B)' = A' \cup B'. \qquad (1)$$

SOLUTION:

Observe that if $A \subset B$, then $A' \subset B'$. Since $A \subset A \cup B$ and $B \subset A \cup B$, we obtain

$$A' \subset (A \cup B)' \quad B' \subset (A \cup B)' \qquad (2)$$

Thus

$$A' \cup B' \subset (A \cup B)'. \qquad (3)$$

Now we shall prove the inverse inclusion, $(A \cup B)' \subset A' \cup B'$. Suppose, $a \notin A' \cup B'$, then open sets 0_1 and 0_2 exist, such that

$$a \in 0_1 \quad \text{and} \quad a \in 0_2 \tag{4}$$

$$0_1 \cap A \subset \{a\}, \quad 0_2 \cap B \subset \{a\}. \tag{5}$$

Since sets 0_1 and 0_2 are open, $0_1 \cap 0_2$ is open and $a \in 0_1 \cap 0_2$. We obtain

$$(0_1 \cap 0_2) \cap (A \cup B) = (0_1 \cap 0_2 \cap A) \cup (0_1 \cap 0_2 \cap B)$$

$$\subset (0_1 \cap A) \cup (0_2 \cap B) \subset \{a\} \cup \{a\} = \{a\}. \tag{6}$$

Hence

$$a \notin (A \cup B)' \quad \text{and} \quad (A \cup B)' \subset A' \cup B'. \tag{7}$$

● **PROBLEM 11–34**

1. Find the interior of the sets

$A = \{^1/_n : n \in N\}$

$B = [0, 1]$

2. Prove that
$$Int(A) = X - \overline{(X - A)} \tag{1}$$

SOLUTION:

DEFINITION OF THE INTERIOR

The interior $Int(A)$ of $A \subset X$ is the largest open set contained in A, that is
$$Int(A) = \bigcup \{0 : (0 \text{ open}) \wedge (0 \subset A)\}. \tag{2}$$

∎

1. $Int(A) = \phi.$

   $Int(B) = (0, 1)$

2. We shall apply
$$A \subset B \Leftrightarrow X - B \subset X - A. \tag{3}$$

Therefore, if $0 \subset A$, then

$$X - A \subset X - 0 = D. \tag{4}$$

Since 0 is an open set, D is a closed set. We obtain

$$Int(A) = \bigcup \{X - D : (D \text{ closed}) \wedge (X - A \subset D)\} =$$

$$= X - \bigcap \{D : (D \text{ closed}) \wedge (X - A \subset D)\} = X - \overline{(X - A)}. \tag{5}$$

● **PROBLEM 11-35**

Find the boundary of the sets

$A = \{^1/_n : n \in N\}$

$B = (0, 1]$

Show that

$$\mathcal{F}r\,(A) = \overline{A} - Int(A). \tag{1}$$

SOLUTION:

DEFINITION OF BOUNDARY

The boundary of a set $A \subset X$ is denoted by $\mathcal{F}r(A)$ and defined by

$$\mathcal{F}r(A) = \overline{A} \cap \overline{(X - A)}. \tag{2}$$

■

We have

$$\overline{A} = A \cup \{0\} \quad \text{and} \quad \overline{(X - A)} = R.$$

Thus

$$\mathcal{F}r(A) = A \cup \{0\} \quad \overline{B} = [0,1]$$

$$\overline{(X - B)} = (-\infty, 0] \cup [1, \infty) \tag{3}$$

Hence

$$\mathcal{F}r(B) = \overline{B} \cap \overline{(X - B)} = \{0, 1\}. \tag{4}$$

We have

444

$$\mathcal{F}r(A) = \overline{A} \cap \overline{(X - A)}. \tag{5}$$

But

$$Int(A) = X - \overline{(X - A)}. \tag{6}$$

Also note that

$$X - (X - B) = B \tag{7}$$

Hence

$$\mathcal{F}r(A) = \overline{A} \cap X - [X - \overline{(X - A)}] =$$

$$= \overline{A} \cap [X - Int(A)] = \overline{A} - Int(A) \tag{8}$$

● **PROBLEM 11–36** DENSE

Show that the following statements are equivalent:

1. D is dense in X.

2. If G is a closed set and $D \subset G$, then $G = X$.

3. Each nonempty basic open set in X contains an element of D.

4. The complement of D has empty interior.

SOLUTION:

DEFINITION OF A DENSE SET

Set $D \subset X$ is dense in X if $\overline{D} = X$.

■

1. ⇒ 2.
Set D is dense in X, $\overline{D} = X$. $D \subset G$, then $\overline{D} \subset \overline{G}$, but $\overline{G} = G$, hence $X \subset G$
and

$$X = G. \tag{1}$$

2. ⇒ 3.
Let $0 \neq \phi$ be open and $0 \cap D = \phi$, then

445

$$D \subset X - 0 \neq X \tag{2}$$

in contradiction with 2 because $X - 0$ is closed.

3. $\Rightarrow$ 4.

Suppose $Int(X - D) = \phi$. Set $Int(X - D)$ is open, hence, there is a nonempty basic set

$$A \subset Int(X - D). \tag{3}$$

But

$$Int(X - D) \subset X - D. \tag{4}$$

Then $A \subset X - D$ and A contains no points of D.

4. $\Rightarrow$ 1.

$$Int(X - D) = X - \overline{[X - (X - D)]} = X - \overline{D} = \phi. \tag{5}$$

Thus

$$\overline{D} = X.$$

● **PROBLEM 11–37**

Set X is given. To promote it to the topological space (X, T), we have to define a family T of its subsets which satisfies certain conditions. There are many ways of defining (or finding) a topology on X. For example, a function f, which assigns its closure to every subest of X determines a topology

$$f : A \to \overline{A}. \tag{1}$$

Without getting into details, we want the closure $f(A)$ of the set to satisfy these conditions:

$$A \subset \overline{A} \qquad \text{(See Problem 11–28)}$$

$$\overline{\phi} = \phi, \overline{\overline{A}} = \overline{A}, \overline{A \cup B} = \overline{A} \cup \overline{B} \qquad \text{(See Problem 11–30)}$$

Prove the following:

THEOREM

Let X represent a set and $f : P(X) \to P(X)$ a function, such that

1. $f(\phi) = \phi$.

2. For each A, $A \subset f(A)$.

3. $f \circ f(A) = f(A)$ for each A.

4. $f(A \cup B) = f(A) \cup f(B)$ for each A, B.

The family

$$T = \{X - f(A) : A \in P(X)\} \qquad (2)$$

is a topology and $\overline{A} = f(A)$ for each A.

∎

Prove this theorem.

SOLUTION:

Note that

$$(A \subset B) \Rightarrow (f(A) \subset f(B)). \qquad (3)$$

Indeed if $A \subset B$, then $A \cup B = B$ and $f(A \cup B) = f(A) \cup f(B) = f(B)$. Now we will show that T is a topology .

I. $\phi, X \in T$.
    Since $X \subset f(X)$, we get $X = f(X)$. Hence $\phi \in T$. Since $X - f(\phi) = X, X \in T$.

II. Suppose $X - f(A)$ and $X - f(B)$ belong to T. Then

$$[X - f(A)] \cap [X - f(B)] = X - [f(A) \cup f(B)] = X - f(A \cup B) \in T. \qquad (4)$$

Thus, intersection of any finite family of sets of T is a member of T.

III. Let $S = \bigcup_{\alpha} X - f(A_\alpha)$.We will show that for some $U \in P(X)$, $S = X - f(U)$. We have

$$S = X - \bigcap_{\alpha} f(A_\alpha) \qquad (5)$$

and

$$X - S = \bigcap_{\alpha} f(A_\alpha) \subset f(A_\alpha) \qquad (6)$$

for each α.
    By (3) and condition 3, we obtain

$$f(X - S) \subset f \circ f(A_\alpha) = f(A_\alpha) \qquad (7)$$

for each α. Thus

$$f(X - S) \subset \bigcap_{\alpha} f(A_\alpha) = X - S. \qquad (8)$$

from (8) and condition 2, we obtain
    Prove the following convenient way of describing F_σ and G_δ sets:

447

$$X - S = f(X - S). \tag{9}$$

Hence

$$S = X - f(X - S) = X - f(U). \tag{10}$$

The union of any family of members of T is again a member of T.

By using T as the topology on X, we will show that $\overline{A} = f(A)$. Indeed, $\overline{A} \subset f(A)$ because each $f(A)$ is closed in T and $A \subset f(A)$, we obtain $\overline{A} \subset \overline{f(A)} = f(A)$. Similarly $f(A) \subset \overline{A}$. Since $X - \overline{A}$ is open in T, for some B, $f(B) = \overline{A}$. Since

$$A \subset \overline{A} \tag{11}$$

we obtain

$$f(A) \subset f(\overline{A}) = f \circ f(B) = f(B) = \overline{A}. \tag{12}$$

That completes the proof.

● **PROBLEM 11-38**

Show that in R the closed interval $[a, b]$ is an F_γ set and also a G_δ set. Show that the set of rational numbers in R is F_γ.

SOLUTION:

DEFINITION OF F_σ AND G_δ SETS

Set F is called an F_σ set if it is the union of at most countably many closed sets. A set G is called a G_δ set if it is the intersection of at most countably many open sets.

∎

A closed interval is a closed set, hence, it is an F_σ. On the other hand,

$$[a,b] = \bigcap_{n=1}(a - \tfrac{1}{n}, b + \tfrac{1}{n}) \tag{1}$$

$[a, b]$ is the intersection of countably many open sets. Hence, $[a, b]$ in R is a G_δ.

The set of rational numbers Q is countable and since each point in R is a closed set, Q is the union of countably many closed sets. Thus, Q is an F_γ.

● **PROBLEM 11-39**

Prove the following convenient way of describing F_σ and G_δ sets:

1. If F is an F_γ, then there is a non-decreasing sequence of closed sets
$$F_1 \subset F_2 \subset F_3 \subset \ldots \text{ where } F = \bigcup_1^\infty F_n. \tag{1}$$

2. If G is a G_δ, then there is a non-increasing sequence of open sets
$$G_1 \supset G_2 \supset G_3 \ldots \text{ where } G = \bigcap_1^\infty G_n. \tag{2}$$

SOLUTION:

1. F is an F_γ. Hence
$$F = \bigcup_1^\infty A_n, A_n \text{ closed.} \tag{3}$$

F is the union of at most countably many closed sets. Let
$$F_1 = A_1, F_2 = A_1 \cup A_2, F_3 = A_1 \cup A_2 \cup A_3, \ldots,$$

$$\ldots, F_n = A_1 \cup \ldots \cup A_n. \tag{4}$$

Then sets F_n are closed and
$$\bigcup_1^\infty A_n = \bigcup_1^\infty F_n, F_1 \subset F_2 \subset F_3 \ldots \tag{5}$$

2. Similarly, if G is a G_δ, then
$$G = \bigcap_1^\infty B_n \tag{6}$$

where B_n are open sets. Let
$$G_1 = B_1, G_2 = B_1 \cap B_2, \ldots, G_n = B_1 \cap \ldots \cap B_n. \tag{7}$$

Then each G_n is open and
$$G_1 \supset G_2 \supset G_3 \supset \ldots \bigcap_1^\infty B_n = \bigcap_1^\infty G_n. \tag{8}$$

● **PROBLEM 11-40**

Prove that:

1. The countable union and finite intersection of F_γ sets is an F_σ.

2. The countable intersection and finite union of G_δ sets is a G_δ.

SOLUTION:

1. Consider the countable union of F_ρ sets F_σ

$$F_i = \bigcup_{k=1}^{\infty} F_{i,k} \tag{1}$$

where $F_{i,k}$ are closed sets. The countable union of F_i's is

$$\bigcup_{i=1}^{\infty} \left[\bigcup_{k=1}^{\infty} F_{i,k} \right] = \bigcup_{i,j} \left\{ F_{i,j} : (i,j) \in N \times N \right\}. \tag{2}$$

Set $N \times N$ is countable and $F_{i,j}$ are closed sets, hence, the union is an F_γ. The finite intersection is

$$\bigcap_{i=1}^{n} \bigcup_{k=1}^{\infty} F_{i,k} = \left[\bigcup_{k=1}^{\infty} F_{1,k} \right] \cap \left[\bigcup_{k=1}^{\infty} F_{2,k} \right] \cap \dots \cap \left[\bigcup_{k=1}^{\infty} F_{n,k} \right] =$$

$$= \bigcup \left\{ F_{1,k_1} \cap \dots \cap F_{n,k_n} : (k_1, \dots, k_n) \in N \times \dots \times N \right\}. \tag{3}$$

Each set $F_{1,k_1} \cap \dots \cap F_{n,k_n}$ is closed and (3) is a countable union.

2. Let

$$G_i = \bigcap_{k=1}^{\infty} G_{i,k} \; ; i = 1, 2, 3, \dots \tag{4}$$

where $G_{i,k}$ are open sets, be a family of G_δ sets. The countable intersection is

$$\bigcap_{i=1}^{\infty} \bigcap_{k=1}^{\infty} G_{i,k} = \bigcap \left\{ G_{i,k} : (i,k) \in N \times N \right\} \tag{5}$$

Sets $G_{i,k}$ are open and $N \times N$ is a countable set. Hence, (5) is a G_δ set. Consider now the finite union of G_i's

$$\bigcup_{i=1}^{n} \bigcap_{k=1}^{\infty} G_{i,k} = \bigcap \{ G_{1,k_1} \cup G_{2,k_2} \cup \dots \cup G_{n,k_n} :$$

$$(k_1, k_2, \dots, k_n) \in N \times N \times \dots \times N \}. \tag{6}$$

(6) is a G_δ set.

● **PROBLEM 11-41**

1. Show that the complement of an F_γ is a G_σ.

2. Show that the complement of a G_δ is an F_γ.

SOLUTION:

1. Let F be an F_γ set, then

450

$$F = \bigcup_{n=1}^{\infty} F_n \tag{1}$$

where F_n are closed sets. The complement of F is

$$X - F = X - \bigcup_{n=1}^{\infty} F_n. \tag{2}$$

By applying DeMorgan's rule, we find

$$X - F = \bigcap_{n=1}^{\infty} (X - F_n). \tag{3}$$

Since each F_n is closed, $X - F_n$ is open for each n. Thus $X - F$ is the intersection of at most countably many open sets.

2. Let G be a G_δ set. Then

$$G = \bigcap_{n=1}^{\infty} G_n \tag{4}$$

where G_n are open sets. The complement of G is

$$X - G = X - \bigcap_{n=1}^{\infty} G_n = \bigcup_{n=1}^{\infty} (X - G_n) \tag{5}$$

where $X - G_n$ are closed sets. Hence $X - G$ is an F_σ set.

● **PROBLEM 11-42**

Show that each F_σ and each G_δ is a Borel set.

SOLUTION:

DEFINITION OF A σ-RING

A nonempty family of sets

$$Q \subset P(X) \tag{1}$$

is called a σ-ring if

1. $A \in Q \Rightarrow X - A \in Q$

2. $\forall n : A_n \in Q \Rightarrow \bigcup_{1}^{\infty} A_n \in Q.$

■

DEFINITION OF BOREL SETS

If (X, T) is a topological space, then a unique smallest σ-ring B containing the topology T of X exists, that is $T \subset B$. Family of sets B is called the family of Borel sets in X.

■

Let F denote an F_σ–set, then

$$F = \bigcup_1^\infty F_n \qquad (2)$$

where F_n are closed sets. The sets $X - F_n$ are open sets. Since B is a σ-ring such that $T \quad B$, each open set belongs to B. Hence

$$\forall n : X - F_n \in B \qquad (3)$$

$$(X - F_n \in B) \Rightarrow (X - (X - F_n) \in B) \qquad (4)$$

Thus

$$\forall n : F_n \in B \quad \text{and} \quad F = \bigcup_1^\infty F_n \in B \qquad (5)$$

If G is a G_δ–set then

$$G = \bigcap_1^\infty G_n \qquad (6)$$

where G_n are open sets. But

$$\bigcap_1^\infty G_n = X - \left[\bigcup_1^\infty (X - G_n) \right] \qquad (7)$$

Each G_n is an open set, therefore

$$G_n \in B \Rightarrow X - G_n \in B \Rightarrow \bigcup_1^\infty (X - G_n) \in B \Rightarrow$$

$$\Rightarrow X - \left[\bigcup_1^\infty (X - G_n) \right] \in B \Rightarrow G \in B. \qquad (8)$$

● **PROBLEM 11–43**

Show that the countable union, countable intersection, and the difference of Borel sets is a Borel set.

SOLUTION:

Let B denote the family of Borel sets in X. B can be defined as the intersection of all σ–rings containing T.

It is easy to verify that the intersection of any family of σ-rings is a σ-ring. Thus, B is a σ-ring.

$$\left(B_i \in B \text{ for } i = 1, 2, 3, ... \right) \Rightarrow \left(\bigcup_1^\infty B_i \in B \right) \qquad (1)$$

Because B is a σ-ring.

$$\left(B_i \in B \text{ for } i = 1, 2, ... \right) \Rightarrow \left(X - B_i \in B \text{ for } i = 1, 2, ... \right) \Rightarrow$$

$$\Rightarrow \left(\bigcup_1^\infty (X - B_i) \in B \right) \Rightarrow \left(X - \bigcup_1^\infty (X - B_i) = \bigcap_1^\infty B_i \in B \right). \qquad (2)$$

$$(A_1, A_2 \in B) \Rightarrow (A_1, X - A_2 \in B) \Rightarrow$$

$$\Rightarrow (A_1 \cap (X - A_2) = A_1 - A_2 \in B). \qquad (3)$$

● **PROBLEM 11–44**

Show that in the Euclidean space R^n there are sets that are not Borel sets.

SOLUTION:

cf P.426
EnD oF 11-14

We shall apply the following:

THEOREM

Let (X, T) be a topological space and B be the family of Borel sets in X. Then

$$\text{card } (B) \le \text{card } (T)^{\text{card N}} \qquad (1)$$

■

Topological space R^n has a countable basis. Hence, the cardinal number of the Euclidean topology of R^n is $2^{\text{card } N}$

$$\text{card } (T) = 2^{\text{card } N}. \qquad (2)$$

From (1) and (2) we find

$$\text{card } (B) \le 2^{\text{card } N}. \qquad (3)$$

Since

$$\text{card } (P(R^n)) = 2^{\text{card } R} \qquad (4)$$

we conclude that there are sets in R^n that are not Borel sets.

● **PROBLEM 11–45**

Let (R, T) denote Euclidean space and X the subset of R

$$X = (0, 1] \cup \{2\}. \tag{1}$$

Find the induced topology T_X.

SOLUTION:

DEFINITION OF SUBSPACE

Let (X, T) denote a topological space and $Y \subset X$. The induced topology T_Y on Y is defined by

$$T_Y =: \{Y \cap T_\alpha : T_\alpha \in T\}. \tag{2}$$

The space (Y, T_Y) is called a subspace of (X, T).

∎

It is easy to show that T_Y defined in (2) is a topology on Y. Consider (R, T). T is the family of all open sets in R. By definition

$$T_X = \{[(0, 1] \cup \{2\}] \cap T_\alpha : T_\alpha \in T\} \tag{3}$$

where T_α are open sets.

The family T_X consists of

1. $\{2\}$

2. all open intervals contained in $(0, 1]$

3. all intervals of the form $]a, 1]$ where $a \in (0, 1)$.

Interval $(0, 1]$ belongs to T_X, hence it is an open set. Since $\{2\} \in T_X$, interval $(0, 1]$ is a closed set in X.

● PROBLEM 11–46

Let (X, T) denote a space and (Y, T_Y) a subspace. Show that if $\{B_\omega : \omega \in \Omega\}$ is a basis for T, then $\{Y \cap B_\omega : \omega \in \Omega\}$ is a basis for T_Y.

SOLUTION:

Let (X, T) denote a topological space and $\{B_\omega : \omega \in \Omega\}$ be its basis. Then each open set in X is the union of members of $\{B_\omega\}$. The induced topology T_Y is defined by

$$T_Y = \{Y \cap T_\alpha\}. \qquad (1)$$

Suppose A is an open set in Y, $A \in T_Y$. Then

$$A = Y \cap T_\alpha \qquad (2)$$

for some α. But $\{B_\omega\}$ is a basis for T,

$$T_\alpha = U B_\omega. \qquad (3)$$

Thus

$$A = Y \cap [U B_\omega] = U(Y \cap B_\omega) \qquad (4)$$

Therefore $\{Y \cap B_\omega : \omega \in \Omega\}$ is a basis for T_Y.

● PROBLEM 11–47

Let (X, T) denote a topological space and (Y, T_Y) be a subspace. Show that a set $A \subset Y$ is T_Y-closed if and only if $A = Y \cap D$, where D is T-closed.

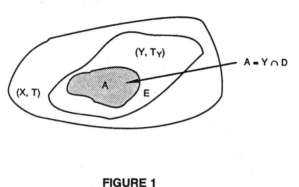

FIGURE 1

SOLUTION:

We will show that the closed sets in Y are the intersections of Y and the closed sets in X. Suppose $A \subset Y$ is closed in Y, then $A = Y - E$, where E is open in Y. Therefore

$$E = Y \cap G \qquad (1)$$

where $G \in T$. Hence

$$A = Y - E = Y - Y \cap G = Y \cap (X - G). \qquad (2)$$

Since $G \in T$, $X - G$ is T-closed.

Now, suppose

$$A = Y \cap D \tag{3}$$

where D is T-closed. Then

$$Y - A = Y - Y \cap D = Y \cap (X - D) \tag{4}$$

where $X - D$ is T-open and $Y - A$ is T_Y-open. Thus A is T_Y-closed.

● PROBLEM 11–48

1. (X, T) is a topological space and (Y, T_Y) is its subspace. In Problem 11–45, we showed that sets, open in a subspace need not be open in the entire space. Prove the following:

THEOREM

Let (X, T) denote a space and (Y, T_Y) its subspace. If $A \subset Y$ is open (closed) in Y, and Y is open (closed) in X, then A is open (closed) in X.

2. Is a subspace of a subspace a subspace of the entire space?

SOLUTION:

1. Suppose A is open in Y. Then

$$A = Y \cap G \tag{1}$$

where G is open in X. Since Y is open in X, $Y \cap G$ is open in X. The same reasoning holds for closed sets.

2. Yes. Because if

$$Z \subset Y \subset X \tag{2}$$

and T_{YZ} is the topology of Z with respect to Y, then

$$T_{YZ} = T_Z \tag{3}$$

Where T_Z is the topology of Z with respect to X. We shall prove (3). Let

$$U \in T_{YZ} \tag{4}$$

then

$$U = Z \cap V \tag{5}$$

where $V \in T_Y$. But

$$V = Y \cap Q \tag{6}$$

where $Q \in T$. Then

$$U = Z \cap V = Z \cap Y \cap Q = Z \cap Q. \tag{7}$$

Hence $U \in T_Z$ and $T_{YZ} \subset T_Z$.

The converse inclusion is obvious.

CHAPTER 12

CONTINUITY, HOMEOMORPHISMS, AND TOPOLOGICAL EQUIVALENCE

CONTINUITY

CONTINUITY AT A POINT

PIECEWISE DEFINITION OF MAPS

OPEN AND CLOSED FUNCTIONS

HOMEOMORPHIC SPACES

IDENTIFICATION SPACES

Let (X, T_X) and (Y, T_Y) represent topological spaces, such that

$X = \{a, b, c, d\}$

$T_X = \{\phi, X, \{a\}, \{a, b\}, \{a, b, c\}\}$

$Y = \{x, y, z, w\}$

$T_Y = \{\phi, Y, \{y\}, \{y, z, w\}$

Which of the functions depicted in Figure 1 is continuous?

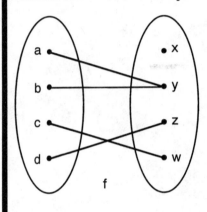

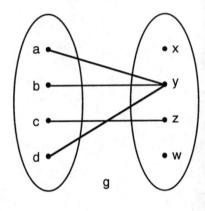

FIGURE 1

SOLUTION:

DEFINITION OF CONTINUOUS FUNCTIONS

Let (X, T_X) and (Y, T_Y) represent topological spaces. A function $f : X \rightarrow Y$ is called continuous if the inverse image of each open set in Y is an open set in X. That is, if $f : X \rightarrow Y$ is continuous, then

$$f^{-1} : T_Y \rightarrow T_X. \tag{1}$$

■

Function f is continuous because f^{-1} maps each open set in T_Y into an open set in T_X.

$$f^{-1}(\phi) = \phi \in T_X$$

$$f^{-1}(Y) = X \in T_X$$

$$f^{-1}(\{y\}) = \{a, b\} \in T_X$$

460

$$f^{-1}(\{y, z, w\}) = X \in T_X. \tag{2}$$

Function g is not continuous, because

$$\{y\} \in T_Y \text{ but } g^{-1}(\{y\}) = \{a, b, d\} \notin T_X. \tag{3}$$

● PROBLEM 12-2

Show that the identity function

$$f: (X, T) \rightarrow (X, T_1) \tag{1}$$

where $\forall\, x \in X : f(x) = x$, is continuous if and only if T is finer than T_1, that is, if

$$T_1 \subset T. \tag{2}$$

SOLUTION:

Suppose f is continuous then

$$A \in T_1 \Rightarrow f^{-1}(A) \in T \tag{3}$$

But $f^{-1}(A) = A$. Hence $T_1 \subset T$.

Now suppose $f : (X, T) \rightarrow (X, T_1)$ is the identity function, $f(x) = x$ and suppose $T_1 \subset T$.

Let $A \in T_1$. Then $f^{-1}(A) = A \in T$ because $T_1 \subset T$. Thus, f is continuous.

∎

Consider the sequence of identity functions

$$(X, T_1) \xrightarrow{\text{identity}} (X, T_2) \xrightarrow{\text{identity}} (X, T_3) \xrightarrow{\text{identity}} (X, T_4) \longrightarrow (\ldots$$

To ensure continuity of the functions, we must have

T_1 is finer than T_2
T_2 is finer than T_3
T_3 is finer than T_4.

● PROBLEM 12-3

1. Show that

$$f: (X, P(X)) \rightarrow (Y, T_Y) \tag{1}$$

is always continuous.

461

2. Show that

$$f: (X, T_X) \rightarrow (Y, \{\phi, Y\}) \tag{2}$$

is always continuous.

SOLUTION:

1. Let us take any open subset in T_Y, $A \in T_Y$. Then $f^{-1}(A)$ is an open set in X because $T_X = P(X)$, i.e., each subset of X is an open set.

2. There are only two open sets in Y, ϕ and Y.

$$f^{-1}(\phi) = \phi \tag{3}$$

which is an open set in X, $\phi \in T_x$. Similarly

$$f^{-1}(Y) = X \in T_X. \tag{4}$$

Each topology in X contains ϕ and X. Hence f defined by (2) is always continuous.

● **PROBLEM 12–4**

Suppose

$$f: (X, T_X) \rightarrow (Y, T_Y) \tag{1}$$

is not a continuous function. Show that the same function

$$f: (X, T'_X) \rightarrow (Y, T'_Y) \tag{2}$$

is also not continuous if

$$T'_X \text{ is coarser than } T_X \text{ (that is } T'_X \subset T_X)$$

and

$$T'_Y \text{ is finer than } T_Y \text{ (that is } T_Y \subset T'_Y).$$

SOLUTION:

Since

$$f: (X, T_X) \rightarrow (Y, T_Y)$$

is not continuous, there is an open set $D \in T_Y$, such that

$$f^{-1}(D) \notin T_X. \tag{3}$$

Consider function (2). Since $D \in T_Y$ and $T_Y \subset T'_Y$, we have

$$D \in T'_Y. \tag{4}$$

Hence $f^{-1}(D) \notin T_X$ and since $T'_X \subset T_X$

$$f^{-1}(D) \notin T'_X. \tag{5}$$

Function defined in (2) is not continuous.

Note that the same will be true if we replace T'_X by T_X in equation (2).

● PROBLEM 12-5

Show that the function

$$f : R \rightarrow R$$

defined by

$$f(x) = \begin{cases} x & \text{if } x \le 1 \\ x + 1 & \text{if } x > 1 \end{cases} \tag{1}$$

is not continuous if R is equipped with the Euclidean topology but becomes continuous when R has the upper limit topology.

SOLUTION:

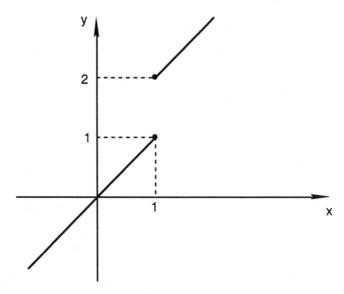

FIGURE 1

The Euclidean topology consists of all open intervals and the unions of open intervals.

Let $A = (0, {}^3/_2)$. Then

$$f^{-1}(A) = f^{-1}(0, {}^3/_2) =]0, 1].$$ (2)

The inverse image of an open set is not an open set. Hence f is not continuous in the Euclidean topology.

The upper limit topology on R consists of unions of open-closed intervals, that is intervals of the form $]a, b]$.

It is easy to verify that for each $]a, b]$

$$f^{-1}[(a, b]] =]c, d].$$ (3)

The inverse image of an open set is an open set, and the function f is continuous.

● PROBLEM 12-6

Function $f : X \to Y$ is continuous if and only if the inverse image of every closed subset of Y is a closed subset of X.

Prove it.

SOLUTION:

$$(f : X \to Y \text{ is continuous}) \Leftrightarrow$$

$$\Leftrightarrow (\forall D \subset Y : D \text{ closed in } Y f^{-1}(D) \text{ closed in } X)$$

$\Rightarrow$ Suppose $f : X \to Y$ is continuous and $D \subset Y$ is any closed subset of Y. Then $Y - D$ is an open subset of Y and

$$f^{-1}(Y - D)$$ (1)

is an open subset of X.

But

$$f^{-1}(Y - D) = X - f^{-1}(D).$$ (2)

Hence $f^{-1}(D)$ is closed.

$\Leftarrow$ Suppose for every $D \subset Y$, D is closed in Y, $f^{-1}(D)$ is closed in X. Let A represent an open subset of Y. Then $Y - A$ is closed in Y and $f^{-1}(Y - A)$ is closed in X. Since

$$f^{-1}(Y - A) = X - f^{-1}(A).$$ (3)

Hence $f^{-1}(A)$ is a open subset of X and f is a continuous function.

Let

$$f : X \to Y$$

$$g : Y \to Z \tag{1}$$

be continuous functions.

Show that the composition function

$$g \circ f : X \to Z \tag{2}$$

is also continuous.

SOLUTION:

Let A represent an open subset of Z. Then $g^{-1}(A)$ is an open subset of Y because g is continuous.

Also $f^{-1}[g^{-1}(A)]$ is an open subset of X, since f is a continuous function.

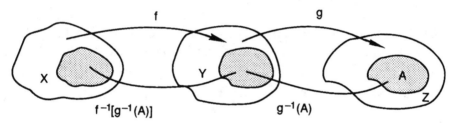

FIGURE 1

But

$$(g \circ f)^{-1}(A) = f^{-1}[g^{-1}(A)]. \tag{3}$$

Thus $(g \circ f)^{-1}(A)$ is an open set in X whenever A is an open set in Z. Hence, $g \circ f$ is a continuous function.

● **PROBLEM 12-8**

Let

$$f : (X, T_X) \to (Y, T_Y). \tag{1}$$

Prove that

465

$$(f \text{ continuous}) \Leftrightarrow \begin{pmatrix} \forall\, f(x) \in Y \quad \forall\, V \ni f(x) \\ \exists\, U \ni x \text{ such that } f(U) \subset V \end{pmatrix}$$

SOLUTION:

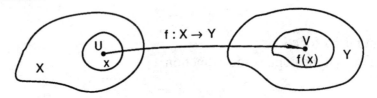

FIGURE 1

$\Rightarrow$ Suppose f is continuous. Take any $f(x) \in Y$ and any neighborhood V of $f(x)$. Then

$$f(x) \in V \text{ and } V \in T_Y. \tag{2}$$

Since f is continuous $f^{-1}(V) \in T_X$.
  Also $x \in f^{-1}(V)$. Hence $U = f^{-1}(V)$.

$\Leftarrow$ Let W represent any open subset of Y. Suppose $x \in f^{-1}(W)$, then $f(x) \in W$. Thus W is a neighborhood of $f(x)$. An open set $U \subset X$ exists, such that $x \in U$ and

$$f(U) \subset W. \tag{3}$$

But

$$U \subset f^{-1}(W). \tag{4}$$

Hence, for each $x \in f^{-1}(W)$

$$x \in U \subset f^{-1}(W). \tag{5}$$

where $U \in T_X$.
  Therefore $f^{-1}(W)$ is the union of open subsets of X and thus, an open subset.

● **PROBLEM 12-9**

Let

$$f : (X, T_X) \to (Y, T_Y) \tag{1}$$

and let $B = \{B_\alpha\}$ denote a basis for T_Y. Function f is continuous if and only if for each $B_\alpha \in B, f^{-1}(B_\alpha)$ is an open subset of X.

SOLUTION:

⇒ Suppose f is continuous. Since each $B_\alpha \in B$ is an open subset of Y,

$$f^{-1}(B_\alpha) \in T_X. \tag{2}$$

⇐ Suppose for each $B_\alpha \in B, f^{-1}(B_\alpha)$ is an open set in X. Let $A \subset Y$ denote an open subset in Y. Since B is the basis

$$A = \bigcup_\alpha B_\alpha. \tag{3}$$

Then

$$f^{-1}(A) = f^{-1}\left(\bigcup_\alpha B_\alpha\right) = \bigcup_\alpha f^{-1}(B_\alpha). \tag{4}$$

Each $f^{-1}(B_\alpha)$ is an open set. Hence, $f^{-1}(A)$ is an open set as a union of open sets. Function f is continuous.

● **PROBLEM 12-10**

Let

$$f: (X, T_X) \rightarrow (Y, T_Y) \tag{1}$$

and let $S = \{S_\alpha\}$ denote a subbasis for the topology T_Y.
Prove that f is continuous if and only if the inverse of every $S_\alpha \in S$ is an open subset of X.

SOLUTION:

If (Y, T_Y) is a topological space, then a class S of open subsets $S \subset T_Y$ is a subbasis for T_Y if and only if finite intersections of members of S form a basis for T_Y.

⇒ Suppose f is continuous. Then the inverse of all open sets are open. Hence, since S_α are open, sets $f^{-1}(S_\alpha) \in T_X$.

⇐ Suppose for every $S_\alpha \in S \; f^{-1}(S_\alpha) \in T_X$. Let $A \in T_Y$. Since $\{S_\alpha\}$ is a subbasis,

$$A = \bigcup_\alpha (S_{\alpha_1} \cap \ldots \cap S_{\alpha_k}). \tag{2}$$

We have

$$f^{-1}(A) = f^{-1}\left[\bigcup_\alpha (S_{\alpha_1} \cap \ldots \cap S_{\alpha_k})\right] =$$

467

$$= \bigcup_{\alpha} f^{-1}(S_{\alpha_1} \cap \ldots \cap S_{\alpha_k}) =$$

$$= \bigcup_{\alpha} \left[f^{-1}(S_{\alpha_1}) \cap \ldots \cap f^{-1}(S_{\alpha_k}) \right]. \tag{3}$$

Since all sets $f^{-1}(S_{\alpha_1}), \ldots, f^{-1}(S_{\alpha_k})$ are open, so is the set $f^{-1}(A)$. Hence, function f is continuous.

Let $\{T_\alpha\}$ represent a family of topologies on X. Suppose

$$f : X \to Y \tag{1}$$

is continuous with respect to each topology T_α. Show that f is continuous with respect to the topology

$$T = \bigcap_{\alpha} T_\alpha. \tag{2}$$

SOLUTION:

First note that intersection of any family of topologies is a topology.

Let A denote an open subset of Y, $A \subset Y$. Consider set $f^{-1}(A)$. Since f is continuous with respect to each topology T_α, the set $f^{-1}(A)$ belongs to each T_α, $f^{-1}(A) \in T_\alpha$. Hence

$$f^{-1}(A) \in \bigcap_{\alpha} T_\alpha = T. \tag{3}$$

Therefore f is continuous with respect to T.

Prove this theorem

$$(f : X \to Y \text{ is continuous}) \Leftrightarrow (\forall \ A \subset X : f(\overline{A}) \subset \overline{f(A)}).$$

SOLUTION:

$\Rightarrow$ Suppose $f : X \to Y$ is continuous. Since

$$f(A) \subset \overline{f(A)} \tag{1}$$

we obtain

$$A \subset f^{-1}[f(A)] \subset f^{-1}[\overline{f(A)}].$$ (2)

Set $\overline{f(A)}$ is closed and since f is continuous $f^{-1}[\overline{f(A)}]$ is closed as well

$$f^{-1}[\overline{f(A)}] = \overline{f^{-1}[\overline{f(A)}]}.$$ (3)

Hence, from (2) and (3), we find

$$A \subset \overline{A} \subset f^{-1}[\overline{f(A)}].$$ (4)

Thus

$$f(\overline{A}) \subset f[f^{-1}[\overline{f(A)}]] = \overline{f(A)}.$$ (5)

$\Leftarrow$ Suppose for any $A \subset X, f(\overline{A}) \subset \overline{f(A)}$. Let B denote any closed subset of Y, $B \subset Y$ and let

$$f^{-1}(B) = A.$$ (6)

Then

$$f(\overline{A}) = f[\overline{f^{-1}(B)}] \subset \overline{f[f^{-1}(B)]} = \overline{B} = B.$$ (7)

Hence

$$\overline{A} \subset f^{-1}[f(\overline{A})] \subset f^{-1}(B) = A$$ (8)

Since $A \subset \overline{A}$, we obtain from (8)

$$A = \overline{A}.$$ (9)

Thus, the inverse image of any closed subset of Y is a closed subset of X. Therefore, $f : X \to Y$ is continuous, by Problem 12–6.

● **PROBLEM 12–13**

Let

$$f : (X, T_X) \to (Y, T_Y)$$ (1)

represent a continuous function. Show that

$$f_A : (A, T_A) \to (Y, T_Y)$$ (2)

where $f_A = f / A$ is the restriction of f to A, is also continuous.

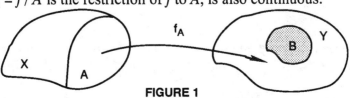

FIGURE 1

SOLUTION:

Note that, for any $B \subset Y$

$$f_A^{-1}(B) = f^{-1}(B) \cap A. \tag{3}$$

Suppose B is an open subset of Y, $B \in T_Y$. Since f is continuous

$$f^{-1}(B) \in T_X. \tag{4}$$

By definition of the induced topology

$$f^{-1}(B) \cap A \in T_A. \tag{5}$$

Hence

$$f^{-1}(B) \cap A = f_A^{-1}(B) \in T_A. \tag{6}$$

and so f_A is continuous.

● **PROBLEM 12-14**

Prove the following:

THEOREM

If $f : (X, T_X) \rightarrow (Y, T_Y)$ is a function and $X = A \cup B$ and $f|A$ and $f|B$ are both continuous (where A and B are topological subspaces of X), then if A and B are both closed or both open, f is continuous. ■

SOLUTION:

If A and B are not both closed or both open, then f does not have to be continuous.

For example, let

$$A = (0, 1) \quad B = [1, 2) \quad X = (0, 2)$$

and let $f|A = 0$ and $f|B = 1$. Then both $f|A$ and $f|B$ are continuous, but f is not continuous.

Suppose both A and B are closed. Let $D \subset Y$ denote any closed subset of Y. We will show that if $D \in T_Y$ then

$$f^{-1}(D) \in T_X. \tag{1}$$

$(f|A)^{-1}(D)$ is closed in A and $(f|B)^{-1}(D)$ is closed in B since both $f|A$ and $f|B$ are continuous. Since A and B are closed $(f|A)^{-1}(D)$ and $(f|B)^{-1}(D)$ are closed subsets of X.

470

From

$$A \cup B = X \tag{2}$$

we obtain

$$f^{-1}(D) = (f|A)^{-1}(D) \cup (f|B)^{-1}(D). \tag{3}$$

Equation (3) is the union of two closed subsets of X, hence $f^{-1}(D)$ is a closed subset of X.

Therefore, f is continuous.

● **PROBLEM 12-15**

Use the theorem of Problem 12–14 to show that the function

$$f: R \rightarrow R$$

$$f(x) = \begin{cases} x & \text{if } x \le 0 \\ 0 & \text{if } x \ge 0 \end{cases} \tag{1}$$

is continuous. R is equipped with the Euclidean topology.

SOLUTION:

Remember that if $f: X \rightarrow Y$ and $A \subset X$, then $f|A$ is the restricted function $f|A : A \rightarrow Y$, where $f|A(x) = f(x)$ for each $x \in A$.

Consider the sets

$$A = \{x \in R : x \le 0\} \quad \text{and} \quad B = \{x \in R : x \ge 0\}. \tag{2}$$

Then

$$R = A \cup B \tag{3}$$

and both A and B are closed.

Both functions $f|A$ and $f|B$ are continuous. By the theorem of Problem 12–14, function f is continuous.

● **PROBLEM 12-16**

Let R^2 (coordinate plane) be a topological space with Euclidean topology. Show that any rotation is continuous.

471

SOLUTION:

Each point of R^2 can be described by its polar coordinates (r, α). Rotation of R^2 about the origin $(0, 0)$ does not change r and transforms α into

$$\alpha_0 + \alpha$$

where α_0 is the angle of rotation. Thus

$$R_{\alpha_0}(r, \alpha) = (r, \alpha + \alpha_0). \tag{1}$$

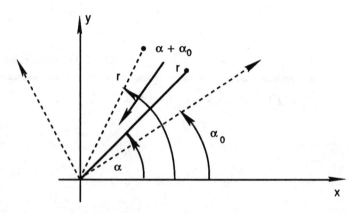

FIGURE 1

Denoting by R_{α_0} rotation through angle α_0, we obtain

$$R_{\alpha_0}^{-1} = R_{-\alpha_0} \tag{2}$$

where $R_{-\alpha_0}$ is the rotation through $-\alpha_0$. Any rotation preserves congruences. Hence, if D is the interior of a square, then $R_{\alpha_0}^{-1}(D)$ is also the interior of a square. The family of interiors of squares forms a basis for R^2 with Euclidean topology.

By Problem 12–9, we conclude that any rotation is continuous.

● **PROBLEM 12–17**

Show that the projection mappings from the plane R^2 into the line R are continuous with respect to Euclidean topology.

SOLUTION:

Both projections can be defined as follows:

472

$$P_x, P_y : R^2 \rightarrow R$$

$$P_x (x, y) = x$$

$$P_y (x, y) = y \qquad (1)$$

Consider $P_y (x, y) = y$. The inverse of any open interval (a, b) is an infinite open strip

$$A = \{(x, y) : a < y < b\}.$$

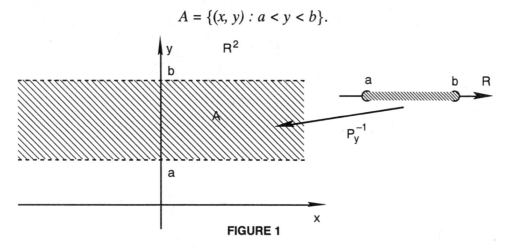

FIGURE 1

The family of open intervals forms a basis for R with the usual topology.

The inverse of any element of the basis is an open set. Hence, by Problem 12–9, the projection P_y is a continuous function. Similarly, we can show that P_x is continuous.

● **PROBLEM 12–18**

Let (X, T_X) represent a topological space and $\{a\}$ a singleton set, which is an open subset of X, $\{a\} \in T_X$. Show that any function

$$f : X \rightarrow Y \qquad (1)$$

where (Y, T_Y) is any topological space, is continuous at $a \in X$.

SOLUTION:

DEFINITION OF CONTINUITY AT A POINT

A function $f : X \rightarrow Y$ is continuous at $x_0 \in X$ if for each neighborhood $W(f(x_0))$ in Y, a neighborhood $V(x_0)$ in X exists, such that $f(V(x_0)) \subset W(f(x_0))$.

■

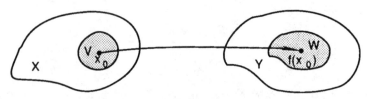

FIGURE 1

Let B represent any open set in Y containing $f(a)$.

$$f(a) \in B. \tag{2}$$

The set $\{a\}$ is an open set in X,

$$\{a\} \in T_X \text{ and } a \in \{a\}. \tag{3}$$

Also, since $f(a) \in B$

$$f(\{a\}) \subset B \tag{4}$$

Thus, f is continuous at $a \in X$.

● **PROBLEM 12-19**

Consider the topological space (X, T) where

$$X = \{a, b, c, d\} \tag{1}$$

and

$$T = \{\phi, X, \{a\}, \{a, b\}, \{a, b, c\}\}. \tag{2}$$

Show that the function $f : X \rightarrow X$, depicted in the diagram, is continuous at $b \in X$ but not continuous at $c \in X$.

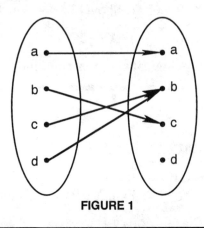

FIGURE 1

474

SOLUTION:

From the diagram

$$f(b) = c \tag{3}$$

The only open sets containing c are $\{a, b, c\}$ and X. We have

$$f^{-1}(\{a, b, c\}) = X \tag{4}$$

and

$$f^{-1}(X) = X. \tag{5}$$

Thus, the inverse of any open set containing $f(b)$ is an open set containing b, and f is continuous at b.

From the diagram

$$f(c) = b \tag{6}$$

Consider an open set containing $f(c)$

$$\{a, b\} \ni b = f(c), \tag{7}$$

Then

$$f^{-1}(\{a, b\}) = \{a, c, d\} \tag{8}$$

Set $\{a, c, d\}$ does not contain any open set containing c.

Hence, f is not continuous at $c \in X$.

● **PROBLEM 12-20**

1. Show that a function

$$f : (X, T_X) \rightarrow (Y, T_Y) \tag{1}$$

is continuous if and only if it is continuous at every point $a \in X$.

2. Show that if (1) is continuous at $a \in X$, then the restriction of f to A, where $a \in A \subset X$, is also continuous at a.

SOLUTION:

1. Suppose f is continuous. Let $a \in X$ denote any point, and let $B \subset Y$ denote an open subset of Y, such that $f(a) \in B$, $B \in T_Y$. Then $f^{-1}(B) \in T_X$ and $a \in f^{-1}(B)$. Thus, f is continuous at $a \in X$. Now, suppose f is continuous at every point $a \in X$. Let $A \subset Y$ denote an open set.

475

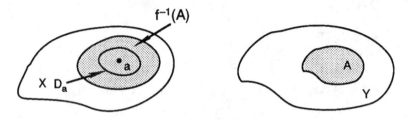

FIGURE 1

For every $a \in f^{-1}(A)$ an open set $D_a \subset X$ exists, such that

$$a \in D_a \subset f^{-1}(A). \tag{2}$$

Thus

$$f^{-1}(A) = \bigcup_a D_a \tag{3}$$

where $a \in f^{-1}(A)$. Set (3) is open as a union of open sets, Hence, f is continuous.

2. Suppose $D \subset Y$ is an open subset containing $f(a)$, see Figure 2.

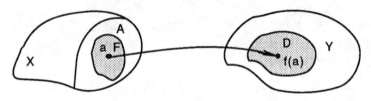

FIGURE 2

Since f is continuous at $a \in X$, there is an open subset F of X, such that

$$a \in F \subset f^{-1}(D). \tag{4}$$

Thus

$$a \in A \cap F \subset A \cap f^{-1}(D) = f_A^{-1}(D) \tag{5}$$

where

$$f_A = f|A. \tag{6}$$

But

$$A \cap F \in T_A. \tag{7}$$

Hence f_A is continuous at $a \in X$ with respect to topology T_A.

476

Show that if a function

$$f : X \rightarrow Y \qquad (1)$$

is continuous at $a \in X$, then it is sequentially continuous at $a \in X$.

SOLUTION:

DEFINITION OF SEQUENTIAL CONTINUITY

A function $f : X \rightarrow Y$ is sequentially continuous at a point $a \in X$ if for every sequence (a_n) in X,

$$(a_n \rightarrow a) \Rightarrow (f(a_n) \rightarrow f(a)) \qquad (2)$$

■

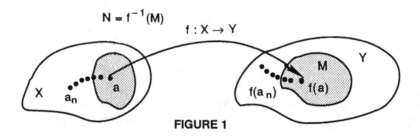

FIGURE 1

Suppose $f : X \rightarrow Y$ is continuous at $a \in X$, and let (a_n) denote a sequence convergent to a, $a_n \rightarrow a$.

We will show that any neighborhood M of $f(a)$, $f(a) \in M \in T_Y$ contains all but a finite number of the elements of the sequence $(f(a_n))$.

We have

$$f(a) \in M \in T_Y \qquad (3)$$

Function f is continuous at $a \in X$. Hence, $f^{-1}(M) = N$ is a neighborhood of a. N contains all but a finite number of elements of (a_n). Then

$$a_n \in N \Rightarrow f(a_n) \in N. \qquad (4)$$

Hence, N contains all but a finite number of elements of $f(a_n)$ and

$$f(a_n) \rightarrow f(a). \qquad (5)$$

● **PROBLEM 12-22**

Give an example of a neighborhood-finite family in R.

SOLUTION:

DEFINITION OF A NEIGHBORHOOD-FINITE FAMILY

A collection $\{A_\omega : \omega \in \Omega\}$ of sets in a topological space (X, T) is called neighborhood-finite if each point of X has a neighborhood $N_0(x)$, such that

$$N_0(x) \cap A_\omega \neq \emptyset \tag{1}$$

for at most finitely many indices ω.

■

Consider the family $\{A_p\}$,

$$A_p = [p, p + 1] \tag{2}$$

where p is an integer.

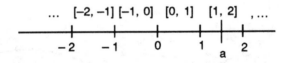

FIGURE 1

Let $a \in R$ denote any point in R. Then, we can always find a sufficiently small neighborhood U of $a \in R$, which intersects with only one element of $\{A_p\}$ (if a is not an integer). If a is an integer, then a sufficiently small neighborhood U of a would intersect with two members of $\{A_p\}$.

Note that a family $\{A_\omega\}$ may be neighborhood-finite, even though each A_α intersects infinitely many other A_β. For example in R_+, we define $\{A_n\}$

$$A_n = \{x : X > n\}. \tag{3}$$

● PROBLEM 12–23

Here is a useful theorem concerning coverings of the space.

THEOREM

Let $\{A_\omega : \omega \in \Omega\}$ represent a family of sets that forms a covering of the space X, that is, $X = \bigcup_\omega A_\omega$. Assume that one of these two conditions holds:

1. all sets A_ω are open, or

2. all sets A_ω are closed, and form an neighborhood-finite family. Then

478

$$\begin{pmatrix} B \subset X \text{ is open} \\ \text{(or closed)} \end{pmatrix} \Leftrightarrow \begin{pmatrix} \text{all } B \cap A_\omega \text{ are open} \\ \text{(or closed) is the subspace } A_\omega \end{pmatrix}.$$

∎

Prove this theorem for the case when condition 2 is true.

SOLUTION:

⇒ This follows immediately from the definition of induced topology.

⇐ Suppose each $B \cap A_\omega$ is closed in the closed A_ω.

The case when each $B \cap A_\omega$ is open in the closed A_ω is similar. Since a subspace of a subspace is a subspace of the entire space X *and each* $B \cap A_\omega$ is closed in the closed A_ω it follows that $B \cap A_\omega$ is closed in X.

$$B \cap A_\omega = \overline{B \cap A_\omega}. \tag{1}$$

Since $\{A_\omega\}$ is neighborhood-finite, so is $\{B \cap A_\omega\}$. Therefore (remember, $\{A_\omega\}$ is a covering)

$$B = \bigcup_\omega \{B \cap A_\omega\} \tag{2}$$

is closed in X.

Here we applied the following:

THEOREM

If $\{A_\omega : \omega \in \Omega\}$ is an neighborhood-finite family in X, then for each $\Lambda \subset \Omega$, the set

$$\bigcup_\lambda \{\overline{A}_\lambda : \lambda \in \Lambda\} \tag{3}$$

is closed in X.

● **PROBLEM 12-24**

The situation in which a continuous function is defined piecewise appears in analysis. For example, a function which is continuous on a segment $[0, n] \subset R$ is defined for each $[k, k + 1]$ separately in such a way, that the adjacent functions agree on the common end points of the segments $[0, 1], [1, 2], \ldots, [n - 1, n]$.

Partial definitions of functions are formulated in this theorem.

THEOREM

Let (X, T) represent a topological space and $\{A_\omega : \omega \in \Omega\}$ its covering. One of two conditions is true:

1. all sets A_ω are open, or

2. all sets A_ω are closed and form a neighborhood-finite family.

For each $\omega \in \Omega$ function

$$f_\omega : A_\omega \to Y \tag{1}$$

is continuous and, such that

$$f_\alpha | A_\alpha \cap A_\beta = f_\beta | A_\alpha \cap A_\beta \tag{2}$$

for each $\alpha, \beta \in \Omega$.

Then a unique continuous function exists

$$f : X \to Y \tag{3}$$

such that for each $\omega \in \Omega$

$$f | A_\omega = f_\omega. \tag{4}$$

∎

Prove this theorem.

SOLUTION:

For each $x \in X$, we define

$$f(x) = f_\alpha(x) \tag{5}$$

where $\alpha \in \Omega$ is any index, such that

$$x \in A_\alpha. \tag{6}$$

The definition is unique, because if $x \in A_\alpha$ and $x \in A_\beta$ then

$$f(x) = f_\alpha(x) = f_\beta(x) \tag{7}$$

Since $f_\alpha | A_\alpha \cap A_\beta = f_\beta | A_\alpha \cap A_\beta$. Hence (5) defines a function which is unique.

Function f is continuous. Let $U \subset Y$ be open. Then

$$f^{-1}(U) \cap A_\omega = f_\omega^{-1}(U) \tag{8}$$

is open in A_ω for each ω.

Thus, by Problem 12–23, $f^{-1}(U)$ is open in X and f is continuous.

● **PROBLEM 12-25**

Show that the function

$$f : R \rightarrow R$$

such that

$$f(x) = x^2 \qquad (1)$$

is not open.

SOLUTION:

Observe that if f is a continuous function, then the inverse image of every open set is an open set and the inverse image of every closed set is a closed set. It will be useful to define the following group of functions.

DEFINITION OF AN OPEN (CLOSED) FUNCTION

A function

$$f : X \rightarrow Y \qquad (2)$$

is called an open (closed) function, if the image of every open (closed) set is open (closed).

∎

Consider an open interval $(-1, 1)$. Function $f(x) = x^2$ maps $(-1, 1)$ into

$$f((-1, 1)) = [0, 1) \qquad (3)$$

which is not an open set. Hence, f defined by (1) is not open.

● **PROBLEM 12-26**

Let $f : X \rightarrow Y$ denote a function and B represent a basis for a topological space (X, T).

Show that, if for every $B_\alpha \in B$ $f(B_\alpha)$ is open in Y, then f is an open function.

SOLUTION:

Suppose A is an open subset of X. Then

$$A = \bigcup_\alpha B_\alpha \, , \, B_\alpha \in B \qquad (1)$$

481

since B is the basis for X. Hence

$$f(A) = f(\bigcup_{\alpha} B_{\alpha}) = \bigcup_{\alpha} f(B_{\alpha}). \qquad (2)$$

By hypothesis, each $f(B_{\alpha})$ is open in Y. Therefore $f(A)$ is open, as a union of open sets, and f is an open function.

As a matter of fact two properties are equivalent:

1. f is an open function.

2. f maps each member of a basis for X to an open set in Y.

● PROBLEM 12–27

Let

$$p : R^2 \to R \qquad (1)$$

represent the projection mapping of R^2 into the x-axis, i.e.,

$$p(x, y) = x. \qquad (2)$$

1. Show that p is an open function.

2. Show that p is not a closed function.

SOLUTION:

1. The family of all open discs in R^2 forms a basis for R^2. We assume that R^2 is equipped with the Euclidean topology.

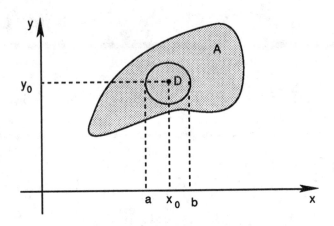

FIGURE 1

Note that the projection of any disc D

$$D = \{(x, y) : (x - x_0)^2 + (y - y_0)^2 < r^2\} \qquad (3)$$

is an open interval

$$p(D) = (a, b) \qquad (4)$$

where (a, b) is an open interval. According to Problem 12–26, we conclude that p, defined by (2), is an open function.

2. Consider the shaded region shown in Figure 2.

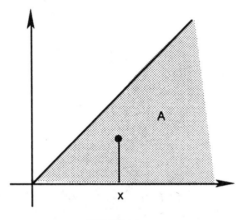

FIGURE 2

$$A = \{(x, y) : x \geq 0, 0 \leq y \leq x\}. \qquad (5)$$

Set A is closed, while its projection $p(A) = [0, \infty)$ is not closed. Thus, p is not a closed function.

● **PROBLEM 12-28**

Prove the following theorem:

THEOREM

Let $f : X \to Y$ represent a closed function and let D denote any subset of Y. Let U denote any open set, such that

$$f^{-1}(D) \subset U \subset X. \qquad (1)$$

Then an open V exists, such that $D \subset V$ and $f^{-1}(V) \subset U$.

■

See Figure 1.

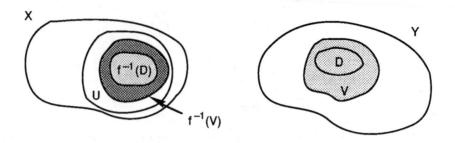

FIGURE 1

SOLUTION:

Let us define

$$V = Y - f(X - U). \qquad (2)$$

By hypothesis

$$f^{-1}(D) \subset U. \qquad (3)$$

Therefore

$$D \subset V.$$

Indeed, suppose $y \in D$, then $f^{-1}(y) \subset f^{-1}(D)$. Hence, $f^{-1}(y) \subset U$ and $y \in f(U)$. Thus, $y \notin f(X - U)$ and $y \in Y - f(X - U)$. Finally

$$y \in V \qquad (4)$$

By hypothesis, f is closed and U is an open subset of X. Then $X - U$ is closed and $f(X - U)$ is closed. Therefore $V = Y - f(X - U)$ is an open subset of Y. Now we prove that $f^{-1}(V) \subset U$.

$$f^{-1}(V) = X - f^{-1}[f(X - U)] \subset X - (X - U) = U \qquad (5)$$

A similar theorem exists for open functions.

THEOREM

Let $f : X \twoheadrightarrow Y$ denote an open function and let D represent any subset of Y. Let W represent any closed set, such that

$$f^{-1}(D) \subset W.$$

Then a closed set P exists, such that $D \subset P$ and

$$f^{-1}(P) \subset W.$$

■

> Prove
>
> $$(f : X \to Y \text{ is a closed function}) \Leftrightarrow (\forall A \subset X : \overline{f(A)} \subset f(\overline{A})).$$

SOLUTION:

$\Rightarrow$ Suppose f is a closed function. Let A represent any subset of X. Then $\overline{A}$ is a closed set and $f(\overline{A})$ is closed.

Since

$$f(a) \subset f(\overline{A}) \tag{1}$$

we have

$$\overline{f(A)} \subset \overline{f(\overline{A})} = f(\overline{A}). \tag{2}$$

$\Leftarrow$ Suppose, for each set $A \subset X, \overline{f(A)} \subset f(\overline{A})$.

Let A represent a closed set, $\overline{A} = A$. We have

$$f(A) \subset \overline{f(A)} \subset f(\overline{A}) = f(A). \tag{3}$$

Therefore

$$f(A) = \overline{f(A)} \tag{4}$$

so that set $f(A)$ is closed and function $f : X \to Y$ is closed.

> Use the function
>
> $$f(x) = \frac{x}{|x| + 1} \tag{1}$$
>
> to show that the spaces R' and $(-1, +1)$ are homeomorphic.

SOLUTION:

DEFINITION OF HOMEOMORPHISM

A continuous bijective (that is, one-to-one and onto) function $f : X \to Y$, such that $f^{-1} : Y \to X$ is also continuous, is called a homeomorphism and denoted by

$$f : X \cong Y. \tag{2}$$

Two spaces X, Y, denoted by $X \cong Y$, are homeomorphic (or of the same topological type), if there is a homeomorphism $f : X \cong Y$.

■

First of all, observe that

$$f(x) = \frac{x}{|x| + 1}$$

where $f : R \to (-1, +1)$ is one-to-one and onto. Both functions f and f^{-1} are continuous. Hence, f, defined by (1), is a homeomorphism.

This result can be generalized to n-dimensional space

$$f : R^n \to B(0, 1) \tag{3}$$

where $B(0, 1)$ is the unit ball. Then $x = (x_1, x_2, \ldots, x_n)$ and

where

$$|x| = \sqrt{x_1^2 + x_2^2 + \ldots + x_n^2}$$

$$f(x) = \frac{x}{|x| + 1} \tag{4}$$

Hence, R^n is homeomorphic to its unit ball $B(0, 1)$

$$R^n \cong B(0, 1). \tag{5}$$

● **PROBLEM 12-31**

Show that the homeomorphism relation is reflexive, symmetric and transitive.

SOLUTION:

Relation is reflexive. For any topological space (X, T), the identity mapping $f(x) = x$

$$f : X \to X \tag{1}$$

is a homeomorphism.

Symmetry. If $X \cong Y$, then a homeomorphism exists, such that

$$f : X \to Y. \tag{2}$$

We define $g = f^{-1}$, where

$$g : Y \to X. \tag{3}$$

Mapping g is a homeomorphism. Hence, $Y \cong X$.

We will show that the homeomorphism relation is transitive.

Suppose $X \cong Y$ and $Y \cong Z$. Then $f : X \to Y$ and $g : Y \to Z$ are homeomorphisms.

Mapping

$$g \circ f : X \to Z \tag{4}$$

where $g \circ f(x) = g[f(x)]$ is a homeomorphism, since both g and f are homeomorphisms.

Relation $\cong$ (X is homeomorphic to Y), defined in any family of topological spaces, is an equivalence relation. Hence, any family of topological spaces can be partitioned into disjoint classes of topologically equivalent spaces.

From now on, we will be more concerned with these classes of topologically equivalent spaces, than with the individual topological spaces.

● PROBLEM 12-32

Show that an area is not a topological property.

SOLUTION:

In Problem 12–31, we defined the classes of topologically equivalent spaces. Now we shall investigate the properties which are common for all members of the same class.

DEFINITION OF TOPOLOGICAL PROPERTIES

A property P of sets is called topological or a topological invariant if, whenever a topological space (X, T) has this property, then every space homeomorphic to (X, T) has property P.

■

That is, property P is common for all members of a class of topologically equivalent spaces.

Consider the function

$$f : R^2 \to R^2 \tag{1}$$

defined by

$$f : (x, y) \to (2x, y), \tag{2}$$

where f is a homeomorphism. It is easy to see that f transforms a unit square into a rectangle, as shown in Figure 1.

Hence, a figure of area one is transformed into a figure of area two.

Area is not a topological property. Similarly, it can be shown that length and volume are not topological properties.

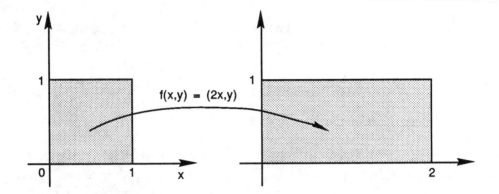

$f(x,y) = (2x,y)$

FIGURE 1

Show that each of the following conditions is necessary and sufficient for a one-to-one mapping, f, to be a homeomorphism:

1. $f(\overline{A}) = \overline{f(A)}$ for every $A \subset X$. (1)

2. $f^{-1}(\overline{B}) = \overline{f^{-1}(B)}$ for every $B \subset Y$. (2)

SOLUTION:

We shall apply the following:

THEOREM

$$(f \text{ is continuous}) \Leftrightarrow (f(\overline{A}) \subset \overline{f(A)} \text{ for every } A \subset X) \quad (3)$$

$$(f \text{ is continuous}) \Leftrightarrow \left(\overline{f^{-1}(B)} \subset f^{-1}(\overline{B}) \text{ for every } B \subset Y\right). \quad (4)$$

Suppose f is a homeomorphism, then f is continuous and for every $A \subset X$

$$f(\overline{A}) \subset \overline{f(A)} \quad (5)$$

Also $f^{-1} : Y \to X$ is continuous and by (4), we obtain

$$\overline{f(A)} \subset f(\overline{A}). \quad (6)$$

Note that $(f^{-1})^{-1} = f$. From (5) and (6), we obtain

$$f(\overline{A}) = \overline{f(A)}. \quad (7)$$

Suppose, for every $A \subset X$, $f(\overline{A}) = \overline{f(A)}$. Then

488

$$f(\overline{A}) \subset \overline{f(A)} \qquad (8)$$

and by (3), we conclude that f is continuous.

Also, for every $A \subset X$

$$\overline{f(A)} = f^{-1}[f^{-1}(A)] \subset f^{-1}[f^{-1}(\overline{A})] = f(\overline{A}). \qquad (9)$$

Hence, $f^{-1} : Y \to X$ is continuous and f is a homeomorphism.

Similarly, we show that condition (2) is necessary and sufficient for a one-to-one function to be homeomorphic.

● **PROBLEM 12–34**

Suppose f maps X onto Y, $f : X \to Y$, where X is a T_1–space.

Show that a necessary and sufficient condition for f to be a homeomorphism is:

1. $\overline{A} = f^{-1}(\overline{f(A)})$ for every $A \subset X$ or $\qquad (1)$

2. $(x \in \overline{A}) \equiv (f(x) \in \overline{f(A)})$ $\qquad (2)$

SOLUTION:

DEFINITION OF A T₁-SPACE

A topological space (X, T) is called a T_1-space, if each single element set is closed, that is,

$$\overline{\{a\}} = \{a\} \text{ for each } a \in X.$$

■

For example, each metric space is a T_1-space.

We shall prove : for $f : X \to Y$, where X is a T_1-space

$$(f \text{ is a homeomorphism}) \Leftrightarrow (\overline{A} = f^{-1}(\overline{f(A)}) \text{ for every } A \subset X). \qquad (3)$$

$\Rightarrow$ Since f is a homeomorphism, it is thus one-to-one. From Problem 12–33, equation (1),

$$f(\overline{A}) = \overline{f(A)} \qquad (4)$$

or

$$f^{-1}[f(\overline{A})] = \overline{A} = f^{-1}[\overline{f(A)}]. \qquad (5)$$

489

$\Leftarrow$ We have to show that f is one-to-one. Suppose $f(a) = f(b)$. Then

$$\overline{\{a\}} = f^{-1}[\overline{f(a)}] = f^{-1}[\overline{f(b)}] = \overline{\{b\}}.\tag{6}$$

Since (X, T) is a T_1-space,

$$a = b.\tag{7}$$

Thus, f is one-to-one. Then from (1),

$$f(\overline{A}) = \overline{f(A)}\tag{8}$$

for every $A \subset X$.

According to Problem 12–33, f is a homeomorphism.

Conditions (1) and (2) are equivalent. From (2), we conclude that every property, expressed in terms of the operation $\overline{A}$ and of operations of set theory and of logics, is topological. We can briefly say that if a point $a \in X$ (or a set, family of sets, etc.) has a given property with respect to the space (X, T), then $f(a)$ has the same property with respect to Y, provided that $f : X \rightarrow Y$ is a homeomorphism.

● **PROBLEM 12–35**

1. Show that two closed intervals, $[a, b]$ and $[c, d]$, are homeomorphic.

2. Show that an open interval $(-1, 1)$ and the real line are homeomorphic.

3. Show that the surface of the sphere with one point removed is homeomorphic to the plane.

SOLUTION:

1. Suppose $a < b$ and $c < d$. Define $f(x)$ by

$$f(x) = \frac{d - c}{b - a} x + \frac{bc - ad}{b - a}\tag{1}$$

f is a homeomorphism, which maps the first interval onto the second. Hence, the two closed intervals are homeomorphic. To show that, one can also use the drawing shown in Figure 1.

2. Function $f : (-1, 1) \rightarrow R$

$$f(x) = \tan \frac{\pi x}{2}\tag{2}$$

is a homeomorphism. Hence, $(-1, 1)$ and R are homeomorphic.

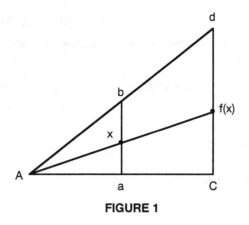

FIGURE 1

3. Consider the surface

$$x^2 + y^2 + (z - 1)^2 = 1 \qquad (3)$$

with the point $(0, 0, 2)$ removed. (See Figure 2).

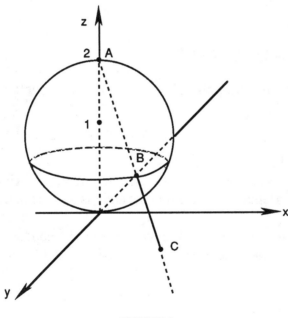

FIGURE 2

Draw a line from A through the point B on the surface of the sphere and through point C on the plane xy.

Each point B on the surface of the sphere is mapped on the plane $z = 0$. This mapping is one-to-one, onto and continuous; the inverse mapping also has the same properties. Thus, a sphere with one point removed and a plane are homeomorphic.

THEOREM

Let $f : X \to Y$ denote a bijection. Then the following properties are equivalent:

1. f is a homeomorphism.

2. $f(\overline{A}) = \overline{f(A)}$ for every $A \subset X$.

3. f is continuous and open.

4. f is continuous and closed.

Prove it.

SOLUTION:

1. $\Leftrightarrow$ 2. This was proved in Problem 12-33.

1. $\Leftrightarrow$ 3. f is a homeomorphism, hence $f^{-1} : Y \to X$ is continuous. Therefore, for each open $A \subset X$,

$$f^{-1}[f^{-1}(A)] = f(A) \tag{1}$$

and $f(A)$ is open in Y.

Suppose f is continuous and open, then the image of each open set in X is an open set in Y. Hence, f^{-1} is continuous and f is a homeomorphism.

3. $\Leftrightarrow$ 4. Note, that if $f : X \to Y$ is bijective, then the conditions that f is open and f is closed, are equivalent. Suppose f is open and $A \subset X$ is closed. Then set B is open

$$A = X - B \tag{2}$$

and

$$f(A) = f(X - B) = f(X) - f(B) = Y - f(B) \tag{3}$$

Hence, since $f(B)$ is open, the set $f(A)$ is closed. Thus, f is closed.

The following theorem helps to determine a given function as being a homeomorphism:

THEOREM 1

Let $f : X \to Y$ and $g : Y \to X$ be continuous and, such that

$$f \circ g = 1_Y \qquad (1)$$

and

$$g \circ f = 1_X \qquad (2)$$

then f is a homeomorphism and

$$f^{-1} = g. \qquad (3)$$

∎

Prove this theorem.

SOLUTION:

First let us prove this.

THEOREM 2

Let $f : X \to Y$ and $g : Y \to X$, such that

$$g \circ f = 1_X \qquad (4)$$

then f is one-to-one and g is onto.

∎

The map of f is one-to-one, since

$$(f(x) = f(y)) \Rightarrow (x = g \circ f(x) = g \circ f(y) = y). \qquad (5)$$

Function g is onto, since for any $x \in X$,

$$x = g[f(x)]. \qquad (6)$$

By applying Theorem 2 and conditions (1) and (2), we conclude that both f and g are bijective. In addition to that, $f^{-1} = g$. Thus, f and f^{-1} are continuous and f is a homeomorphism.

● **PROBLEM 12–38**

Let

$$f : (X, T_X) \to (Y, T_Y) \qquad (1)$$

denote a homeomorphism and let (A, T_A) represent a subspace of (X, T_X). Let

$$f_A : (A, T_A) \to (B, T_B) \tag{2}$$

be a restriction of f to A, $f_A = f \mid A$ and let

$$B = f(A)$$

and let T_B be a topology induced on B.
  Show that f_A is a homeomorphism.

SOLUTION:

  Since f is a bijection (that is, one-to-one and onto), $f_A : A \to B$ is also a bijection. The restriction of a continuous function is also a continuous function. Hence, f_A is continuous.
  Let $U \subset A$ be T_A open, then

$$U = A \cap V \tag{3}$$

where $V \in T_X$. Since f is one-to-one,

$$f(A \cap V) = f(A) \cap f(V) \tag{4}$$

Thus,

$$f_A(U) = f(U) = f(A) \cap f(V) = B \cap f(V) \tag{5}$$

Since f is open and $V \in T_X$, $f(V) \in T_Y$. Therefore,

$$B \cap f(V) \in T_B \tag{6}$$

and so, f_A is open.

● **PROBLEM 12-39**

Show that the subsets X and Y of the plane R^2 with the Euclidean topology, shown in Figure 1, are not homeomorphic. The topologies of X and Y are the usual induced topologies.

$$X = \{x : d(x, a_1) = 1 \text{ or } d(x, a_2) = 1\} \tag{1}$$

$$Y = \{y : d(y, a_3) = 1\} \tag{2}$$

$$a_1 = (0, 1) \quad a_2 = (0, -1) \quad a_3 = (3, 0).$$

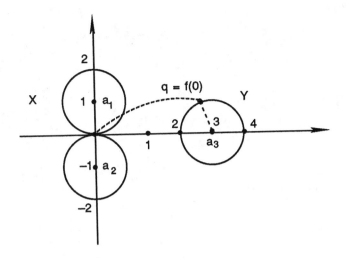

FIGURE 1

SOLUTION:

Suppose a homeomorphism

$$f : X \to Y \tag{3}$$

exists. Let us denote

$$f(0) = q \tag{4}$$

$$X_1 = X - \{0\}$$

$$Y_1 = Y - \{q\}. \tag{5}$$

By Problem 12–38, we conclude that

$$f_1 : X_1 \to Y_1 \tag{6}$$

is also a homeomorphism, with respect to the induced topologies.

Set Y_1 is connected by setting

$$q = (3 + \cos \beta_0, \sin \beta_0), \tag{7}$$

we define

$$g : (0, 2\pi) \to Y_1 \tag{8}$$

by

$$g(\beta) = (3 + \cos(\beta_0 + \beta), \sin(\beta_0 + \beta)). \tag{9}$$

Function g is homeomorphic interval $(0, 2\pi)$ is connected, and thus, Y_1 is connected.

To obtain a contradiction, we will show that X_1 is not connected.

The sets

$$A = \{(x, y) : x > 0\}$$

$$B = \{(x, y) : x < 0\} \tag{10}$$

are both open in R^2. Hence,

$$A_1 = X_1 \cap A$$

$$B_1 = X_1 \cap B \tag{11}$$

are open subsets of X_1, such that

$$A_1, B_1 = \phi, \quad A_1 \cap B_1 = \phi, \quad A_1 \cup B_1 = X_1 \tag{12}$$

space X_1 is not connected. Hence, X_1 and Y_1 are not homeomorphic, because connectedness is a topological property.

● **PROBLEM 12-40**

A topological space, (X, T), is given, along with an equivalence relation R. Construct the identification topology on $X/_R$ (sometimes called the quotient topology).

SOLUTION:

If X is a set and R is an equivalence relation on X, then R determines a partition of X into equivalence classes.

Two elements, $x, y \in X$, belong to the same class, if and only if, $x \, R \, y$ (x is in R-relation to y). The set of equivalence classes is denoted by $X/_R$.

Let us define a mapping

$$f : X \rightarrow X/_R \tag{1}$$

by

$$f(x) = [x] \tag{2}$$

where $[x]$ is the equivalence class, such that $x \in [x]$.

f is called the identification mapping (or quotient mapping). See Figure 1.

Let us introduce a topology on $X/_R$.

A subset U of $X/_R$ is open, if and only if, $f^{-1}(U)$ is open in X. In Problem 12-41, we will show that this is, indeed, a topology. This topology is called the identification topology or quotient topology.

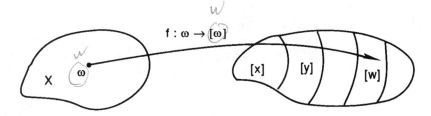

FIGURE 1

● **PROBLEM 12-41**

Prove that the identification topology, defined in Problem 12–40, is indeed, topology.

SOLUTION:

Let us denote the collection of all open sets of $X/_R$ by T_R.

1. $$f^{-1}(\phi) = \phi \text{ and } f^{-1}(X/_R) = X \tag{1}$$

(X, T) is a topological space, hence, $\phi, X \in T$. Thus,

$$\phi \in T_R \text{ and } X/_R \in T_R \tag{2}$$

2. Let $E, F \in T_R$, then $f^{-1}(E)$ and $f^{-1}(F)$ are open sets in X. Thus,

$$f^{-1}(E) \cap f^{-1}(F) \in T. \tag{3}$$

But

$$f^{-1}(E) \cap f^{-1}(F) = f^{-1}(E \cap F) \in T. \tag{4}$$

Hence, $E \cap F$ is an open set in $X/_R$

$$E \cap F \in T_R \tag{5}$$

3. Let $\{E_\alpha\}$ represent a family of open sets in $X/_R$. Then each set

$$f^{-1}(E_\alpha) \in T \tag{6}$$

and

$$\bigcup_\alpha f^{-1}(E_\alpha) \in T. \tag{7}$$

We have

$$\bigcup_\alpha f^{-1}(E_\alpha) = f^{-1}\left(\bigcup_\alpha E_\alpha\right) \in T \tag{8}$$

and $\bigcup_\alpha E_\alpha$ is an open subset of $X/_R$.

The collection of open sets of $X/_R$ forms a topology on $X/_R$.

The closed interval [0, 1] is equipped with the absolute value topology. An equivalence relation is defined by:

1. 0 is equivalent to 1

2. every other element of the segment [0, 1] is equivalent only to itself.

The equivalence classes are {0, 1} and {x} for $0 < x < 1$.
    Show that the identification space defined is homeomorphic to a circle.

SOLUTION:

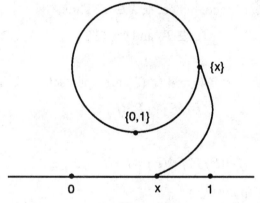

FIGURE 1

The endpoints 0 and 1 of the segment [0, 1] become a single point {0, 1} of the new topological space. Hence, we wrap the segment [0, 1] around a circle of a radius of $1/_{2\pi}$ and obtain a continuous function from [0, 1] onto a circle. The inverse function is also continuous.

The identification space, obtained from [0, 1], is homeomorphic to a circle.

CHAPTER 13

SEPARATION AXIOMS

Let $X = \{0, 1\}$. Show that X, with the indiscrete topology, is not a T_0-space. Give an example of topology T, for which (X, T) is a T_0-space.

SOLUTION:

DEFINITION OF T_0-SPACE

The space (X, T) is said to be a T_0-space, if for any two distinct $a, b \in X$, there is a neighborhood of at least one, which does not contain the other.

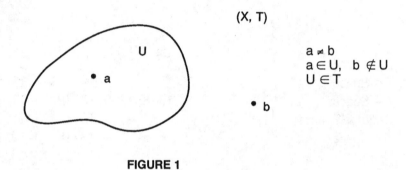

(X, T)

$a \neq b$
$a \in U, \quad b \notin U$
$U \in T$

FIGURE 1

With the indiscrete topology $T = \{\phi, X\}$ one cannot separate 0 from 1 or 1 from 0. Consider the set $X = \{0, 1\}$ with topology $T = \{\phi, X, \{0\}\}$. Then, for the two distinct points 0 and 1, an open set $\{0\}$ exists, such that

$$0 \in \{0\} \text{ but } 1 \notin \{0\}.$$

Hence, (X, T) is a T_0-space. Note that each metric space (X, d) is a T_0-space.

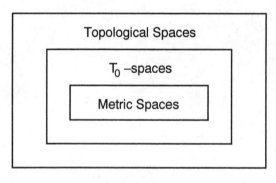

FIGURE 2

● PROBLEM 13-2

Explain why, in general, a pseudometric space is not T_0.

SOLUTION:

A pseudometric on X is a function $D : X \times X \to R_+$, such that

1. $D(x, y) = D(y, x)$

2. $D(x, y) \le D(x, z) + D(z, y)$

for all $x, y, z, \in X$.

Thus, it is possible that for two distinct points $a, b \in X$,

$$D(a, b) = 0.$$

Any neighborhood of a contains b and any neighborhood of b contains a.

Hence, a pseudometric space is not a T_0-space. Points $a, b \in X$ are distinct, but such that, $D(a, b) = 0$ cannot be separated.

● PROBLEM 13-3

Show that every metric space is a T_1-space.

SOLUTION:

We shall start with the

DEFINITION OF T_1-SPACE

A topological space (X, T) is called a T_1-space, if every single element set is closed, that is,

$$\forall a \in X \quad \{a\} = \overline{\{a\}}. \tag{1}$$

Note, that in a metric space (X, d), the condition for a set A to be closed, can be expressed by the implication

$$\left(\begin{array}{c} \lim_{n \to \infty} x_n = x \\ x_n \in A \end{array} \right) \Rightarrow (x \in \overline{A}).$$

501

Since lim $a = a$, we have

$$\overline{\{a\}} = \{a\}.$$

Thus, every metric space is a T_1-space (see Figure 1).

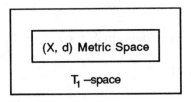

FIGURE 1

There are topological spaces which are not T_1-spaces. For example, space X, containing two points $X = \{a, b\}$ with topology $T = \{\phi, X\}$, is not a T_1-space.

We showed before, that every metric space can be regarded as a topological space. Figure 2 illustrates the results.

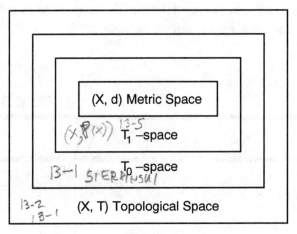

FIGURE 2

From the definitions, it is clear that any T_1-space is also a T_0-space.

● PROBLEM 13-4

Show that a topological space (X, T) is a T_1-space iff for any pair of distinct points $a, b \in X$, the open sets $G, H \in T$ exist, such that

$$a \in G, \quad b \notin G \quad \text{and} \quad b \in H, \quad a \notin H.$$

See Figure 1.

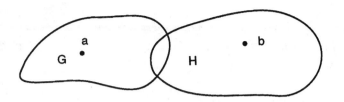

FIGURE 1

SOLUTION:

Defn (c) #13-3/P.50

Suppose (X, T) is a T_1-space. Then, for any $x \in X$, $\{x\}$ is a closed set. Let $a, b \in X$ and $a \neq b$. The sets $X - \{a\}$ and $X - \{b\}$ are open, and

$$a \in X - \{b\} \quad \text{and} \quad b \notin X - \{b\}$$

$$b \in X - \{a\} \quad \text{and} \quad a \notin X - \{a\}.$$

Conversely, suppose $x \in X$. We shall show that $\{x\}$ is closed, i.e., $X - \{x\}$ is open. Let $y \in X - \{x\}$, then $y \neq x$ and an open set H_y exists, such that

$$y \in H_y \quad \text{and} \quad x \in H_y.$$

Thus,

$$y \in H_y \subset X - \{x\} \text{ and } X - \{x\} = \bigcup_{y \neq x} H_y.$$

Since all H_y are open sets, $X - \{x\}$ is open and $\{x\}$ is closed, $\{x\} = \overline{\{x\}}$.

● PROBLEM 13-5

Let X represent a finite set. Prove that the only topology on X, which makes X into a T_1-space, is the discrete topology.

SOLUTION:

Let X represent any finite set and T a topology on X, such that (X, T) is a T_1-space.

A space (X, T) is t_1, iff every one-point subset of X is closed. Since the union of two closed subsets is a closed subset, we conclude that all subsets of X are closed.

Therefore, all subsets of X are open and T is the discrete topology.

Let (X, T) denote a T_1-space. Show that the following conditions are equivalent:

1. $a \in X$ is an accumulation point of A.

2. every open set containing a contains an infinite number of points of A.

Thm 17.9 P 99 MUNKRES

SOLUTION:

1. $\Rightarrow$ 2.

arbitrary

Suppose $a \in X$ is an accumulation point of A, and G is an open set $a \in G$, containing only a finite number of points of A different from a. Then

$$a \notin B = \{a_1, a_2, ..., a_n\} := A \cap [G - \{a\}].$$

B is a finite subset of a T_1-space, hence, it is closed and $X - B$ is open. Let

$$H = (X - B) \cap G. \quad a \in X - B \, \& \, a \in G$$

Then H is open, $a \in H$ and H do not contain any points of A different from a. Hence, a is not an accumulation point of A.

2. $\Rightarrow$ 1.

By definition of an accumulation point. *Defn P. 441 @ TOP*

Prove, that if (X, T) and (Y, T') are homeomorphic and (X, T) is a T_1-space (or T_0-space), then so is (Y, T').

SOLUTION:

Let f denote a homeomorphism

$$f : X \rightarrow Y$$

and X be a T_1-space. A space (X, T) is T_1, if and only if every one-point subset of X is closed.

Let y represent any point of Y, $y \in Y$. The set $f^{-1}(y)$ is a one-point subset of X and since X is T_1, the set of $\{f^{-1}(y)\}$ is closed.

Since $f : X \rightarrow Y$ is a homeomorphism, it maps closed sets into closed sets. Therefore, for any $y \in Y$

$$\{y\} = \overline{\{y\}}.$$

Thus, (Y, T') is a T_1-space. Similarly, we can show that, if X is T_0, then so is Y.

● **PROBLEM 13-8**

Prove the following theorem:

THEOREM

Each T_2-space is a T_1-space.

SOLUTION:

DEFINITION OF T_2-SPACE (or a Hausdorff space or a separated space)

A topological space is a Hausdorff space if, for each pair of points $a \neq b$, two disjoint open sets A and B exist, such that

$$a \in A, \quad b \in B \quad A \cap B = \phi. \tag{1}$$

It is easy to see that each metric space is a T_2-space. Now, we shall show that each T_2-space is also a T_1-space. Suppose (X, T) is a T_2-space. Let $a \in X$ represent a given point. For each $x \in X, x \neq a$

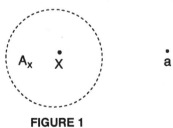

FIGURE 1

there is an open A_x, excluding the dotted circle in Figure 1, such that $x \in A_x$ and $a \notin A_x$. Thus

$$X - \{a\} = \bigcup_{x \neq a} A_x \tag{2}$$

and $X - \{a\}$ is an open set as a union of a family of open sets A_x.

Hence, $\{a\}$ is closed and X is a T_1-space.

505

Find a T_1–space which is not a T_2–space.

    Show that the properties of being a T_1– and a T_2–space are hereditary.

SOLUTION:

Consider the set X, consisting of 0 and all points $1/_n$, for $n = 1, 2, 3, \ldots$

$$X = \{0, 1, 1/_2, 1/_3, 1/_4, \ldots\} \tag{1}$$

We define the topology T on X: Sets containing 1 are open, if and only if they are complements of finite sets.

Sets which do not contain the point 1 are open, when they are open in the sense of the usual topology of real numbers.

Hence, each open set containing 0 is infinite.

Therefore, points 0 and 1 cannot be separated by two open disjoint sets. The space is T_1–space, but not T_2.

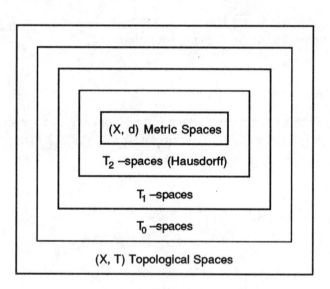

FIGURE 1

Show that the set X with the order topology is a T_2–space.

SOLUTION:

Let X represent any set totally ordered by $<$, $(X, <)$. Let S represent the family of subsets of X of the form

$$\{x : x < a\} \quad \text{or} \quad \{x : a < x\}$$

for all $a \in X$.

The family S forms a subbasis for a topology T on X, called the order topology induced by $<$. We shall show that (X, T) is a T_2-space.

Let $a, b \in X$ represent two distinct points. Since X is totally ordered, either $a < b$ or $b < a$. Suppose $a < b$. There are two possibilities now:

1. An element $c \in X$ exists, such that

$$a < c < b.$$

Then

$$\{x : x < c\} \quad \text{and} \quad \{x : c < x\}$$

are two disjoint neighborhoods of a and b respectively.

2. No $c \in X$ exists, such that

$$a < c < b.$$

Then

$$\{x : x < b\} \quad \text{and} \quad \{x : a < x\}$$

are disjoint neighborhoods of a and b respectively.

● PROBLEM 13-11

If

$$f : (X, T) \rightarrow (Y, T') \tag{1}$$

is onto and one-to-one, f^{-1} is continuous and X is a T_2-space, prove that Y is a T_2-space.

SOLUTION:

Let y_1 and y_2 represent any two distinct points of Y. Since f is one-to-one and onto, two distinct points of X exist, such that

$$x_1, x_2 \in X$$

$$f^{-1}(y_1) = x_1, \quad f^{-1}(y_2) = x_2.$$

(X, T) is a Hausdorff space, therefore, there are two open sets $U_1, U_2 \subset X$, such that

$$x_1 \in U_1, \quad x_2 \in U_2, \quad U_1 \cap U_2 = \phi.$$

Since f is bijective,

$$f(U_1) \subset Y, \quad f(U_2) \subset Y$$

$$f(U_1) \cap f(U_2) = \phi$$

Now, since f^{-1} is continuous, the function $(f^{-1})^{-1} = f$ maps open sets into open sets. Hence, $f(U_1), f(U_2) \in T'$ are open sets.

$$y_1 \in f(U_1), \quad y_2 \in f(U_2)$$

We conclude that (Y, T') is a T_2–space.

If two spaces are homeomorphic and one of them is a T_2–space, then so is the other.

● PROBLEM 13–12

Prove the theorems:

1. $\left(\begin{array}{c} (X,T) \text{ a Hausdorff space} \\ A \text{ a finite subset of } X \end{array} \right) \Rightarrow (A \text{ is closed}).$

2. $\left(\begin{array}{c} (X,T) \text{ a Hausdorff space} \\ A \subset X, x \text{ is a cluster point of } A \\ U \text{ a neighborhood of } X \end{array} \right) \Rightarrow (U \cap A \text{ is infinite}).$

SOLUTION:

1. Let $x \in X$. We shall show that $\{x\}$ is closed. Indeed, let

$$y \in X - \{x\}$$

then a neighborhood U of y exists, such that

$$x \in U.$$

Hence,

$$U \subset X - \{x\}$$

and the set $X - \{x\}$ is open. Thus, $\{x\}$ is closed.

Any subset $\{x_1, \ldots x_n\}$ of a Hausdorff space is closed.

2. Suppose, on the contrary, that $U \cap A$ is finite. Then

$$U \cap [A - \{x\}]$$

is closed. Hence

$$U - [U \cap (A - \{x\}]$$

is open.

But

$$x \in U - [U \cap (A - \{x\})] = U - (A - \{x\}).$$

Then $U - (A - \{x\})$ is a neighborhood of x.

Since x is a cluster point of A, then

$$\{[U - (A - \{x\})] - \{x\}\} \cap A \neq \phi.$$

Contradiction!

● **PROBLEM 13-13**

Show, that if (X, T) is a Hausdorff space, then every convergent sequence in X has a unique limit.

SOLUTION:

Let (x_n) denote a convergent sequence with two limits a, b, such that $a \neq b$.

Since (X, T) is a Hausdorff space, the open sets U_1 and U_2 exist, such that

$$a \in U_1, \quad b \in U_2, \quad U_1 \cap U_2 = \phi.$$

But (x_n) converges to a. Thus,

$$\exists k_1 \ \forall n > k_1 \ x_n \in U_1.$$

Similarly,

$$\exists k_2 \ \forall n > k_2 \ x_n \in U_2.$$

But the sets U_1 and U_2 are disjoint. Contradiction. Hence, $a = b$.

Give an example of a regular space, which is not a T_1-space.

SOLUTION:

DEFINITION OF REGULAR SPACE

A topological space (X, T), is said to be regular if, given any closed sub-set $F \subset X$ and any point $x \in X$, such that $x \notin F$, there are open sets U and V, such that

$$F \subset U, \quad x \in V, \quad \text{and} \quad U \cap V = \phi \tag{1}$$

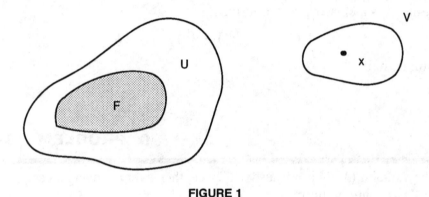

FIGURE 1

Consider the set $X = \{a, b, c\}$ with topology $T = \{\phi, X, \{a\}, \{b, c\}\}$. Note that the closed sets are $\phi, X, \{a\}, \{b, c\}$.

The topological space (X, T) is a regular space. But (X, T) is not a T_1-space. Set $\{c\}$ which is finite is not closed $\{c\} \neq \overline{\{c\}}$.

Let (Y, T_Y) denote the subspace of (X, T). Let $y \in Y$ and $A \subset Y$. Show, that if y does not belong to the T_Y-closure of A, then y does not belong to the T-closure of A.

SOLUTION:

By definition,

$$T_Y\text{-closure of } A = \overline{A} \cap Y \tag{1}$$

where $\overline{A}$ is the T–closure of A.

But

$$y \in Y. \qquad (2)$$

Therefore, if

$$y \notin \overline{A} \cap Y \qquad (3)$$

then

$$y \notin \overline{A}. \qquad (4)$$

Note, that an equivalent definition of regular space is: a topological space (X, T) is called regular if, for every point x and every closed set F, such that $x \notin F$, an open set G exists, such that

$$x \in G \quad \text{and} \quad \overline{G} \cap F = \phi.$$

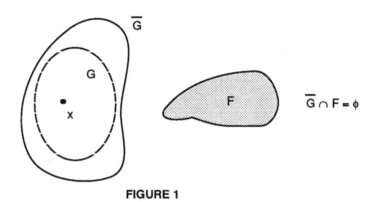

FIGURE 1

● PROBLEM 13–16

Show that every subset of a regular space is regular, i.e., the property of being a regular space is hereditary.

SOLUTION:

Let (X, T) denote a regular space and (A, T_A) its subspace. See Figure 1. Let $y \in A$ and let F represent a T_Y–closed subset of A, such that $y \notin F$. Hence, by Problem 13–15,

$$y \notin \overline{F}$$

where $\overline{F}$ is the T–closure of F. Space (X, T) is regular. Two open sets G and H exist, such that

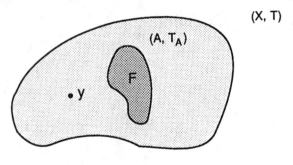

FIGURE 1

$$\overline{F} \subset G, \quad y \in H, G \cap H = \phi.$$

But

$$G \cap A \text{ and } H \cap A \text{ are } T_A\text{–open}$$

subsets of A.

   $y \in A \cap H$ because $y \in A$ and $y \in H$. Sets $A \cap G$ and $A \cap H$ are disjoint because $G \cap H = \phi$.

   Also, since

$$F \subset A \text{ and } F \subset \overline{F} \subset G \Rightarrow F \subset A \cap G.$$

Thus, (A, T_A) is also regular.

● **PROBLEM 13-17**

Show that a space (X, T) is regular, if and only if given any $x \in X$ and any neighborhood U of x, $x \in U \in T$, there is a neighborhood V of x, such that

$$\overline{V} \subset U.$$

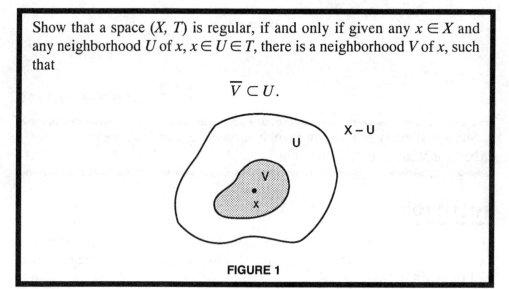

FIGURE 1

SOLUTION:

Suppose (X, T) is regular. Let U represent the neighborhood of x. Set $X - U$

is closed and $x \notin X - U$. Hence, open sets V and Z exist, such that

$$x \in V, \quad X - U \subset Z, \quad V \cap Z = \phi.$$

Since $X - U \subset Z$, we have $X - Z \subset U$. Also, since $V \cap Z = \phi$, we have

$$V \subset X - Z \subset U.$$

Hence,

$$V \subset \overline{V} \subset X - Z \subset U.$$

Now, suppose $x \in X$ and U are any neighborhood of x. Then, a neighborhood V of x exists, such that

$$\overline{V} \subset U.$$

Let $x \in X$ and F represent any closed subset of X, such that $x \notin F$.
    $X - F$ is a neighborhood of x. Then V exists, such that, V is open, $x \in V$

$$\overline{V} \subset X - F.$$

Set $X - \overline{V}$ is open and $F \subset X - \overline{V}$, also V is an open subset which contains x. Since $V \subset \overline{V}$

$$V \cap (X - \overline{V}) = \phi.$$

Thus, V and $X - \overline{V}$ are the sets that we are looking for.
    (X, T) is regular.

● PROBLEM 13-18

Show that every T_3–space is also a T_2–space.

SOLUTION:

DEFINITION OF A T_3–SPACE

A regular T_1–space is called a T_3–space.

■

Let (X, T) denote a T_3–space. We shall show that (X, T) is also a Hausdorff space.
    Let $a, b \in X$ represent distinct points. Space (X, T) is a T_1–space, therefore, $\{a\}$ is a closed set. Since a and b are distinct,

$$b \in \{a\}$$

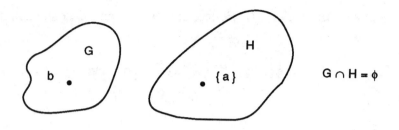

FIGURE 1

Since (X, T) is a regular space, two disjoint open sets G and H exist, such that

$$b \in G \quad \text{and} \quad \{a\} \subset H$$

Hence, a and b belong to two disjoint open sets G and H. (X, T) is a Hausdorff space (T_2-space).

● **PROBLEM 13–19**

Show that any metric space (X, d) is a T_3-space.

SOLUTION:

Let (X, d) denote a metric space and U represent any neighborhood of $x \in X$. There is a ball $B(x, r)$, such that

$$x \in B(x, r) \subset U. \tag{1}$$

Let us take any number r'

$$0 < r' < r. \tag{2}$$

Then

$$B(x, r') \subset B(x, r) \tag{3}$$

and

$$\overline{B(x, r')} = \{y : d(x, y) \leq r'\} \subset B(x, r) \subset U. \tag{4}$$

Hence, for any $x \in X$ and any neighborhood U of x, there is a neighborhood $B(x, r')$ of x, such that

$$\overline{B(x, r')} \subset U.$$

According to Problem 13–17, we conclude that (X, d) is regular. Since (X, d) is also T_1, any metric space is a T_3-space.

514

We shall define a topology on the set of real numbers R by giving an open neighborhood system. Let $x \neq 0$, then P_x is the family of all open intervals which contain x. For $x = 0$, we define P_0 as the family of all sets of the form $]- a, a[- \{1/_n\}$, where n is a positive integer. Show that (R, T) defined above is T_2, but not T_3.

SOLUTION:

First, we show that (R, T) is T_2. Let $x \in R$ and $y \in R$, $x \neq y$. If both x and y are different from 0, then

$$U =]x - \frac{|x - y|}{2}, x + \frac{|x - y|}{2}[$$

$$V =]y - \frac{|x - y|}{2}, y + \frac{|x - y|}{2}[$$

Bot U and V are open sets and

$$x \in U, \quad y \in V, \quad U \cap V = \phi.$$

If one point is 0, i.e., $x = 0$, then

$$U =]- \frac{|y|}{2}, \frac{|y|}{2}[- \{\frac{1}{n}\}.$$

Hence, (R, T) is a Hausdorff space (i.e., a T_2-space).

Now, we shall show that (R, T) is not T_3. Let $x = 0$ ad $F = \{1/_n\}$, where n is a positive integer. The set F is closed.

Let V represent any neighborhood of $x = 0$

$$V =]- a, a[- \{1/_n\}.$$

No open set U exists, such that $F \subseteq U$ and $V \cap U = \phi$.

Hence, (R, T) is not regular and, therefore, not T_3.

Show that the property of being T_3 is hereditary, that is, that any subspace of a T_3-space is a T_3-space.

SOLUTION:

Let Y denote a subspace of a T_3-space (X, T). Let F be closed in Y and

$x \in Y$ and $x \notin F$.

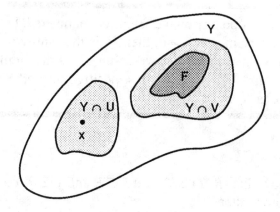

FIGURE 1

Since F is closed in Y,

$$F = Y \cap F'$$ (1)

where F' is a closed subset of X. Then $x \notin F'$. Space X is regular, hence, the open sets U and V in X exist, such that

$$x \in U, \quad F' \subset V, \quad U \cap V = \phi.$$ (2)

The sets $Y \cap U$ and $Y \cap V$ are open in Y and

$$x \in Y \cap U, \quad F \subset Y \cap V, \quad (Y \cap U) \cap (Y \cap V) = \phi.$$ (3)

Hence, space (Y, T_y) is regular. Since (Y, T_y) is a subspace of a T_1-space (X, T), it is also T_1. Therefore, (Y, T_y) is a regular and a T_1-space, i.e., a T_3-space.

● **PROBLEM 13-22**

Show that any space (X, T), containing more than one point with the indiscrete topology, is normal.

SOLUTION:

DEFINITION OF NORMAL SPACE

A topological space (X, T) is said to be normal if, given any two disjoint closed sets F_1 and F_2 in X, there are disjoint open sets U and V, such that

$$F_1 \subset U \quad \text{and} \quad F_2 \subset V$$

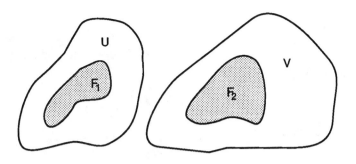

FIGURE 1

The indiscrete topology consists of two sets X and ϕ.

$$T = \{X, \phi\}.$$

Hence, the only closed sets are X and ϕ because $X - \phi = X$ and $X - X = \phi$.

Thus, there are no non-empty disjoint closed subsets of X. The space is normal.

It is easy to show that the space with discrete topology is normal. In this topology, each set is closed and open.

● **PROBLEM 13–23**

Prove this theorem:

THEOREM

A topological space (X, T) is normal iff, for every closed set F and every open set H containing F, an open set U exists, such that

$$F \subset U \subset \overline{U} \subset H. \tag{1}$$

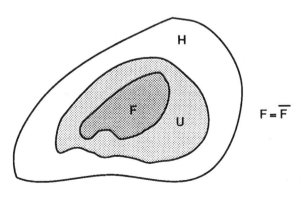

FIGURE 1

517

SOLUTION:

$\Rightarrow$ Let (X, T) denote a normal space. Let F represent a closed set and H an open set, such that

$$F \subset H \tag{2}$$

The set $X - H$ is closed and

$$F \cap (X - H) = \phi. \tag{3}$$

F and $X - H$ are two disjoint closed sets. Hence, the open sets U and U' exist, such that

$$F \subset U, \quad X - H \subset U', \quad U \cap U' = \phi. \tag{4}$$

Since

$$U \cap U' = \phi, \text{ we have } U \subset X - U' \tag{5}$$

also, since

$$X - H \subset U', \text{ we have } X - U' \subset H. \tag{6}$$

Set $X - U'$ is closed, hence, we conclude

$$F \subset U \subset \overline{U} \subset X - U' \subset H. \tag{7}$$

$\Leftarrow$ Let F_1 and F_2 denote disjoint closed sets. Then

$$F_1 \subset X - F_2 \tag{8}$$

and $X - F_2$ is open. An open set exists, such that

$$F_1 \subset U \subset \overline{U} \subset X - F_2. \tag{9}$$

But, since $\overline{U} \subset X - F_2$, we must have $F_2 \subset X - \overline{U}$. Also since $U \subset \overline{U}$, we have $U \cap (X - \overline{U}) = \phi$.

Thus, since $X - \overline{U}$ is open $F_1 \subset U$, $F_2 \subset X - \overline{U}$ and $U \cap (X - \overline{U}) = \phi$ where U and $X - \overline{U}$ are open sets.

● **PROBLEM 13-24**

Show that every metric space is normal.

SOLUTION:

Let (X, d) denote a metric space. Metric d induces the topology T on X. To show that (X, T) is normal, we shall apply the separation axiom.

THEOREM (SEPARATION AXIOM)

Let A_1 and A_2 represent closed disjoint subsets of a metric space (X, d). Then the disjoint open sets U_1 and U_2 exist, such that

$$A_1 \subset U_1 \quad \text{and} \quad A_2 \subset U_2$$

■

From this, we conclude that every metric space is normal.

● **PROBLEM 13-25**

Consider the set $X = \{a, b, c\}$ with the topology

$$T = \{\phi, X, \{a\}, \{b\}, \{a, b\}\}. \tag{1}$$

Is (X, T) a T_1–space, a regular space, or a normal space?

SOLUTION:

The closed sets are

$$X, \phi, \{b, c\}, \{a, c\}, \{c\} \tag{2}$$

Not every singleton subset of X is closed. For example,

$$\{a\} \neq \overline{\{a\}}. \tag{3}$$

Hence, the space (X, T) is not T_1. Also, (X, T) is not a regular space. Take a closed subset $\{c\}$ and $a \notin \{c\}$. Then, the only open set, which contains $\{c\}$ is the whole space X, which contains a. We shall show that (X, T) is a normal space.

Let F_1 and F_2 represent disjoint closed subsets of X. Then one of them, say F_1 is the empty set ϕ.

The sets ϕ and X are disjoint open sets and

$$\phi = F_1 \subset \phi; \quad F_2 \subset X.$$

Thus, (X, T) is a normal space.

● **PROBLEM 13-26**

Show that every T_4–space is also a T_3–space. See Figure 1.

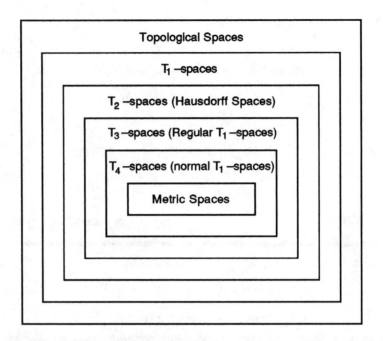

FIGURE 1

SOLUTION:

DEFINITION OF A T_4-SPACE

A normal space which is also a T_1-space is called a T_4-space. ∎

Let (X, T) denote a T_4-space. Hence, (X, T) is normal and T_1. Suppose F is a closed subset of X and $a \notin F$. Since (X, T) is T_1, the singleton set $\{a\}$ is closed. Sets F and $\{a\}$ are closed and disjoint. Since (X, T) is normal, the open sets U_1 and U_2 exist, such that

$$\{a\} \subset U_1, \quad F \subset U_2, \quad U_1 \cap U_2 = \phi.$$

Therefore (X, T) is regular and T_3.

The diagram illustrates the relationship between different kinds of topological spaces, defined in this chapter.

● **PROBLEM 13–27**

Prove that if Y is a closed subset of a T_4-space (X, T), then the subspace (Y, T_Y) is also T_4-space.

520

SOLUTION:

Since every subspace of a T_1-space is T_1 and (X, T) is T_1 also, Y is a T_1-space. Since Y is closed, a subset F of Y is closed in Y, if and only if F is closed in X. Hence, if F_1 and F_2 are disjoint closed subsets of Y, they are also disjoint closed subsets of X.

Thus, the open sets U_1 and U_2 exist, such that

$$F_1 \subset U_1, \quad F_2 \subset U_2 \text{ and } U_1 \cap U_2 = \phi.$$

Then

$$F_1 \subset U_1 \cap Y, \quad F_2 \subset U_2 \cap Y,$$

and $U_1 \cap Y$ and $U_2 \cap Y$ are disjoint subsets of Y, open in Y. Since (Y, T_Y) is T_1 and normal, it is T_4.

● **PROBLEM 13–28**

Suppose that the space (X, T) is homeomorphic to the space (Y, T'), and that X is T_4. Prove that Y is T_4.

SOLUTION:

We have already shown that if a space X is homeomorphic to a space Y and X is T_1, then Y is T_1.

Now, suppose X is a normal space and $f : X \to Y$ is a homeomorphism.

Let G_1 and G_2 represent any two disjoint closed subsets of Y. Then $f^{-1}(G_1)$

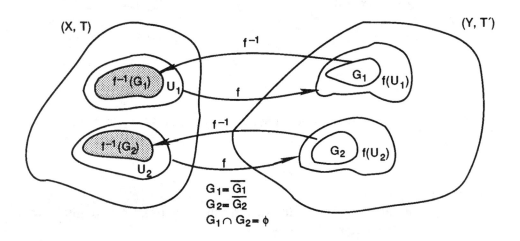

FIGURE 1

521

and $f^{-1}(G_2)$ are disjoint closed subsets of X. Since X is normal, there are two disjoint open sets U_1 and U_2 in X, such that

$$f^{-1}(G_1) \subset U_1, \quad f^{-1}(G_2) \subset U_2$$

The sets $f(U_1)$ and $f(U_2)$ are open in Y and

$$G_1 \subset f(U_1), \quad G_2 \subset f(U_2)$$

$$f(U_1) \cap f(U_2) = \phi.$$

Hence, (Y, T') is normal and T_4.

● PROBLEM 13-29

Describe the notion of the extension of a continuous function defined on a topological subspace.

SOLUTION:

The problem of extension of a function is one of the central and most difficult questions in all of topology. Suppose (X, T) is a topological space and (Y, T_Y) is its topological subspace.

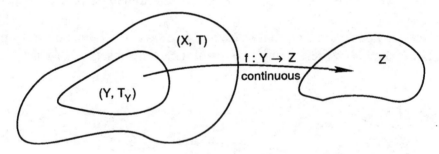

FIGURE 1

Let $f : Y \to Z$ denote a continuous function from Y into some space (Z, T'). The question is: does a continuous function $F : X \to Z$ exist, such that

$$F(y) = f(y) \text{ for each } y \in Y,$$

that is, $F \,|\, Y = f$?

In most cases the answer to this question is unknown. Only normal spaces have certain extension properties.

Here is one of the most important theorems in topology:

URYSOHN'S LEMMA

If a topological space (X, T) is normal, then, given any disjoint closed non-empty subsets A and B of X, there is a continuous function $f : X \to Z$, $Z = [0, 1]$, (Z has the absolute value topology), such that

$$\text{for every } a \in A, f(a) = 0$$

and $$\text{for every } b \in B, f(b) = 1.$$

∎

Prove that the converse of Urysohn's lemma is true.

SOLUTION:

Suppose (X, T) has the property described in Urysohn's lemma. Let A and B represent closed disjoint non-empty subsets of X. There is a continuous function $f : X \to [0, 1]$, such that for all $a \in A$, $f(a) = 0$ and for all $b \in B$, $f(b) = 1$.

Sets

$$U = \{x : 0 \le x < \tfrac{1}{2}\}$$

$$U' = \{x : \tfrac{1}{2} < x \le 1\}$$

are disjoint open subsets of $[0, 1]$. Since f is continuous, $f^{-1}(U)$ and $f^{-1}(U')$ are disjoint open subsets of X, such that

$$A \subset f^{-1}(U)$$

$$B \subset f^{-1}(U').$$

Hence, (X, T) is normal.

Let (X, T) denote a T_4–space which contains more than one point. Show that a non-constant continuous function $f : X \to [0, 1]$ exists.

SOLUTION:

Space X contains at least two points, for example, x and y. Since (X, T) is a T_4-space, it must be normal and T_1. For any T_1-space,

$$\{x\} = \{\bar{x}\}$$

$$\{y\} = \{\bar{y}\}$$

That is, any one-point subset is closed. The sets $\{x\}$ and $\{y\}$ are disjoint closed non-empty subsets of X.

Therefore, since X is normal, a continuous function

$$f : X \to [0, 1]$$

exists, such that

$$f(x) = 0 \quad \text{and} \quad f(y) = 1.$$

Function f is a continuous non-constant function from X into $[0, 1]$.

● **PROBLEM 13-32**

Prove this generalization of Urysohn's lemma:

Let A and B represent disjoint closed non-empty subsets of a normal space (X, T). Then a continuous function

$$f : X \to [a, b]$$

exists, such that

$$f(x) = a \quad \text{for all} \quad x \in A$$

$$f(x) = b \quad \text{for all} \quad x \in B.$$

SOLUTION:

Since (X, T) is a normal space, we can apply Urysohn's lemma (Problem 13–30).

Function f', which is continuous exists, such that

$$f' : X \to [0, 1]$$

and for all $x \in A$, $f(x) = 0$ and for all $x \in B$, $f(x) = 1$.

Consider function f, defined by

$$f(x) = a + (b - a) f'(x).$$

Since f' is continuous, so is f. Furthermore,

$$f(x) = \begin{cases} a & \text{for all } x \in A \\ b & \text{for all } x \in B \end{cases}.$$

19-7

● **PROBLEM 13-33**

Show that Urysohn's lemma is a special case of Tietze's extension theorem.

TIETZE'S EXTENSION THEOREM

Let (X, T) denote a T_4-space and Y any closed subset of X. If f is any continuous function $f : Y \to R$ (where R is the space of real numbers with the absolute value topology), then there is a continuous extension F of f, from X into R

$$F : X \to R.$$

■

SOLUTION:

Let (X, T) denote a T_4-space and let A and B represent disjoint closed non-empty subsets of X. We define

$$Y = A \cup B \quad \text{and} \quad f : Y \to R$$

by $f(x) = 0$ for all $x \in A$ and $f(x) = 1$ for all $x \in B$.

Function f is continuous. Hence, according to Tietze's theorem, it can be extended.

● **PROBLEM 13-34**

Let (X, T) denote any T_4-space and A represent any closed subset of X. Show that for any continuous function

$$f : A \to R^n$$

a continuous extension F of f exists, such that

$$F : X \to R^n$$

Note that R^n is an n-dimensional Cartesian product of R.

See Figure 1.

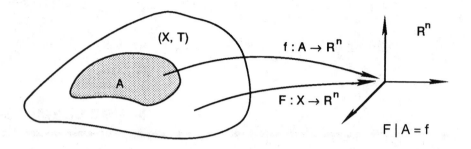

FIGURE 1

SOLUTION:

f is a continuous function of A into R^n. Thus, it can be written as

$$f(x) = (f_1(x), f_2(x), \ldots, f_n(x)).$$

Each of the functions $f_1, \ldots, f_n$ is continuous and maps A into R. According to Tietze's extension theorem, for each of the functions f_i, $i = 1, \ldots, n$, there exists a continuous extension F_i of f_i, $i = 1, 2, \ldots, n$, from X into R. We define

$$F(x) = (F_1(x), \ldots, F_n(x)).$$

Since all functions $F_1, F_2, \ldots, F_n$ are continuous, so is F. Function F is a continuous extension of f from X into R^n.

● **PROBLEM 13-35**

> Let R denote the space of real numbers with the absolute value topology and Z the set of integers, $Z \subset R$, with subspace topology. Let R^2 represent the plane with product topology and $Z^2 \subset R^2$ the set of all pairs (m, n), where m and n are integers, with subspace topology.
>
> Using this "scenario," prove the existence of a continuous function
>
> $$F : R \to R^2.$$

SOLUTION:

Consider any function

$$f : Z \to Z^2.$$

In both cases, the subspace topology is the discrete topology. Hence, any function from Z into Z^2 is continuous. Sets Z and Z^2 have the same cardinal number. It is possible to find function $f : Z \to Z^2$, which is one-to-one and

526

continuous.

According to Problem 13–34, f has a continuous extension F from R into R^2.

According to Problem 13–34,

● **PROBLEM 13–36**

Consider the class of real-valued functions

$$F = \{f_1(x) = \sin x, f_2(x) = \sin 2x, ..., f_n(x) = \sin nx, ...\}$$

defined in R. Show that F does not separate points of R.

SOLUTION:

DEFINITION

A class $F = \{f_\omega : \omega \in \Omega\}$ of functions from X into Y is said to separate points if, for any pair of distinct points $a, b \in X$, a function $f \in F$ exists, such that $f(a) \neq f(b)$.

■

Consider a pair of distinct points 0 and π.

$$\sin 0 = \sin 2 \cdot 0 = \sin 3 \cdot 0 = ... \qquad = 0$$

$$\sin \pi = \sin 2\pi = \sin 3\pi = ... \qquad = 0$$

Hence class F does not separate points.

● **PROBLEM 13–37**

Let $C(X, R)$ denote the class of all real-valued continuous functions, defined on a topological space (X, T). Show that, if the class $C(X, R)$ separates points, then (X, T) is a Hausdorff space.

SOLUTION:

Let $a, b \in X$ represent distinct points. Since $C(X, R)$ separates points, a continuous function

$$f : X \rightarrow R$$

527

exists, such that $f(a) \neq f(b)$.

There are two open disjoint subsets U_1 and U_2 of R, such that

$$f(a) \in U_1 \quad \text{and} \quad f(b) \in U_2.$$

Since f is a continuous function, sets $f^{-1}(U_1)$ and $f^{-1}(U_2)$ are open and disjoint.

$$a \in f^{-1}(U_1)$$

$$b \in f^{-1}(U_2).$$

Hence, (X, T) is a Hausdorff space (T_2-space).

● PROBLEM 13–38

Prove that a completely regular space is regular.

SOLUTION:

DEFINITION OF COMPLETELY REGULAR SPACE

A topological space (X, T) is completely regular if, for any closed subset F of X and any $a \in X$, such that $a \notin F$, a continuous function $f : X \rightarrow [0, 1]$ exists, such that for every

$$x \in F, \quad f(x) = 1 \quad \text{and} \quad f(a) = 0.$$

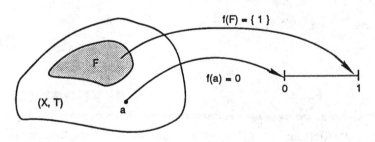

FIGURE 1

Now we show that completely regular space is also regular.

Let F represent a closed subset of X and $a \in X$ a point which does not belong to F.

By hypothesis, a continuous function

$$f : X \rightarrow [0, 1]$$

exists, such that $f(F) = \{1\}$ and $f(a) = 0$. An interval $[0, 1]$ is a Hausdorff space. Hence, two open disjoint subsets U_1 and U_2 of $[0, 1]$ exist, such that

$$0 \in U_1 \quad \text{and} \quad 1 \in U_2.$$

Since f is continuous, $f^{-1}(U_1)$ and $f^{-1}(U_2)$ are open. These subsets are disjoint and such that

$$a \in f^{-1}(U_1), \quad F \subset f^{-1}(U_2).$$

Hence, (X, T) is regular.

Show that the property of being a completely regular space is hereditary, i.e., every subspace of a completely regular space is completely
regular.

SOLUTION:

Let (X, T) denote a completely regular space and let (Y, T_Y) be its subspace.

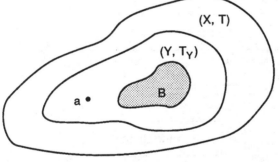

FIGURE 1

Let B be closed in Y and a represent a point in Y, such that $a \notin B$. Then

$$B = Y \cap D$$

where D is a closed set in X. Since $a \in Y$ and $a \notin B$, we have

$$a \notin D.$$

Since (X, T) is completely regular, a continuous function $f : X \rightarrow [0, 1]$ exists,
such that

$$f(a) = 0 \quad \text{and for each } x \in D \quad f(x) = 1.$$

Then $f \mid Y$ is a continuous function.

$$f \mid Y : Y \rightarrow [0, 1]$$

such that $f\,|\,Y(a) = 0$ and for each $x \in B, f\,|\,Y(x) = 1$.

Thus, (Y, T_y) is a completely regular space.

● **PROBLEM 13–40**

Consider the drawing of Problem 13–26. Where would you locate on this drawing the Tychonoff spaces?

SOLUTION:

DEFINITION OF TYCHONOFF SPACE

A completely regular T_1–space is called a Tychonoff space.

∎

In Problem 13–38, we proved that a completely regular space is also a regular space.

Hence, only a Tychonoff space is a T_3–space.

Any normal T_1–space (i.e., T_4–space) is a Tychonoff space by virtue of Urysohn's lemma.

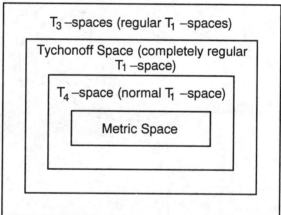

FIGURE 1

● **PROBLEM 13–41**

Prove the following:
THEOREM

The class $C(X, R)$ of all real-valued continuous functions on a completely regular T_1–space X separates points.

∎

530

SOLUTION:

Let a and b represent any distinct points in X. Since X is T_1, the set $\{a\}$ is closed. Points a and b are distinct, hence,

$$b \notin \{a\}.$$

Space (X, T) is completely regular. Hence, a real-valued continuous function f on X exists, such that

$$f(b) = 0 \quad \text{and} \quad f(a) = 1.$$

This function separates points a and b

$$f(a) \neq f(b).$$

CHAPTER 14

CARTESIAN PRODUCTS

Let (X_1, T_1) and (X_2, T_2) denote topological spaces.

DEFINITION OF OPEN SETS IN $X_1 \times X_2$

A set $A \subset X_1 \times X_2$ is called open in $X_1 \times X_2$ iff it is the union of Cartesian products $G \times H$, where $G \in T_1$, and $H \in T_2$ ∎

Prove the following:

THEOREM

The Cartesian product of two topological spaces is a topological space. ∎

SOLUTION:

From the definition, we conclude that the family of all sets $G \times H$ is a base of $X_1 \times X_2$, where $G \in T_1$ and $H \in T_2$.

From the identity

$$(G_1 \times H_1) \cap (G_2 \times H_2) = (G_1 \cap G_2) \times (H_1 \cap H_2) \tag{1}$$

we see that, if $G_1, G_2 \in T_1$ and $H_1, H_2 \in T_2$, then the intersection $(G_1 \times H_1) \cap (G_2 \times H_2)$ of open sets $G_1 \times H_1$ and $G_2 \times H_2$ is an open set.

Since

$$\left(\bigcup_\alpha G_\alpha\right) \times \left(\bigcup_\beta H_\beta\right) = \bigcup_{\alpha,\beta} (G_\alpha \times H_\beta) \tag{2}$$

we conclude that the union of an arbitrary family of open sets in $X_1 \times X_2$ is open.

Thus, the Cartesian product of two topological spaces is a topological space.

● PROBLEM 14-2

Show that in R^2 the usual topology agrees with the definition of Problem 14-1.

SOLUTION:

Let A denote any open set in R^2. It can be easily shown that the collection of open squares with the sides parallel to the x and y axes forms a basis for

RECTANGLES

R^2 space.

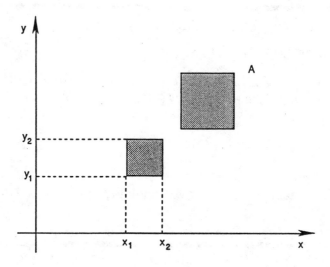

FIGURE 1

Thus

$$A = \bigcup_\alpha S_\alpha, \qquad (1)$$

Rectangle

where each S_α is an open square with the sides parallel to the x and y axes. Since the squares' boundaries are open as indicated by the dashed lines, the boundaries are not included in S_α.

Observe that each square can be represented as the Cartesian product of two open intervals

$$S_\alpha =]\, x_1, x_2\, [\, \times\,]\, y_1, y_2\, [\qquad (2)$$

Hence, the collection of sets of the type $]\, x_1, x_2\, [\, \times\,]\, y_1, y_2\, [$ forms a basis of R^2.

● **PROBLEM 14-3**

Prove that

THEOREM

If $\{A_\alpha\}$ is a base of X_1, and $\{B_\beta\}$ is a base of X_2, then $\{A_\alpha \times B_\beta\}$ is a base of $X_1 \times X_2$.

BUT NOT EVERY BASIS IS A CARTESIAN PRODUCT ■

SOLUTION:

Suppose $W \subset X_1 \times X_2$ is an open set in $X_1 \times X_2$. Then, by Problem 14-1,

W is the union of Cartesian products $E_\gamma \times F_\delta$ where E_γ and F_δ are open subsets of X_1 and X_2 respectively. Thus

$$W = \cup(E_\gamma \times F_\delta) \tag{1}$$

Since $\{A_\alpha\}$ is a base of X_1 and $\{B_\beta\}$ is a base of X_2, we have for each E_γ

$$E_\gamma = \cup A_{\alpha, \gamma} \tag{2}$$

and for each F_δ

$$F_\delta = \cup B_{\beta, \delta} \tag{3}$$

where $A_{\alpha, \gamma} \in \{A_\alpha\}$, $B_{\beta, \delta} \in \{B_\beta\}$. Hence

$$E_\gamma \times F_\delta = (\cup A_{\alpha, \gamma}) \times (\cup B_{\beta, \delta}) = \cup (A_{\alpha, \gamma} \times B_{\beta, \delta}). \tag{4}$$

Therefore

$$W = \cup(E_\gamma \times F_\delta) = \cup (A_\alpha \times B_\beta). \tag{5}$$

The set of $\{A_\alpha \times B_\beta\}$ is the base of $X_1 \times X_2$.

● **PROBLEM 14–4**

Show that

THEOREM

The projections

$$\pi_1 : X \times Y \rightarrow X; \quad \pi_1(x, y) = x \tag{1}$$

$$\pi_2 : X \times Y \rightarrow Y; \quad \pi_2(x, y) = y \tag{2}$$

are continuous mappings. ∎

SOLUTION:

Let A be an open subset of X. Then

$$\pi_1^{-1}(A) = A \times Y. \tag{3}$$

Since both A and Y are open sets, $\pi_1^{-1}(A)$ is an open set and π_1 is a continuous mapping. Similarly we show that π_2 is a continuous mapping, since for any open subset B of Y

$$\pi_2^{-1}(B) = X \times B \tag{4}$$

$X \times B$ is an open set.

This result holds for the generalized Cartesian product of any family of topological spaces. cf, 14-M/P.543

Let

$$f: T \rightarrow X \times Y \qquad (1)$$

where

$$f(t) = (f_1(t), f_2(t)). \qquad (2)$$

Show that f is continuous iff f_1 and f_2 are continuous. Also, f is continuous at t_0 iff f_1 and f_2 are continuous at t_0.

SOLUTION:

Suppose that f is continuous at t_0. Since

$$f_1(t) = \pi_1[f(t)] \qquad (3)$$

by Problem 14–4, we conclude that f_1 is continuous at t_0.

Now, suppose that f_1 and f_2 are continuous at t_0. Let $A \subset X \times Y$ be open and let

$$t_0 \in f^{-1}(A). \qquad (4)$$

We shall show that

$$t_0 \in Int \, f^{-1}(A). \qquad (5)$$

We can apply the following property:
If S is a subbase of Y and $f: X \rightarrow Y$ and if $f^{-1}(D)$ is open for each $D \in S$, then f is continuous.

Thus, we can assume that A belongs to a subbase of $X \times Y$. Setting

$$A = G \times Y \qquad (6)$$

we obtain

$$f^{-1}(A) = f_1^{-1}(G). \qquad (7)$$

Therefore

$$t_0 \in f_1^{-1}(G). \qquad (8)$$

But f_1 is continuous at t_0, thus

$$t_0 \in Int \, f_1^{-1}(G) = Int \, f^{-1}(A). \qquad (9)$$

Let

$$f : X \rightarrow Y$$

$$g : W \rightarrow Z \qquad (1)$$

where X, Y, W and Z are topological spaces.
   Show that the product mapping

$$h = f \times g : X \times W \rightarrow Y \times Z \qquad (2)$$

is continuous iff f and g are continuous.

SOLUTION:

We have

$$h(x, w) = (f(x), g(w)). \qquad (3)$$

Let w_0 represent a given element of W. Then

$$h_1(x) =: h(x, w_0) = (f(x), g(w_0)) \qquad (4)$$

and we obtain

$$f = \pi_1 \text{ o } h_1. \qquad (5)$$

We shall apply the following.

THEOREM

A continuous mapping of two variables is continuous with respect to each variable.

■

Hence, if h is continuous, then so is f. By the same argument g is continuous.
   Now, suppose that f and g are continuous. Then, by applying

$$h^{-1}(A \times B) = f^{-1}(A) \times g^{-1}(B) \qquad (6)$$

we conclude that h is continuous.

● **PROBLEM 14-7**

We shall investigate the properties which are invariants of Cartesian multiplication.

Prove:

THEOREM

The product of two closed sets is a closed set.

∎

SOLUTION:

Let $A \subset X$ be closed in X and $B \subset Y$ be closed in Y. We have

$$(X \times Y) - (A \times B) = [(X - A) \times Y] \cup [X \times (Y - B)]. \qquad (1)$$

Thus, the set $(X \times Y) - (A \times B)$ is the union of two open sets and is therefore an open set.

Similarly it can be shown that if both spaces are T_2–spaces, then their Cartesian product is a T_2–space.

Also regularity is invariant under Cartesian multiplication. But normality is not invariant under Cartesian multiplication.

● **PROBLEM 14–8**

Show that the Cartesian product of two T_2–spaces is a T_2–space.

SOLUTION:

Suppose that X and Y are T_2–spaces. Let $z_1 \in X \times Y$ and $z_2 \in X \times Y$, such that

$$z_1 \neq z_2 \qquad (1)$$

$$z_1 = (x_1, y_1) \qquad z_2 = (x_2, y_2) \qquad (2)$$

Since $z_1 \neq z_2$, we must have either

$$x_1 \neq x_2 \qquad (3)$$

or

$$y_1 \neq y_2 \qquad (4)$$

Suppose that $x_1 \neq x_2$. Since X is a T_2–space, two open sets A_1 and A_2 exist, such that,

$$A_1 \subset X, \ A_2 \subset X \qquad (5)$$

$$x_1 \in A_1, x_2 \in A_2 \tag{6}$$

$$A_1 \cap A_2 = \phi. \tag{7}$$

Therefore

$$z_1 \in A_1 \times Y \quad z_2 \in A_2 \times Y. \tag{8}$$

The sets $A_1 \times Y$ and $A_2 \times Y$ are open and

$$(A_1 \times Y) \cap (A_2 \times Y) = \phi. \tag{9}$$

Hence the set $X \times Y$ is a T_2–space.

● PROBLEM 14-9

Show that the diagonal $D \subset X \times X$ is homeomorphic to X.

SOLUTION:

DEFINITION OF A DIAGONAL

The set

$$D = \{(x, y) : x = y\} \tag{1}$$

is the diagonal of

$$X^2 = X \times X$$

■

Consider the projection

$$\pi_1 : X \times X \to X$$

$$\pi_1 (x, y) = X \tag{2}$$

In Problem 14–4 we proved that the projection is a continuous mapping. Hence, the required homeomorphism is the projection π_1.

$$X \simeq D. \tag{3}$$

Prove the following:

THEOREM

If X is a T_2-space, then the diagonal is closed in $X \times X = X^2$. ∎

SOLUTION:

We shall show that $X^2 - D$ is an open set. That is, we have to show that for any point

$$(x, y) \in X^2 - D \qquad (1)$$

there are two open sets A and B, such that

$$x \in A, \quad y \in B \qquad (2)$$

and

$$A \times B \subset X^2 - D. \qquad (3)$$

Since X is a T_2-space and $x \neq y$, there are two open sets A and B, such that

$$x \in A \quad \text{and} \quad y \in B \qquad (4)$$

and

$$A \cap B = \phi. \qquad (5)$$

Therefore

$$(A \times B) \cap D = \phi \qquad (6)$$

that is

$$A \times B \subset X^2 - D. \qquad (7)$$

Note that the following is also true: if the diagonal of $X \times X$ is closed, then X is a T_2-space.

1. Show that if

$$f : X \to Y \qquad (1)$$

is continuous, then the graph of f

$$G = \{(x, y) : y = f(x)\} \tag{2}$$

is homeomorphic to X.

2. Also show that if Y is a T_2–space, then G is closed in $X \times Y$.

SOLUTION:

1. Consider the mapping

$$h(x) = (x, f(x)). \tag{3}$$

$h(x)$ is a homeomorphism of X onto G. Thus $X \cong G$.

2. Let us define a mapping

$$p(x, y) = (f(x), y). \tag{4}$$

Then

$$[p(x, y) \in D] \equiv [f(x) = y] \equiv [(x, y) \in G] \tag{5}$$

Thus

$$G = p^{-1}(D). \tag{6}$$

Since p is continuous and D is closed, we conclude that G is closed.

● **PROBLEM 14–12**

Construct the subbase of the generalized Cartesian product.

SOLUTION:

Consider the family of topological spaces $\{X_\omega : \omega \in \Omega\}$. The generalized Cartesian product is defined by

$$Z = \underset{\omega \in \Omega}{\times} X_\omega \tag{1}$$

The αth coordinate of $f \in Z$, $\pi_\alpha(f)$ is defined by

$$\pi_\alpha(f) = f(\alpha) \tag{2}$$

then

$$\pi_\alpha : Z \to X_\alpha \tag{3}$$

is the projection of Z on X_α.

The topology in Z (called Tychonow topology) is defined as follows:

DEFINITION

A subbase of Z is defined as the family of sets of the form

$$H_{\alpha,G} = \pi_\alpha^{-1}(G) = \{f : f(\alpha) \in G\} \tag{4}$$

where G is open in X_α.

∎

Observe that $H_{\alpha,G}$ is the product of G and of all spaces X_β, where $\beta \neq \alpha$.

● **PROBLEM 14-13**

Show that if $A_\omega \subset X_\omega$ for each $\omega \in \Omega$, then

$$\underset{\omega}{\times} A_\omega = \underset{\omega}{\times} \overline{A}_\omega. \tag{1}$$

That is, the Cartesian product of closed sets is always closed.

SOLUTION:

Let $(x_\omega) \in \underset{\omega}{\times} X_\omega$ be, such that

$$(x_\omega) \in \overline{\underset{\omega}{\times} A_\omega}. \tag{2}$$

We shall show that for every ω

$$x_\omega \in \overline{A}_\omega \tag{3}$$

and hence

$$(x_\omega) \in \underset{\omega}{\times} \overline{A}_\omega. \tag{4}$$

Let $x_\omega \in U_\omega$, where U_ω is open in X_ω then

$$(x_\omega) \in\, < U_\omega >. \tag{5}$$

By $< U_\alpha >$ we denote the "slice" of $\underset{\omega}{\times} X_\omega$ where each factor is X_ω except the αth, which is U_α. We have

$$\phi \neq \underset{\omega}{\times} A_\omega \cap < U_\omega > = (U_\omega \cap A_\omega) \times [X\{A_\beta : \beta \neq \omega\}].$$

Hence

$$U_\omega \cap A_\omega \neq \phi \quad \text{and} \quad x_\omega \in \overline{A}_\omega.$$

The converse inclusion is proved in the same manner by reversing the steps.

Let $\{X_\omega : \omega \in \Omega\}$ represent any family of topological spaces. Show that for each fixed $\beta \in \Omega$,

$$\pi_\beta : \underset{\omega}{\times} X_\omega \to X_\beta$$

is continuous and open.

$cf\ 14\text{-}4$
$p\ 535\text{-}6$

SOLUTION:

Let U denote an open set in X_β. Then

$$\pi_\beta^{-1}(U) = \ < U >$$

is open in $\underset{\omega}{\times} X_\omega$. Hence π_β is continuous.
Since

$SLICE\ (14\text{-}13)$

$$\pi_\beta < U_{\omega_1}, \ldots, U_{\omega_n} > = \begin{cases} X_\beta & \text{if } \beta \neq \omega_1, \ldots, \omega_n \\ U_\beta & \text{if } \beta = \omega_k \end{cases}$$

the image of any basic open set is open. Hence π_β is open.

Let $\{X_\omega : \omega \in \Omega\}$ denote a family of topological spaces and let

$$f : Y \to \underset{\omega}{\times} X_\omega . \qquad (1)$$

Prove that

(f is continuous) $\Leftrightarrow$ ($\pi_\omega \circ f$ is continuous for each $\omega \in \Omega$).

SOLUTION:

$\Rightarrow$ Suppose f is continuous. From 14–14 each π_ω is continuous. Hence

$$\pi_\omega \circ f$$

is continuous for each $\omega \in \Omega$.

$\Leftarrow$ Suppose $\pi_\omega \circ f$ is continuous. Then for each element $<U_\alpha>$ of subbasis in $\underset{\omega}{\times} X_\omega$, we have

$$f^{-1} < U_\alpha > = f^{-1}[\pi_\alpha^{-1}(U_\alpha)] = (\pi_\alpha \circ f)^{-1}(U_\alpha).$$

$SLICE$
$< \ >$
$cf\ 14\text{-}13$

Hence $x<U_\alpha>$ is open and f is continuous.

■

For any

$$f : Y \to \underset{\omega}{\times} X_\omega$$

the map

$$\pi_\alpha \circ f : Y \to X_\alpha$$

is called the αth coordinate function of f.

● **PROBLEM 14-16**

Let (X_n, d_n) denote a finite or infinite sequence of metric spaces. Define the metric of the product

$$\underset{k=1}{\overset{n}{\times}} X_k \text{ and } \underset{n=1}{\overset{\infty}{\times}} X_n.$$

SOLUTION:

Let $X_1, X_2, \ldots, X_n$ denote a finite sequence of metric spaces. Let x and y denote points in $\underset{k=1}{\overset{n}{\times}} X_k$. Then

$$x = (x_1, x_2, \ldots, x_n)$$

$$y = (y_1, y_2, \ldots, y_n). \tag{1}$$

The distance between x and y can be defined as follows:

$$d(x,y) = \sqrt{\sum_{k=1}^{n} d_k(x_k, y_k)}. \tag{2}$$

It is easy to show that (2) defines a metric in the Cartesian product $X_1 \times X_2 \times \ldots \times X_n$.

Consider the infinite Cartesian product $X_1 \times X_2 \times \ldots \times X_n \times \ldots$. The distance between points

$$x = (x_1, x_2, \ldots) \quad \text{and} \quad y = (y_1, y_2, \ldots)$$

can be defined by

$$d(x,y) = \sum_{n=1}^{\infty} \frac{1}{2^n} d_n(x_n, y_n). \tag{3}$$

Note that d_n is the metric in X_n.

In Problem 14–16, we defined the product $\times X_n$ of metric spaces and the distance. Hence $\left(\underset{n=1}{\overset{\infty}{\times}} X_n, d \right)$ is a metric space.

Metric d induces topology T_d on $\underset{n=1}{\overset{\infty}{\times}} X_n$. In Problem 14–12 we defined the Tychonow topology, T, as $\underset{n=1}{\overset{\infty}{\times}} X_n$. What is the relation between these topologies? (i.e., which one is weaker, or are they the same).

SOLUTION:

The following theorem answers this question.

THEOREM

A set $A \subset X_1 \times X_2 \times \ldots$ is open in the metric sense if, and only if, A is open in the Tychonow topology.

∎

Hence the topology induced by the metric agrees with the Tychonow topology.

CHAPTER 15

COUNTABILITY PROPERTIES

Let (R, T) denote the set of real numbers with topology induced by the absolute value metric. Show that

$$\{B(x, 2) : x \in R\}$$

i.e., the collection of open balls

$$B(x, 2) = \{y : y \in R, |x - y| < 2\}$$

is an open cover of R.

Give an example of an open subcover and an improvement of $\{B(x, 2) : x \in R\}$.

SOLUTION:

DEFINITION OF OPEN COVER

An open cover of a topological space (X, T) is a collection

$$\{A_\alpha : \omega \in \Omega, A_\alpha \in T\}$$

of open subsets of X, such that

$$X = \bigcup_\alpha A_\alpha.$$

■

$\{B_\beta\}$ is an open subcover of $\{A_\alpha\}$, if

$$\{B_\beta\} \subset \{A_\alpha\}.$$

The set $\{B(x, 2) : x \in R\}$ is an open cover of R because

$$R = \bigcup_x B(x, 2).$$

The set

$$\{B(n, 2) : n \text{ is an integer}\}$$

is an open subcover of R.

An open cover $\{D_\gamma\}$ is said to be an improvement of $\{A_\alpha\}$, if for each D_γ, there is A_α such that

$$D_\gamma \subset A_\alpha.$$

Hence, we can choose

$$\{B(x, 1) : x \in R\}$$

as an improvement of $\{B(x, 2) : x \in R\}$.

Show that every open cover of a closed and bounded interval $A = [a, b]$ is reducible to a finite cover.

SOLUTION:

We shall apply Heine-Borel Theorem:

If $A = [a, b]$ is a closed and bounded interval and $\{U_n\}$ is a class of open sets, such that

$$A \subset \bigcup_n U_n,$$

then one can choose a finite number of open sets $U_{n_1}, U_{n_2}, \ldots, U_{n_k}$, such that

$$A \subset U_{n_1} \cup U_{n_2} \cup \ldots \cup U_{n_k}.$$

The conclusion from the above theorem is that every open cover of interval $[a, b]$ is reducible to a finite cover.

Show that every metric space is a first countable space.

SOLUTION:

DEFINITION OF FIRST COUNTABLE SPACE

A topological space (X, T) is called a first countable space if, for every point $a \in X$, a countable class B_a of open sets containing a exists, such that every open set A containing a also contains a member of B_a. ■

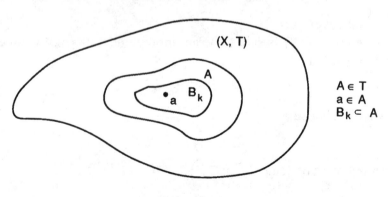

$A \in T$
$a \in A$
$B_k \subset A$

FIGURE 1

548

That is, a topological space (X, T) is a first countable space iff at every point $a \in X$, a countable local base exists.

Let (X, d) denote a metric space and let $a \in X$ be any point. The set of balls

$$\{B(a, 1), B(a, {}^1/_2), B(a, {}^1/_3), B(a, {}^1/_4), ...\}$$

forms a countable local base at $a \in X$.

For every open set A containing a, a ball $B(a, {}^1/_k)$ exists such that

$$a \in B(a, {}^1/_k) \subset A.$$

Hence every metric space is a first countable space.

● PROBLEM 15-4

Show that any subspace (Y, T_Y) of a first countable space (X, T) is also first countable.

SOLUTION:

Let (X, T) be a first countable space and (Y, T_Y) its subspace. Let $a \in Y$.

By hypothesis (X, T) is a first countable space. Hence a countable base B_a in (X, T) exists.

$$B_a = \{B_k : k \in N\}$$

Let us define for each B_k $B'_k = Y \cap B_k$.

Sets B'_k are open in (Y, T_Y) they form a T_Y–local base at $a \in Y$.

The set $\{B'_k : k \in N\}$ is countable. Hence (Y, T_Y) is a first countable space.

● PROBLEM 15-5

Let

$$A_n = \{A_1, A_2, A_3, ...\}$$

be a nested local base at $a \in X$. Let $(a_1, a_2, ...)$ denote a sequence such that

$$a_1 \in A_1, a_2 \in A_2, ..., a_k \in A_k, ...$$

Show that (a_n) converges to a.

549

SOLUTION:

Let U denote an open set containing a.

Set $\{A_1, A_2, \ldots\}$ is a local base at a. Hence, set A_k exists, such that

$$A_k \subset U.$$

Sets $A_1, A_2, A_3, \ldots$ are nested if

$$A_1 \supset A_2 \supset A_3 \supset A_4 \supset \ldots$$

Hence for any $m > k$ we have

$$A_k \supset A_m$$

and since $a_m \in A_m$ we obtain

$$a_m \in A_k \subset U.$$

Thus

$$a_n \to a.$$

● **PROBLEM 15-6**

Show that the property of being a first countable space is a topological property.

SOLUTION:

Suppose topological spaces (X, T) and (Y, T_1) are homeomorphic

$$X \simeq Y.$$

We shall show that if (X, T) is a first countable space, then (Y, T_1) is also a first countable space.

Let

$$f : X \to Y$$

be a homeomorphism and let (X, T) be a first countable space. Let $y \in Y$ and V be an open subset of Y, such that $y \in V$. (See Figure 1).

Since f is a homeomorphism, $f^{-1}(y)$ is a point in X and $f^{-1}(V)$ is an open set in X such that

$$f^{-1}(y) \in f^{-1}(V).$$

(X, T) is a first countable space, hence $x = f^{-1}(y)$ has a countable local base B_x. An open set $B \in B_x$ exists, such that

550

$$x = f^{-1}(y) \in B \subset f^{-1}(V).$$

Thus

$$y \in f(B) \subset V$$

where $f(B)$ is open in (Y, T').

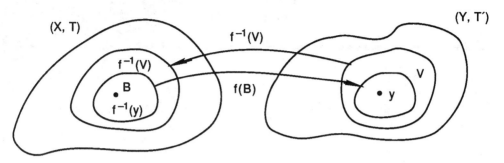

FIGURE 1

● **PROBLEM 15-7**

Show that if

$$A_x = \{A_1, A_2, A_3, \ldots\}$$

is a countable local base at $x \in X$, then a nested local base exists at x.

SOLUTION:

Let

$$B_1 = A_1, B_2 = A_1 \cap A_2, \ldots, B_n = A_1 \cap \ldots \cap A_n.$$

Each of the sets $B_1, B_2, \ldots$ is open and contains x. Also

$$B_1 \supset B_2 \supset B_3 \supset \ldots$$

Now, let U be an open set containing x, then $k \in N$ exists, such that

$$B_k \subset A_k \subset U.$$

Hence

$$\{B_1, B_2, B_3, \ldots\}$$

is a nested local base at $x \in X$.

● PROBLEM 15-8

> Prove this theorem:
>
> **THEOREM**
>
> A function defined on a first countable space (X, T) is continuous at $x \in X$ iff it is sequentially continuous at x.   ∎

SOLUTION:

We shall show that if (X, T) is a first countable space, then

$$f : X \to Y$$

is continuous at $x \in X$, if and only if, for every sequence (a_n) converging to $x \in X$, the sequence $(f(a_n))$ converges to $f(x)$. It has been shown that if f is continuous, then f is sequentially continuous. It remains to be shown that

(f is sequentially continuous) $\Rightarrow$ (f is continuous)

or equivalent

(f is not continuous) $\Rightarrow$ (f is not sequentially continuous).

Let $\{A_1, A_2, \ldots\}$ be a nested local base at $x \in X$.

Suppose f is not continuous. Then, an open set $V \subset Y$ exists, such that

$$f(x) \in V, \quad \text{and for every } n \in N, \; A_n \not\subset f^{-1}(V).$$

Thus, for every $n \in N$, an $a_n \in A_n$ exists, such that $a_n \notin f^{-1}(V)$ and therefore $f(a_n) \notin V$.

According to Problem 15–5, (a_n) converges to x, but

$$f(a_n) \not\to f(x).$$

● PROBLEM 15-9

> Show that the set of real numbers R with the cofinite topology is not a first countable space.

SOLUTION:

Cofinite topology T contains ϕ and the complements of finite sets. Suppose (R, T) is a first countable space. Let $A(1) = \{A_1, A_2, A_3, \ldots\}$ be a countable open local base at $1 \in R$. Each A_n is T–open, hence its complement $R - A_n$ is closed and hence finite.

The set

$$A = \bigcup_n (R - A_n)$$

is the countable union of finite sets. Thus A is countable and R is not countable. A point $a \in R$ exists, such that

$$a \neq 1, \quad a \notin A.$$

We have

$$a \in R - A = R - (\bigcup_n (R - A_n)) = \bigcap_n A_n.$$

Hence $a \in A_n$ for every $n \in N$.

The set $R - \{a\}$ is a T–open as a complement of a finite set.

$$1 \in R - \{a\}$$

because $a \neq 1$.

$A(1)$ is a local base at $1 \in R$. $A_k \in A(1)$ exists, such that

$$A_k \subset R - \{a\}.$$

Hence $a \notin A_k$.

That contradicts the fact that for every $n \in N$, $a \in A_n$.

● **PROBLEM 15–10**

Show that R^2 with the usual topology is a second countable space.

SOLUTION:

DEFINITION OF A SECOND COUNTABLE SPACE

A topological space (X, T) is called a second countable space, if a countable basis B exists for the topology T.

■

Let us define B as a class of open balls in R^2 with rational radii and centers, whose both coordinates are rational numbers. The Cartesian product of three countable sets is a countable set. Hence, B is a countable set. It is easy to see that B is a basis for the usual topology on R^2.

Hence (R^2, T) is a second countable space.

1. Show that (R, T), where R is the set of real numbers and T is the discrete topology, is not the second countable space.

2. Show that: Any second countable space is first countable.

SOLUTION:

1. It is easy to see that if T is a discrete topology, then its base contains all singleton sets.

Set R, and hence, the class of singleton subsets $\{a\}$ of R, is not countable. R with the discrete topology T is not the second countable space.

2. Suppose (X, T) is the second countable space.

Hence, B is a countable base for (X, T). Let B_x consist of all members of B which contain $x \in X$. Then B_x is a countable local base at $x \in X$. Hence, (X, T) is a first countable space.

Show that:

1. Every subspace of a second countable space is second countable.

2. Any second countable space is separable.

SOLUTION:

1. Let B represent a countable base for the second countable space (X, T).

$$B = \{B_n : n \in N\}$$

Then, for any subspace A

$$B_A = \{A \cap B_n : n \in N\}$$

is a countable base for A. Thus, (A, T_A) is a second countable space.

2. **DEFINITION OF SEPARABLE SPACE**

X is said to be separable if X contains a countable dense subset.

■

Let (X, T) denote a second countable space with a countable base B. For each $B_n : n \in N$ select $x(B_n) \in B_n$.

The set

$$\{x(B_n) : B_n \in B\}$$

is a countable subset of X, which is dense.

Hence, (X, T) is separable.

● **PROBLEM 15-13**

Prove the following:

THEOREM (Lindelöf)

If A is a subset of a second countable space (X, T), then every open cover of A is reducible to a countable cover. ∎

SOLUTION:

Let

$$B = \{B_n : n \in N\}$$

denote a countable base for X and let $H = \{H_\alpha : \alpha \in \Omega\}$ denote an open cover of A.

$$A \subset \cup H_\alpha.$$

Hence, for every $x \in A$, $H_\alpha \in H$ exists, such that $x \in H_\alpha$. Since B is a base for X, for every $x \in A$, $B_x \in B$ exists, such that

$$x \in B_x \subset H_\alpha.$$

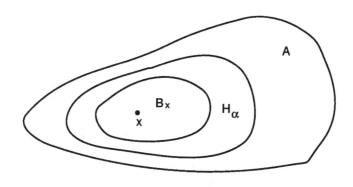

FIGURE 1

555

Thus

$$A \subset \cup \{B_x : X \in A\}$$

The family of sets $\{B_x : x \in A\}$ is a subset of B, hence it is countable

$$\{B_x : x \in A\} = \{B_k : k \in K\}$$

where K is a countable index set. For each $k \in K$, we can choose one set $H_k \in H$, such that

$$B_k \subset H_k.$$

Thus

$$A \subset \bigcup_k B_k \subset \bigcup_k H_k$$

and $\{H_k : k \in K\}$ is a countable subcover of H.

● PROBLEM 15–14

Show that every second countable space is a Lindelöf space.

SOLUTION:

DEFINITION OF A LINDELÖF SPACE

A topological space (X, T) is called a Lindelöf space, if every open cover of X is reducible to a countable cover.

■

We shall prove the following:
THEOREM

If (X, T) is a second countable space, then every base B for X is reducible to a countable base for X.

■

Let (X, T) represent a second countable space. Then X has a countable base

$$B = \{B_n : n \in N\}.$$

Let $H = \{H_\alpha : \alpha \in \Omega\}$ represent any base for X. Then for each $n \in N$

$$B_n = U\{H_\alpha : H_\alpha \in H_n\}$$

where $H_n \subset H$. Hence H_n is an open cover of B_n.

According to Problem 15–13, H_n is reducible to a countable subcover

H'_n.

$$B_n = U\{H_\alpha : H_\alpha \in H'_n\}, \quad H'_n \subset H_n$$

and H'_n is countable.

Note that

$$H' = \{H_\alpha : H_\alpha \in H'_n, n \in N\}$$

is a base for X, since B is a base. Also

$$H' \subset H$$

and H' are countable.

● PROBLEM 15-15

Show that every closed subspace of a Lindelöf space is Lindelöf.

SOLUTION:

Suppose A is a closed subspace of a Lindelöf space (X, T) and $\{H_k\}$ is an open cover of A, $k \in K$. Each H_k is open in A,

$$H_k = A \cap V_k$$

where V_k is open in X.

Since A is closed, $X - A$ is open in X. Hence,

$$\{X - A\} \cup \{V_k : k \in K\}$$

is an open cover of X.

There is a countable subcover

$$\{X - A\} \cup \{V_{k_n} : n = 1, 2, 3, ...\}$$

of

$$\{X - A\} \cup \{V_k : k \in K\}.$$

Then

$$\{V_{k_n} : n = 1, 2, 3, ...\}$$

is a countable open cover of A. Therefore

$$\{A \cap V_{k_n} : n = 1, 2, 3, ...\}$$

is a countable open subcover of $\{H_k : k \in K\}$. Hence, A is a Lindelöf space.

Show that a discrete space X is Lindelöf, if and only if, X is a countable set.

SOLUTION:

Let (X, T) denote a discrete topological space, i.e., T consists of all the subsets of X.

Suppose (X, T) is Lindelöf. Then every open cover of X has a countable subcover. Each singleton set $\{x\}$, $x \in X$ is open in discrete topology. The family of sets

$$\{\{x\} : x \in X\}$$

is an open cover of X

$$X = \bigcup_{x \in X} \{x\}.$$

Since X is Lindelöf, a countable subcover exists

$$\{x_1\}, \{x_2\}, \{x_3\}, \dots.$$

Hence, X is a countable set.

Conversely, if X is a countable set and (X, T) is a discrete space, then X is Lindelöf. Let

$$A = \{A_\alpha : \alpha \in \Omega\}$$

represent an open cover of X.

$$X \subset \bigcup_{\alpha} A_\alpha.$$

For each $x \in X$, $A_x \in A$ exists, such that

$$x \in A_x$$

since X is countable

$$X = \{x_1, x_2, \dots\}$$

we can choose

$$x_1 \in A_1, x_2 \in A_2, \dots$$

Hence, $A_1, A_2, A_3, \dots$ is an open subcover of A. Thus, (X, T) is Lindelöf.

Show that:

1. The real line R with the usual topology is a separable set.

2. The real line R with the discrete topology is not a separable set.

SOLUTION:

1. Let Q denote the set of rational numbers. We have showed that Q is a countable set. Since Q is dense in R, the set of real numbers R with the usual topology is separable.

2. Let T denote the discrete topology on R. Then every subset of R is both closed and open. Therefore, the only dense set in R is R itself. Set R is not countable. Hence, R with the discrete topology is not a separable space.

● PROBLEM 15-18

Show that if a topological space (X, T) is a second countable space, then (X, T) is separable.

SOLUTION:

Suppose (X, T) is a second countable space. The family of sets

$$B = \{B_n : n \in N\}$$

is a countable base of X. For each $n \in N$, we choose a point a_n, such that

$$a_n \in B_n.$$

The set

$$A = \{a_n : n \in N\}$$

is countable.

We shall show that each point

$$x \in X - A$$

is an accumulation point of A, that is, that $\overline{A} = X$.

Let $X \in X - A$ and let U represent an open set containing x. At least one set $B_k \in B$ exists, such that

559

$$a_k \in B_k \subset U.$$

Since $a_k \in A$ and $x \in X-A$,

$$a_k \neq x.$$

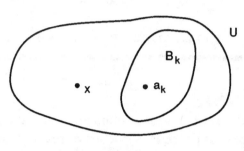

FIGURE 1

We conclude that x is an accumulation point of A, since every open set U containing x also contains a point of A different from x.

● **PROBLEM 15-19**

1. Show that a discrete space X is separable, if and only if, X is countable.

2. Show that (X, T) with the cofinite topology T is separable.

SOLUTION:

1. Let (X, T) denote a discrete space. Then every subset of X is both open and closed. Therefore, the only dense subset of X is X itself. Hence, X contains a countable dense subset, if and only if, X is countable. Thus, a discrete space X is separable iff X is countable.

2. Suppose X is countable. Then, X is a countable dense subset of (X, T).

Now, suppose, X is not countable. Then X contains non-finite, countable, subset A.

In the cofinite topology, the only closed sets are the finite sets and X. Hence, the closure of A is the entire space X, that is

$$\overline{A} = X.$$

Since A is countable, (X, T) is separable.

Show that (R^2, T) is separable, where R^2 is the plane and T is the topology generated by the half-open rectangles.

$$]a, b] \times]c, d] = \{(x, y) : a < x \leq b, c < y \leq d\}$$

SOLUTION:

For any rectangle $]a, b] \times]c, d]$ a point (x_0, y_0) exists with rational coordinates, such that

$$(x_0, y_0) \in]a, b] \times]c, d] .$$

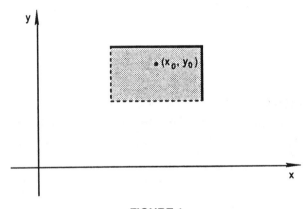

FIGURE 1

Consider the set $A = Q \times Q$ consisting of all points in R^2 with rational coordinates.

Since Q is countable, so is A. The set A is dense in R^2. Hence, (R^2, T) is separable.

Show that any open subspace of a separable space is separable.

SOLUTION:

Let (X, T) represent a separable space and let

$$\{x_n : n \in N\}$$

represent a countable dense subset of (X, T).

We shall show that any open subspace U of X is separable. We define

$$B = U \cap \{x_n : n \in N\}.$$

Let V represent any subset of U which is open in U. Hence V is open in X, since U is open in X. Therefore, V contains some x_n, and hence, a point of B.

We shall apply the following theorem:

THEOREM

A subset A of a topological space (X, T) is dense, if and only if, every open subset of X contains some point of A.

∎

Therefore, B is dense in U. Set U contains a countable dense subset B, and hence, U is separable.

● **PROBLEM 15–22**

Let $B(x, r)$ represent an open sphere in a metric space (X, d), and let

$$d(x, y) < \tfrac{1}{3} r. \tag{1}$$

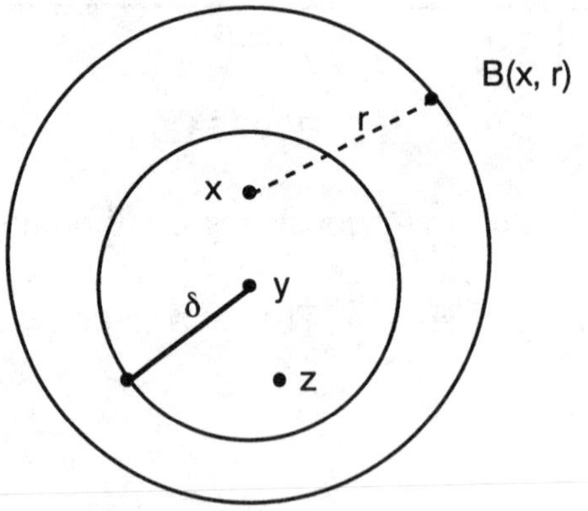

FIGURE 1

Show that if $\tfrac{1}{3} r < \delta < \tfrac{2}{3} r$, then

$$x \in B(y, \delta) \subset B(x, r). \tag{2}$$

SOLUTION:

From (1) and condition $\tfrac{1}{3} r < \delta$ we obtain

$$d(x, y) < \tfrac{1}{3}\, r < \delta. \tag{3}$$

Hence

$$x \in B(y, \delta). \tag{4}$$

Suppose

$$z \in B(y, \delta) \tag{5}$$

then

$$d(y, z) < \delta. \tag{6}$$

According to triangle inequality

$$d(z, x) < d(y, z) + d(y, x). \tag{7}$$

From (7), (6), (1) we obtain

$$d(z, x) < d(y, z) + d(y, x) < \delta + \tfrac{1}{3}\, r < \tfrac{2}{3}\, r + \tfrac{1}{3}\, r = r. \tag{8}$$

Hence

$$z \in B(x, r)$$

and

$$B(y, \delta) \subset B(x, r). \tag{9}$$

● **PROBLEM 15-23**

THEOREM

Every separable metric space is second countable.

∎

Prove it.

SOLUTION:

Let (X, d) denote a separable metric space. Then A is a countable dense subset of X

$$\overline{A} = X.$$

According to S we denote the class of all open spheres with centers in A and with rational radii, that is,

$$S = \{B(a, r) : a \in A, r \in Q\}. \tag{1}$$

Since A and Q are countable sets, S is countable.

We will show that S is a base for the topology on X. That is, that for every open set $U \subset X$ and for every $x \in U$

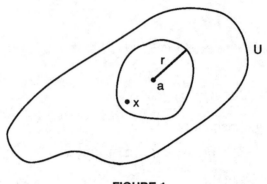

FIGURE 1

there is a sphere $B(a, r) \in S$, such that

$$x \in B(a, r) \subset U. \tag{2}$$

Since $x \in U$, there is an open sphere $B(x, r_0)$, such that

$$x \in B(x, r_0) \subset U. \tag{3}$$

Since A is dense in X, we can always find $a_0 \in A$, such that

$$d(x, a_0) \leq \frac{1}{3} r_0. \tag{4}$$

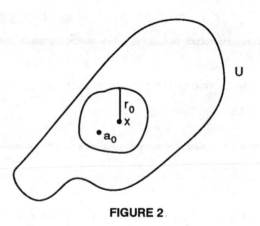

FIGURE 2

Let r_1 denote a rational number, such that

$$\frac{1}{3} r_0 < r_1 < \frac{2}{3} r_0. \tag{5}$$

Then, according to Problem 15–22, we have

$$x \in B(a_0, r_1) \subset B(x, r_0) \subset U. \tag{6}$$

Since $B(a_0, r_1) \in S$ and S are countable, S is a countable base for the topology on X.

Show that the linear space $C[0, 1]$ of all continuous functions on the closed interval $[0, 1]$ with the norm defined by

$$\|f\| = \sup\{|f(x)| : x \in [0, 1]\}$$

is second countable.

SOLUTION:

Let $f \in C[0, 1]$. Then, according to the Weierstrass approximation theorem for any $\varepsilon > 0$, a polynomial $P(x)$ exists with rational coefficients, such that

$$\| f(x) - P(x) \| < \varepsilon$$

that is, for all $x \in [0, 1]$

$$|f(x) - P(x)| < \varepsilon.$$

Hence, the collection of polynomials with rational coefficients is dense in $C[0, 1]$.

The set P of all polynomials with rational coefficients is countable. Hence, $C[0, 1]$ contains a countable dense set, i.e., $C[0, 1]$ which is separable. Since each separable metric space (see Problem 15–23) is second countable, $C[0, 1]$ is second countable.

Show that every Lindelöf metric space is separable.

SOLUTION:

Suppose (X, d) is a Lindelöf space. Let $\alpha > 0$ denote any real positive number, and let A denote a maximal subset of X such that

$$d(a, b) \geq \alpha \quad \text{for every } a, b \in A. \tag{1}$$

The existence of such maximal subset is guaranteed according to Zorn's lemma.

For each $a \in A$ consider $B(a, \alpha/2)$ and

$$D = X - U\{\overline{B(a, \alpha/4)} : a \in A\} \qquad (2)$$

D is an open set.

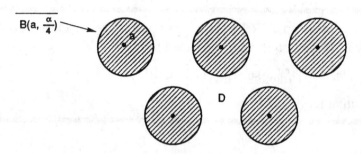

$B(a, \frac{\alpha}{4})$

D

FIGURE 1

Note that

$$\{D\} \cup \{B(a, \alpha/2) : a \in A\} \qquad (3)$$

is a covering of X by open sets. Since X is Lindelöf, a countable subcovering exists.

Suppose we remove $B(a, \alpha/2)$ from the original covering (3) for any $a \in A$, then the remaining sets would not cover X (none of them would contain a). Thus, $\{B(a, \alpha/2) : a \in A\}$ is countable, hence, A is countable.

We can repeat the above construction for each $\alpha = 1/n$; $n = 1, 2, 3, \ldots$ and obtain corresponding maximal subsets $\{A_n\}$.

Let

$$P = \bigcup_{n \in N} \{A_n\} \qquad (4)$$

P is countable as the union of countably many countable sets. We shall show that P is dense in X.

Suppose $x \in X$ and $\beta > 0$. We have to show that there is $y \in P$, such that $d(y, x) < \beta$.

Set

$$n > 1/\beta. \qquad (5)$$

Then there is $y \in A_n$, such that

$$d(y, x) < \beta. \qquad (6)$$

Let U represent any nonempty open subset of X. We choose $x \in U$ and $\beta > 0$, such that

$$B(x, \beta) \subset U.$$

Then $B(x, \beta)$, and hence U, contains an element of P. Therefore P is dense in X. Space (X, d) is separable.

● **PROBLEM 15-26**

Prove the following important theorem:

THEOREM

If (X, d) is a metric space, then the following statements are equivalent:

1. X is separable

2. X is second countable

3. X is Lindelöf.

■

SOLUTION:

In Problem 15–12, we proved that any second countable space is separable, i.e., we proved

$$(2) \Rightarrow (1) \tag{1}$$

In Problem 15–23, we proved that every separable metric space is second countable, i.e.,

$$(1) \Rightarrow (2) \tag{2}$$

Hence

$$(1) \equiv (2).$$

In Problem 15–14, we proved that every second countable space is Lindelöf, i.e.,

$$(2) \Rightarrow (3). \tag{3}$$

In Problem 15–25, we proved that every Lindelöf metric space is separable, i.e.,

$$(3) \Rightarrow (1) \tag{4}$$

From Equations (4) and (2) we obtain

$$(3) \Rightarrow (2). \tag{5}$$

Therefore

$$(1) \equiv (2) \equiv (3). \tag{6}$$

CHAPTER 16

COMPACTNESS

Why is any compact space also a Lindelöf space?

SOLUTION:

DEFINITION OF COMPACT SPACE

A space (X, T) is said to be compact if given any open cover $\{U_\omega\}$, $\omega \in \Omega$, of X, there is a finite subcover of $\{U_\omega\}$, $\omega \in \Omega$.

■

Let (X, T) represent any space and $A \subset X$. An open cover of A is a collection $\{U_\omega\}$, $\omega \in \Omega$ of open sets, such that

$$A \subset \bigcup_\omega U_\omega.$$

Also, $\{U_\omega\}$, $\omega \in \Omega$ is an open cover of A if $\{U_\omega \cap A\}$, $\omega \in \Omega$ is an open cover of the subspace A. A is said to be compact if every open cover of A has a finite subcover. Hence, A is compact if the subspace A is compact.

A space (X, T) is said to be a Lindelöf space if every open cover of X has a countable open subcover.

Therefore, any compact space is also Lindelöf.

Show that $(0, 1)$ is not a compact space with the absolute value topology.

SOLUTION:

The open interval with the absolute value topology is Lindelöf since it is a subspace of a second countable space R. With the absolute value topology, $(0, 1)$ is an example of a space which is Lindelöf, but not compact.

We shall now show that $(0, 1)$ is not compact with the absolute value topology.

Consider the collection of open subsets of $(0, 1)$

$$\{U_n\}, n = 1, 2, 3, 4, \ldots$$

where

$$U_n = \left(\frac{1}{n + 1}, 1\right).$$

This is an open cover of (0, 1) because

$$(0,1) = \bigcup_n \left(\frac{1}{n+1}, 1\right).$$

No finite number of the U_n will cover (0, 1). Suppose, on the contrary that

$$U_{a_1}, U_{a_2}, ..., U_{a_k}$$

cover (0, 1). Let U_{a_l} represent the set with the highest index, then

$$(0,1) = U_{a_1} \cup U_{a_2} \cup ... \cup U_{a_k} = U_{a_l} = \left(\frac{1}{a_l + 1}, 1\right) \neq (0,1).$$

Hence, a contradiction.

We conclude that (0, 1) is not a compact space with the absolute value topology.

● PROBLEM 16–3

1. Use the Heine-Borel theorem to show that every closed and bounded interval [a, b] with the absolute value topology is compact.

2. Show that every finite subset of a topological space (X, T) is compact.

SOLUTION:

1. We quote:

HEINE-BOREL THEOREM

Let [a, b] represent a closed and bounded interval and let $\{U_\alpha\}$ represent a class of open sets, such that

$$[a,b] \subset \bigcup_\alpha U_\alpha.$$

Then one can select a finite number of the open sets, say

$$U_{\alpha_1}, U_{\alpha_2}, ..., U_{\alpha_k}$$

in such a way that

$$[a,b] \subset U_{\alpha_1} \cup U_{\alpha_2} \cup ... \cup U_{\alpha_n}.$$

According to the Heine-Borel theorem, every closed and bounded interval [a, b] on the real line with the absolute value topology is compact.

2. Let $A = \{a_1, a_2, ..., a_n\}$ be a finite subset of a topological space (X, T). Let $\{U_\alpha\}$ be an open cover of A

$$A \subset \bigcup_\alpha U_\alpha.$$

Then each element of A belongs to at least one of the sets $\{U_\alpha\}$. Hence

$$a_1 \in U_{\alpha_1}$$

$$a_2 \in U_{\alpha_2}$$

$$\vdots$$

$$a_n \in U_{\alpha_n}$$

and

$$A \subset U_{\alpha_a} \cup ... \cup U_{\alpha_n}.$$

Any finite subset of a topological space is compact.

● PROBLEM 16-4

Let N represent the set of natural numbers with topology T defined as follows: a subset $U \subset N$ is open if it contains all, but mostly finite elements of N. Show that (N, T) is a compact space.

SOLUTION:

Let $\{U_\omega\}$, $\omega \in \Omega$ represent any open cover of N. We choose any element of $\{U_\omega\}$, say U_{ω_1}.

By definition U_{ω_1} contains all but most finitely elements of N. Let us denote these elements by

$$k_2, k_3, ..., k_m.$$

Since $\{U_\omega\}$ is an open cover of N, an open set U_{ω_2} exists, such that

$$k_2 \in U_{\omega_2}.$$

Similarly, we can find $U_{\omega_3}, ..., U_{\omega_m}$ such that

$$k_3 \in U_{\omega_3}, ..., k_m \in U_{\omega_m}.$$

For an open cover $\{U_\omega\}$, $\omega \in \Omega$ we found a finite subcover

$$\{U_{\omega_1}, U_{\omega_2}, ..., U_{\omega_m}\} \text{ of } \{U_\omega\}.$$

Since $\{U_\omega\}$, $\omega \in \Omega$ was an arbitrary cover of N, the space (N, T) is compact.

1. Show that any infinite subset A of a discrete topological space X is not compact.

2. Prove that a subset of a discrete space is compact, if and only if it is finite.

SOLUTION:

1. Let (X, T) represent a discrete topological space and A represent an infinite subset of A. We shall find an open cover of A which has no finite subcover. Consider the class of singleton sets

$$\{\{a\} : a \in A\}.$$

It is an open cover of A because

$$A = \bigcup_{a \in A} \{a\}$$

and since all subsets of a discrete space are open, each set $\{a\}$ is open. Note that no proper subclass of $\{\{a\} : a \in A\}$ is a cover of A.

Since A is infinite so is $\{\{a\} : a \in A\}$. Thus, the open cover $\{\{a\} : a \in A\}$ of A contains no finite subcover and A is not compact.

2. In Problem 16–3, we proved that every finite subset of a topological space is compact. Hence, a subset of a discrete space is compact, if and only if it is finite.

Show that a continuous image of a compact set is also compact.

SOLUTION:

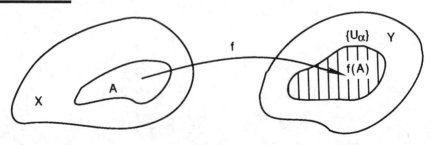

FIGURE 1

Suppose $f : X \rightarrow Y$ is continuous and X and Y are topological spaces and A is a compact subset of X.

Let $\{U_\alpha\}$ represent an open cover of $f(A)$

$$f(A) \subset \bigcup_\alpha U_\alpha.$$

Then, we have

$$A \subset f^{-1}[f(A)] \subset f^{-1}[\bigcup_\alpha U_\alpha] = \bigcup_\alpha f^{-1}(U_\alpha).$$

Since f is continuous, all sets $f^{-1}(U_\alpha)$ are open and $\{f^{-1}(U_\alpha)\}$ is an open cover of A. By hypothesis, A is a compact set, hence $\{f^{-1}(U_\alpha)\}$ is reducible to a finite cover, say

$$f^{-1}(U_{\alpha_1}), f^{-1}(U_{\alpha_2}), \dots, f^{-1}(U_{\alpha_n}).$$

Then

$$A \subset f^{-1}(U_{\alpha_1}) \cup \dots \cup f^{-1}(U_{\alpha_n})$$

and

$$f(A) \subset f[f^{-1}(U_{\alpha_1}) \cup \dots \cup f^{-1}(U_{\alpha_n})] \subset U_{\alpha_1} \cup \dots \cup U_{\alpha_n}.$$

Thus $f(A)$ is compact.

● **PROBLEM 16–7**

Show that topological space (X, T) with the cofinite topology T is compact.

SOLUTION:

Suppose $\{U_\alpha\}$ is an open cover of X. Choose any

$$U \in \{U_\alpha\}.$$

Since T is the cofinite topology, $X - U$ is a finite set, which we can denote as

$$X - U = \{u_1, u_2, \dots, u_n\}.$$

$\{U_\alpha\}$ is a cover of X, therefore, for each of the elements of $\{u_1, u_2, \dots, u_n\}$, at least one set U_{α_i} exists, such that

$$u \in_i U_{\alpha_i} \text{ and } U_{\alpha_i} \in \{U_\alpha\}$$

for $i = 1, 2, \dots, n$.

Thus

$$X - U \subset U_{\alpha_1} \cup \dots \cup U_{\alpha_n}$$

573

and
$$X = U \cup (X - U) = U \cup U_{\alpha_1} \cup \dots \cup U_{\alpha_n}.$$

Hence X is compact.

1. Find an example of a subset of a compact space which is not compact.

2. Prove the following

THEOREM

A closed subset of a compact space is also compact.

■

SOLUTION:

1. Consider the closed interval $[a, b]$ which is compact by the Heine-Borel theorem. Open interval (a, b), which is a subset of $[a, b]$, is not compact (see Problem 16–2).

2. Let $F \subset X$ represent a closed subset of a compact space (X, T) and let
$$\{U_\alpha\}$$
represent an open cover of F,
$$F \subset \bigcup_\alpha U_\alpha.$$

We have
$$X = (X - F) \cup \bigcup_\alpha U_\alpha.$$

Since F is closed, $X - F$ is open and
$$\{U_\alpha, X - F\}$$
is an open cover of X.

By hypothesis, X is compact and the cover $\{U_\alpha, X - F\}$ is reducible to a finite cover, say
$$X = U_1 \cup \dots \cup U_n \cup (X - F)$$

where $U_1, \dots, U_n \in \{U_\alpha\}$.

Since F and $X - F$ are disjoint we obtain

574

$$F \subset U_1 \cup \ldots \cup U_n.$$

Therefore, an open cover $\{U_\alpha\}$ of F is reducible to a finite subcover $\{U_1, U_2, \ldots, U_n\}$.

Set F is compact.

● **PROBLEM 16-9**

Show that the following conditions are equivalent:

1. X is compact

2. For every family $\{F_\alpha\}$ of closed subsets of X,

$$\left(\bigcap_\alpha F_\alpha = \phi \right) \Rightarrow \left(\begin{array}{c} \{F_\alpha\} \text{ contains a finite subclass} \\ \{F_{\alpha_1}, \ldots, F_{\alpha_k}\} \text{ such that} \\ F_{\alpha_1} \cap \ldots \cap F_{\alpha_k} = \phi \end{array} \right).$$

SOLUTION:

1. $\Rightarrow$ 2.

Suppose $\bigcap_\alpha F_\alpha = \phi$, then according to de Morgan's law

$$X = X - \left(\bigcap_\alpha F_\alpha \right) = \bigcup_\alpha (X - F_\alpha)$$

$\{X - F_\alpha\}$ is an open cover of X, because all F_α are closed. But, by hypothesis X is compact, hence a finite subcover exists, say

$$X = (X - F_{\alpha_1}) \cup \ldots \cup (X - F_{\alpha_k})$$

where

$$F_{\alpha_1}, \ldots, F_{\alpha_k} \in \{F_\alpha\}.$$

Again, according to de Morgan's law

$$\phi = X - X = X - [(X - F_{\alpha_a}) \cup \ldots \cup (X - F_{\alpha_k})] = F_{\alpha_1} \cap \ldots \cap F_{\alpha_n}.$$

2. $\Rightarrow$ 1.

Let $\{U_\alpha\}$ represent an open cover of X.

$$X = \bigcup_\alpha U_\alpha.$$

According to de Morgan's law

$$\phi = X - X = \bigcap_\alpha (X - U_\alpha).$$

All sets $X - U_\alpha$ are closed and have an empty intersection. By hypothesis, a subclass of $\{X - U_\alpha\}$ exists, say $(X - U_{\alpha_1}), \ldots, (X - U_{\alpha_k})$, such that

$$(X - U_{\alpha_1}) \cap \ldots \cap (X - U_{\alpha_k}) = \phi.$$

According to de Morgan's law

$$X - [(X - U_{\alpha_1}) \cap \ldots \cap (X - U_{\alpha_k})] = U_{\alpha_1} \cup U_{\alpha_2} \cup \ldots \cup U_{\alpha_k} = X.$$

Thus, X is compact.

● **PROBLEM 16-10**

Show that the family of open intervals

$$\{(0, 1), (0, \tfrac{1}{2}), (0, \tfrac{1}{3}), (0, \tfrac{1}{4}), \ldots\}$$

has the finite intersection property.

SOLUTION:

DEFINITION

A family $\{A_\alpha\}$ of sets is said to have the finite intersection property, if every finite subfamily

$$\{A_{\alpha_1}, \ldots, A_{\alpha_n}\}$$

has a non-empty intersection, that is

$$A_{\alpha_1} \cap \ldots \cap A_{\alpha_n} = \phi.$$

∎

Let $\{(0, a_1), \ldots, (0, a_n)\}$ represent any subfamily of open intervals, then

$$(0, a_1) \cap \ldots \cap (0, a_n) = (0, a) \neq \phi$$

where

$$a = min\,(a_1, a_2, \ldots, a_n).$$

Observe that the intersection of all the members of the family

$$(0, 1) \cap (0, \tfrac{1}{2}) \cap \ldots = \phi$$

is empty.

Prove this theorem:

THEOREM

A topological space (X, T) is compact, if and only if every class $\{F\alpha\}$ of closed subsets of X, which has the finite intersection property (see Problem 16–10), has a non-empty intersection, i.e.,

$$\bigcap_\alpha F_\alpha \neq \phi.$$

∎

SOLUTION:

The implication of Problem 16–9 can be written in the form

$$\left(\bigcap_\alpha F_\alpha \neq \phi\right) \Rightarrow \begin{pmatrix} \exists\, \alpha_1, \alpha_2, \ldots, \alpha_k \text{ such that} \\ F_{\alpha_1} \cap F_{\alpha_2} \cap \ldots \cap F_{\alpha_k} = \phi \end{pmatrix} \qquad (1)$$

Let a and b represent sentences, then

$$(a \Rightarrow b) \equiv (a' \vee b)$$

and

$$(b' \Rightarrow a') \equiv (b \vee a').$$

Let us take the negation of (1)

$$\begin{pmatrix} \forall\, \alpha_1, \alpha_2, \ldots, \alpha_k \text{ such that} \\ F_{\alpha_1} \cap F_{\alpha_2} \cap \ldots \cap F_{\alpha_k} = \phi \end{pmatrix} \Rightarrow \left(\bigcap_\alpha F_\alpha \neq \phi\right). \qquad (2)$$

Since (1) implies (2), the proof is completed.

Let A represent a compact subset of a Hausdorff space (X, T). Show that, if

$$q \in X - A, \qquad (1)$$

then the open sets U and V exist, such that

$$q \in U, \quad A \subset V, \quad U \cap V = \phi. \qquad (2)$$

See Figure 1.

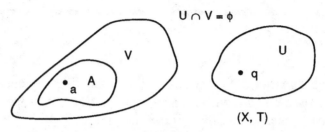

FIGURE 1

SOLUTION:

Choose $a \in A$, since $q \in X - A$, $a \neq q$ and since (X, T) is a Hausdorff space, open sets U_a and V_a exist such that

$$q \in U_a, \quad a \in V_a, \quad U_a \cap V_a = \phi. \tag{3}$$

The family of sets $\{V_a : a \in A\}$ forms an open cover of A

$$A \subset \bigcup_{a \in A} V_a. \tag{4}$$

Since A is compact, a finite subcover $V_{a_1}, V_{a_2}, \ldots, V_{a_k}$ exists, such that

$$A \subset V_{a_1} \cup V_{a_2} \cup \ldots \cup V_{a_k}. \tag{5}$$

Let us define

$$V = V_{a_1} \cup V_{a_2} \cup \ldots \cup V_{a_k} \tag{6}$$

and

$$U = U_{a_1} \cap \ldots \cap U_{a_k}. \tag{7}$$

Obviously U and V are open as the union and finite intersection of open sets.
From (5) and (6) we have

$$A \subset V,$$

from (3) and (7) we have

$$q \in U.$$

We have to show that $U \cap V = \phi$.

$$U \cap V = (U_{a_1} \cap \ldots \cap U_{a_k}) \cap (V_{a_1} \cup V_{a_2} \cup \ldots \cup V_{a_k}). \tag{8}$$

Since

$$U_{a_l} \cap V_{a_l} = \phi \tag{9}$$

for $l = 1, 2, \ldots, k$, we have

$$U \cap V = \phi.$$

Show that any compact subset of a Hausdorff space is closed.

SOLUTION:

Let A represent a compact subset of a Hausdorff space (X, T) and let $q \notin A$.

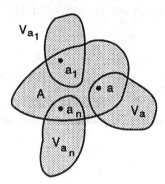 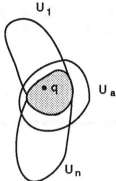

FIGURE 1

We shall show that if A is compact, then $X - A$ is open.

Let $a \in A$; then since (X, T) is Hausdorff, there are neighborhoods U_a and V_a of a and q respectively, such that

$$a \in U_a, \quad q \in V_a, \quad U_a \cap V_a = \phi. \tag{1}$$

The family of sets

$$\{V_a : a \in A\} \tag{2}$$

forms an open cover of A. Since A is compact, there are finitely many

$$a_1, a_2, \ldots, a_n \quad \text{such that} \quad \{V_{a_1}, \ldots, V_{a_n}\} \tag{3}$$

forms on an open cover of A. Let

$$U = U_{q_1} \cap \ldots \cap U_{q_n} \quad \text{and} \quad V = V_{a_1} \cup \ldots \cup V_{a_n}. \tag{4}$$

Then $q \in U$ and $U \cap V = \phi$. Since $q \in X - A$ and $q \in U$ and $U \cap A = \phi$, then $X - A$ is open and A is closed.

Show that a subset of a compact T_2–space (Hausdorff space) is compact, if and only if it is closed.

SOLUTION:

Let (X, T) represent a Hausdorff compact space and $A \subset X$, then

$$(A \text{ is compact}) \Leftrightarrow (A \text{ is closed}).$$

$\Leftarrow$ In Problem 16–8 we proved that any closed subset of a compact space is compact.

$\Rightarrow$ In Problem 16–13 we proved that any compact subset of a Hausdorff space is closed.

● **PROBLEM 16-15**

Prove   *ll X IS NORMAL (?)*

THEOREM

Let A and B represent disjoint compact subsets of a Hausdorff space (X, T). Then open sets U and V exist, such that

$$A \subset U, \quad B \subset V, \quad U \cap V = \phi. \qquad (1)$$

■

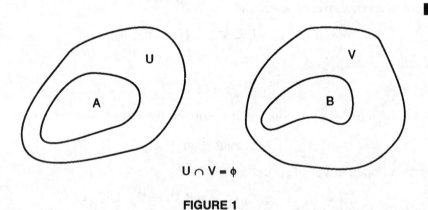

$U \cap V = \phi$

FIGURE 1

SOLUTION:

where is used?
this used?

Let $a \in A$: then, since $A \cap B = \phi$, $a \notin B$. By hypothesis, B is compact and open sets U_a and V_a exist, such that   *cf: 16~13/P579 maybe*

$$a \in U_a, \quad B \subset V_a, \quad U_a \cap V_a = \phi. \qquad (2)$$

The family of sets $\{U_a : a \in A\}$ forms an open cover of A and since A is compact we can select $\{U_{a_1}, U_{a_2}, \dots, U_{a_n}\}$ a finite subcover of A.

Let us denote

580

$$U = U_{a_1} \cup \ldots \cup U_{a_n}$$

$$V = V_{a_1} \cap \ldots \cap V_{a_n}. \tag{3}$$

Then

$$A \subset U \quad \text{and} \quad B \subset V. \tag{4}$$

Both U and V are open sets. For each $k = 1, 2, \ldots, n$

$$U_{a_k} \cap V_{a_k} = \phi. \tag{5}$$

Thus

$$U_{a_k} \cap V = \phi$$

and

$$U \cap V = (U_{a_1} \cup \ldots \cup U_{a_n}) \cap V = \phi. \tag{6}$$

● PROBLEM 16–16

Show that every compact Hausdorff space is normal.

SOLUTION:

Suppose (X, T) is a compact Hausdorff space and F_1 and F_2 are disjoint closed subsets of X.

By Problem 16–8, both sets F_1 and F_2 are also compact.

By Problem 16–15, open subsets U_1 and U_2 exist such that

$$F_1 \subset U_1, \quad F_2 \subset U_2, \quad U_1 \cap U_2 = \phi.$$

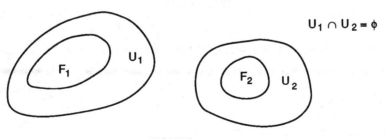

FIGURE 1

We conclude that (X, T) is normal (see definition in Problem 13–22). See Figure 2.

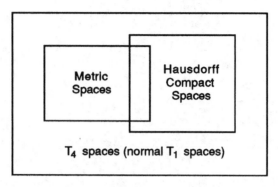

FIGURE 2

● **PROBLEM 16-17**

1. Let A represent a compact subset of a Hausdorff space (X, T).

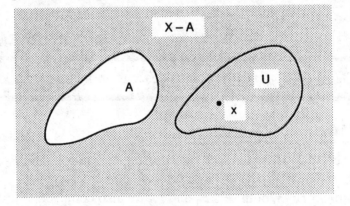

FIGURE 1

Show that if $x \notin A$, then there is an open set U, such that

$$x \in U \subset X - A. \tag{1}$$

2. Prove that a compact subset A of a Hausdorff space (X, T) is closed.

SOLUTION:

1. Open sets U and V exist such that

$$x \in U, \quad A \subset V \quad \text{and} \quad U \cap V = \phi. \tag{2}$$

Therefore,

$$U \cap A = \phi \quad \text{and} \quad x \in U \subset X - A. \tag{3}$$

582

2. We shall prove that $X - A$ is open. Let

$$x \in X - A$$

that is,

$$x \notin A.$$

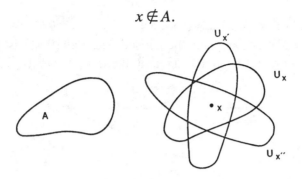

FIGURE 2

Since A is compact, an open set U_x exists, such that

$$x \in U_x \subset X - A. \tag{4}$$

Therefore,

$$X - A = \bigcup_{x \in X - A} U_x \tag{5}$$

and $X - A$ is open, while the union of open sets and A is closed. Compare with Problem 16–13.

● PROBLEM 16-18

How does an equivalence relation lead to the new compact spaces? (Hint: use results of Problem 16–17).

Show that the circle is compact.

SOLUTION:

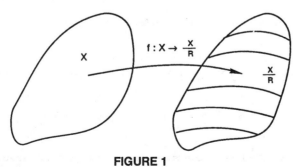

FIGURE 1

Suppose X is a compact space and R is an equivalence relation on X.

The space $X/_R$ is called the identification space. Since the identification mapping f from X onto $X/_R$,

$$f: X \rightarrow X/_R$$

is continuous and the space $X/_R$ is compact.

We have already shown that the closed interval $[0, 1]$ is compact. An equivalence relation between $[0, 1]$ and the circle can be established as follows:

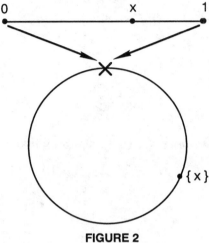

FIGURE 2

Let 0 be equivalent to 1, and every other element of $[0, 1]$ be equivalent only to itself.

By identifying 0 and 1, we obtain a circle. The circle is an identification space derived from $[0, 1]$, hence the circle is compact.

● **PROBLEM 16–19**

Prove:

THEOREM

Let f represent a one-to-one continuous function

$$f: X \rightarrow Y$$

from a compact space X into a Hausdorff space Y. Then X and $f(X)$ are homeomorphic.

■

See Figure 1.

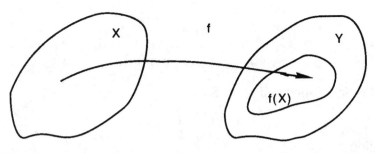

FIGURE 1

SOLUTION:

Function

$$f : X \to f(X) \tag{1}$$

is a continuous bijection (i.e., one-to-one and onto). Hence,

$$f^{-1} : f(X) \to X \tag{2}$$

exists. By hypothesis, f is continuous, so we only have to show that f^{-1} is continuous.

Function f^{-1} is continuous if, for every closed subset F of X,

$$[f^{-1}]^{-1}(F) = f(F) \tag{3}$$

it is a closed subset of $f(X)$.

Let F represent a closed subset of a compact space, then (by Problem 16–8) F is compact. Since F is continuous, $f(F)$ is a compact subset of $f(X)$. $f(X)$ is the Hausdorff space as a subspace of the Hausdorff space Y. Thus, $f(F)$ is closed and f^{-1} is continuous. Therefore,

$$f : X \to f(X) \tag{4}$$

is a homeomorphism.

● PROBLEM 16–20

Let (X, T) represent compact and let (X, T') represent the Hausdorff space.

Prove that

$$(T' \subset T) \implies (T' = T).$$

SOLUTION:

Consider the identity function defined on X:

585

$$f(x) = x$$

$$f : (X, T) \to (X, T').$$

Function f is one-to-one and onto. Since

$$T' \subset T,$$

f is continuous.

Since (X, T) is compact and (X, T') is the Hausdorff space, we conclude that (X, T) and (X, T') are homeomorphic. Therefore

$$T' = T.$$

● **PROBLEM 16-21**

Here is an important theorem which enables us to find more compact spaces.

TYCHONOFF THEOREM

Let $\underset{k}{\times} X_k \ k \in K$ represent the product space of the countable family of non-empty spaces $\{(X_\kappa, T_k) : k \in K\}$. Then

$$\underset{k}{\times} X_k$$

is compact, if and only if each component space is compact.

■

Prove it.

SOLUTION:

Suppose $\underset{k}{\times} X_k$ is compact. For each $l \in K$, the projection map

$$p_l : \underset{k}{\times} X_k \to X_l$$

is continuous and onto.

Therefore X_l is compact for each $l \in K$. Now, suppose each X_k is compact. Let $\{t_j\}, j \in J$ represent any ultranet in $\underset{k}{\times} X_k$ with the kth coordinate of t_j denoted by $t_j(k)$. Then

$$\{p_k(t_j)\} = \{t_j(k)\}, j \in J$$

is an ultranet in X_k and $\{t_j(k)\}$ converges in X_k. Therefore, $\{t_k\}, k \in K$ converges in $\underset{k}{\times} X_k$.

586

A space (X, T) is compact, if and only if every ultranet in X converges. Hence,

$$\underset{k}{\times} X_k$$

is compact.

● **PROBLEM 16–22**

Show that the cylinder, the torus and the cube are compact spaces.

SOLUTION:

We have already proved that the closed interval $I = [0, 1]$ and the circle C are compact.

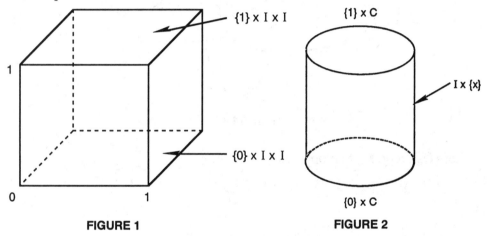

FIGURE 1   FIGURE 2

According to the Tychonoff theorem, the cube $[0, 1] \times [0, 1] \times [0, 1]$ is compact because $[0, 1]$ is compact.

Similarly, we show that the cylinder is compact, see Figure 2.

Since C and $[0, 1]$ are compact, the cylinder $C \times [0, 1]$ is compact.

The torus $C \times C$ is also compact, see Figure 3.

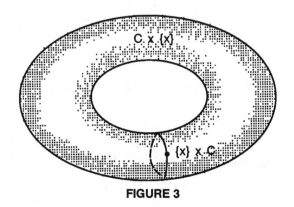

FIGURE 3

587

1. Show that if A is compact and F is closed, then $A \cap F$ is compact.

2. Prove that if $A_1, A_2, \ldots, A_k$ are compact subsets of a topological space X, then $A_1 \cup \ldots \cup A_k$ is also compact.

SOLUTION:

1.

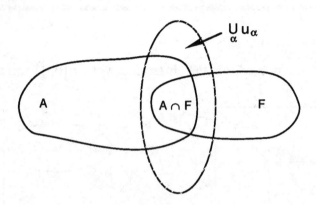

FIGURE 1

Let $\{U_\alpha\}$ represent an open cover of $A \cap F$

$$A \cap F \subset \bigcup_\alpha U_\alpha. \tag{1}$$

Then

$$A \subset \bigcup_\alpha U_\alpha \cup (X - F). \tag{2}$$

and since F is closed $\{\{U_\alpha\}, X - F\}$ is an open cover of A. Set A is compact so we can choose a finite subcover, say

$$A \subset U_{\alpha_1} \cup \ldots \cup U_{\alpha_n} \cup (X - F). \tag{3}$$

Hence

$$A \cap F \subset U_{\alpha_1} \cup \ldots \cup U_{\alpha_n} \tag{4}$$

and $A \cap F$ is compact.

2. Let A and B represent compact subsets of (X, T) and let $\{U_\alpha\}$ represent an open cover of $A \cup B$

$$A \cup B \subset \bigcup_\alpha U_\alpha. \tag{5}$$

Then $\{U_\alpha\}$ is an open cover of A and B, which are compact.

$$A \subset \bigcup_\alpha U_\alpha. \tag{6}$$

We can choose finite subcover

$$A \subset U_{\alpha_1} \cup \ldots \cup U_{\alpha_n} \qquad (7)$$

and

$$B \subset U_{\alpha_{n+1}} \cup \ldots \cup U_{\alpha_{n+k}}. \qquad (8)$$

Hence

$$A \cup B \subset U_{\alpha_1} \cup \ldots \cup U_{\alpha_{n+k}} \qquad (9)$$

and $A \cup B$ is compact.

● **PROBLEM 16–24**

Show that a finite subset A of a topological space (X, T) is sequentially compact.

SOLUTION:

DEFINITION OF SEQUENTIALLY COMPACT SETS

A subset A of a topological space (X, T) is sequentially compact if every sequence in A contains a subsequence which converges to a point in A.

∎

Let A represent a finite subset of (X, T), and let

$$(x_1, x_2, x_3, \ldots) \qquad (1)$$

represent a sequence in A.

Since A is finite, at least one of the elements in A, say x_0, appears an infinite number of times in the sequence. We can choose the subsequence of (1) to be

$$(x_0, x_0, x_0, \ldots) \qquad (2)$$

which converges to the point x_0 belonging to A.

● **PROBLEM 16–25**

Show that a continuous image of a sequentially compact set is sequentially compact.

SOLUTION:

Let

$$f : X \to Y \tag{1}$$

be a continuous function and let A represent a subset of X which is sequentially compact. We will show that $f(A)$ is sequentially compact in Y.

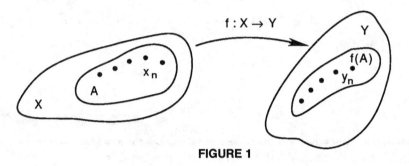

FIGURE 1

Let

$$(y_1, y_2, \ldots) \tag{1}$$

represent a sequence in $f(A)$. Then there exists a sequence in A

$$(x_1, x_2, \ldots) \tag{2}$$

such that

$$f(x_1) = y_1, f(x_2) = y_2, \ldots , f(x_n) = y_n, \ldots .$$

Since A is sequentially compact, sequence (2) contains a subsequence, say

$$(x_{k_1}, x_{k_2}, \ldots) \tag{3}$$

which converges to a point $x_0 \in A$.

Since f is continuous, we have

$$\left((x_{k_1}, x_{k_2}, \ldots) \text{ converges to } x_0 \right) \Rightarrow \left(\begin{array}{l} (f(x_{k_1}), f(x_{k_2}), \ldots) \\ \text{converges to } f(x_0) \end{array} \right).$$

We have $f(x_0) \in f(A)$.

Thus, $f(A)$ is sequentially compact.

● **PROBLEM 16–26**

1. Show that every bounded closed interval $[a, b]$ is countably compact.
2. Show that the open interval (a, b) is not countably compact.

590

SOLUTION:

DEFINITION OF COUNTABLY COMPACT SETS

A subset A of a topological space (X, T) is countably compact, if every infinite subset $B \subset A$ has an accumulation point in A.

∎

1. Let B represent an infinite subset of a closed interval $[a, b]$.

$$B \subset [a, b].$$

We shall apply:

BOLZANO-WEIERSTRASS THEOREM

Every bounded infinite set of real numbers has an accumulation point.

∎

Thus B has an accumulation point x. Since $[a, b]$ is closed, and $B \subset [a, b]$, the accumulation point x of B belongs to $[a, b]$. Hence $[a, b]$ is countably compact.

2. We shall show that the open interval $(0, 1)$ is not countably compact. Consider the infinite subset of $(0, 1)$

$$\{ 1/2, 1/3, 1/4, 1/5, \ldots \}.$$

This subset has only one accumulation point 0 which does not belong to $(0, 1)$.

Hence $(0, 1)$ is not countably compact.

● PROBLEM 16-27

Show that a closed subset of a countably compact space is countably compact.

SOLUTION:

Let (X, T) represent a countably compact space and F be a closed subset of X.

Let A represent any infinite subset of F. A is also an infinite subset of a countably compact space (X, T). Therefore an accumulation point x of A exists, such that

$$x \in X$$

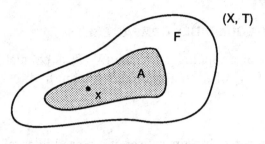

FIGURE 1

Since $A \subset F$, x is also an accumulation point of F. By hypothesis, F is closed, so it contains all its accumulation points. Thus

$$x \in F.$$

Hence, any infinite subset A of a closed set F has an accumulation point $x \in F$. Therefore, F is countably compact.

● **PROBLEM 16-28**

Prove:

$$(X \text{ is compact}) \Rightarrow (X \text{ is countably compact}).$$

SOLUTION:

Suppose (X, T) is compact.
Let A represent a subset of X

$$A \subset X$$

with no accumulation points in X.

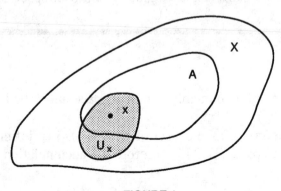

FIGURE 1

Each point $x \in X$ belongs to an open set U_x which contains, at most, one point of A.

Now, consider the family of sets

$$\{U_x : x \in X\} \tag{1}$$

Family (1) consists of open sets and

$$X \subset \bigcup_{x \in X} U_x. \tag{2}$$

Hence (1) is an open cover of X. Since X is compact (by hypothesis), we can select a finite subcover

$$U_{x_1}, U_{x_1}, \ldots, U_{x_n}. \tag{3}$$

Then

$$A \subset X \subset U_{x_1} \cup \ldots \cup U_{x_n}. \tag{4}$$

Since each of the sets (3) contains, at most, one point of A, A is finite. Therefore, every infinite subset of X contains an accumulation point in X, that is, X is countably compact.

● **PROBLEM 16–29**

In Problem 16–28 we showed that if the space is compact, then it is countably compact.

$$((X, T) \text{ compact}) \Rightarrow ((X, T) \text{ countably compact}).$$

Give an example showing that this implication cannot be reversed.

SOLUTION:

We shall give an example of a space which is countably compact, but not compact. Let N represent the set of positive integers with the topology T generated by the sets

$$\{1, 2\}, \{3, 4\}, \{5, 6\}, \ldots$$

and let $A \neq \phi$ represent a non-empty subset of N.

Let $k \in A$, then, if k is even, then $k - 1$ is a limit point of A and if k is odd, then $k + 1$ is a limit point of A.

Hence A has an accumulation point and (N, T) is countably compact. Since

$$\{\{1, 2\}, \{3, 4\}, \{5, 6\}, \ldots\}$$

is an open cover of N which has no finite subcover, (N, T) is not compact.

1. Show that sequential compactness is a topological property.

2. Show that countable compactness is a topological property.

SOLUTION:

1. Let (X, T) and (Y, T') represent homeomorphic topological spaces. Then a homeomorphism exists

$$f : X \rightarrow Y.$$

Suppose X is sequentially compact and let

$$(y_n) \tag{1}$$

be a sequence in Y.
   Then, since f is onto and one-to-one,

$$f^{-1}(y_n) \tag{2}$$

is a sequence in X. Since X is sequentially compact, (2) contains a subsequence, say

$$f^{-1}(y_{k_n}) \tag{3}$$

which converges to a point in X,

$$f^{-1}(y_{k_n}) \rightarrow x \in X. \tag{4}$$

Since f is continuous,

$$f(f^{-1}(y_{k_n})) = (y_{k_n}) \rightarrow f(x) \in Y. \tag{5}$$

Thus, (1) contains a subsequence (y_{k_n}), which converges to a point $f(x)$ in Y. (Y, T') is sequentially compact.

2. Let B represent an infinite subset of Y. Then $f^{-1}(B)$ is an infinite subset of X. X is countably compact. Hence $f^{-1}(B)$ has an accumulation point x in X. Thus B has an accumulation point $f(x)$ in Y and Y is countably compact.

Prove that if a space (X, T) is sequentially compact, then it is countably compact.

$$(X \text{ sequentially compact}) \Rightarrow (X \text{ countably compact}).$$

SOLUTION:

Suppose (X, T) is sequentially compact and A is any infinite subset of X, $A \subset X$. We shall show that A has an accumulation point in A. We can find a sequence

$$(a_n) \quad n \in N$$

in A with distinct terms, i.e., for any $k, l \in N, k \neq l$

$$a_k \neq a_l.$$

Since X is sequentially compact, (a_n) contains a subsequence which converges to a point in X.

$$(a_{k_n}) \to x \in X.$$

Then $x \in X$ is an accumulation point of A and (X, T) is countably compact.

There are some other properties related to compactness, for example, paracompactness, metacompactness, and pseudocompactness.

● **PROBLEM 16-32**

Show that the coordinate plane R^2 with the Pythagorean metric topology is locally compact.

SOLUTION:

DEFINITION OF LOCALLY COMPACT SPACES

A space (X, T) is said to be locally compact if for any $x \in X$ and any neighborhood U of x, there is a compact set A such that

$$x \in Int(A) \subset A \subset U \tag{1}$$

∎

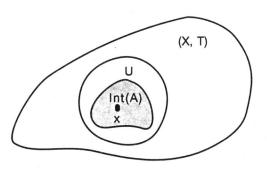

FIGURE 1

595

Plane R^2 is not compact, since it is not bounded.

Let

$$x \in R^2$$

and let U represent any neighborhood of x. Then there exists $r > 0$ such that

$$x \in B(x, r) \subset U. \tag{2}$$

We have

$$B(x, {}^r/_2) \subset \overline{B(x, {}^r/_2)} \subset B(x, r) \subset U. \tag{3}$$

The set

$$\overline{B(x, {}^r/_2)}$$

is a closed and bounded subset of R^2, hence it is compact.

$$x \in B(x, {}^r/_2) \subset \overline{B(x, {}^r/_2)} \subset U. \tag{4}$$

Hence R^2 with the Pythagorean metric topology is locally compact.

● PROBLEM 16–33

Prove that any compact T_2–space (X, T) is locally compact.

SOLUTION:

The criterion for local compactness given in Problem 16–32 is simpler for the T_2–spaces.

DEFINITION

Let (X, T) represent a T_2–space. Then X is locally compact if and only if, given any

$$x \in X,$$

there is a compact set A, such that

$$x \in Int(A).$$

■

One can prove that for T_2–spaces, the existence of one compact subset A of X such that $x \in Int(A)$ guarantees that for any neighborhood U of x, there is a compact set B, such that

$$x \in Int(B) \subset B \subset U.$$

596

Let (X, T) be a compact T_2–space. Then X is a compact neighborhood of any $x \in X$.

● **PROBLEM 16-34**

Let Q represent the subspace of rational numbers in the space R of real numbers with the absolute value topology. Show that Q is not locally compact.

SOLUTION:

Space R is obviously locally compact.
Take any $x \in Q$ and a compact set A such that

$$x \in int(A). \qquad (1)$$

Remember that a subset A of R (or R^n) is compact, if and only if A is bounded and closed.
Then A contains infinitely many elements of Q.

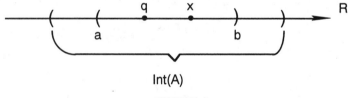

Int(A)

FIGURE 1

We can find rational numbers a and b, such that $(a, b) \subset R$

$$x \in (a, b) \cap Q \subset Int(A). \qquad (2)$$

Let q represent an irrational number, such that

$$q \in (a, b). \qquad (3)$$

We define an open cover of A as follows:

for each $a \in A$,

$$U(a) = \begin{cases} if \quad q < a \quad \{s \in R : a < s\} \\ if \quad q > a \quad \{t \in R : t < a\} \end{cases} \qquad (4)$$

$\{U(a) \cap A : a \in A\}$ is an open cover of A which has not finite subcover. Hence Q is not locally compact.

Prove the following

THEOREM

If (X, T) is a Hausdorff space (i.e., a T_2-space) locally compact, then (X, T) is a regular space

∎

SOLUTION:

We shall apply the results of Problem 13–17. A space (X, T) is regular iff given any $x \in X$ and any neighborhood U of x, there is a neighborhood V of x such that

$$\overline{V} \subset U.$$

Let (X, T) represent a locally compact T_2-space. If $x \in X$, and U is any neighborhood of x, then there is a compact set A, such that

$$x \in Int(A) \subset A \subset U \tag{1}$$

(see Problem 16–32).

Since A is compact, A is closed, hence

$$\overline{Int(A)} \subset A. \tag{2}$$

We define V as

$$V = Int(A) \tag{3}$$

and obtain

$$x \in V \subset \overline{V} \subset U. \tag{4}$$

Since V is a neighborhood of x, (X, T) is regular.

1. Show that an open subspace of a locally compact space is locally compact.

2. Show that a closed subspace of a locally compact T_2-space is locally compact.

SOLUTION:

1. Let U represent an open subspace of a locally compact space (X, T) and $x \in U$.

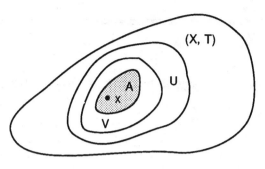

FIGURE 1

Let V represent any neighborhood of x, such that

$$x \in V \subseteq U. \tag{1}$$

Since (X, T) is locally compact, a compact set A exists such that

$$x \in Int(A) \subseteq A \subseteq V. \tag{2}$$

Hence, U is locally compact.

2. Let F represent a closed subspace of X and $x \in F$ and let A represent any compact subset of X, such that

$$x \in Int(A) \subseteq A. \tag{3}$$

A is closed because (X, T) is T_2. The set

$$F \cap A$$

is a closed subset of the compact set A and thus, is compact. We have

$$x \in Int_F(F \cap A) \subseteq F \cap A \subseteq F \tag{4}$$

Where $Int_F(F \cap A)$ denotes $Int(F \cap A)$ in F.

By definition in Problem 16–33, we conclude that since F is T_2, F is locally compact.

● PROBLEM 16–37

In Problem 16–6, we showed that compactness is preserved by continuous functions. One can find examples showing that local compactness is not preserved by continuous functions. Prove the following:

THEOREM

If $f: (X, T) \rightarrow (Y, T')$ is a continuous, open function from a locally compact space (X, T) onto Y, then (Y, T') is also locally compact. ∎

SOLUTION:

A function $f: X \rightarrow Y$ is said to be open, if for any open subset $U \subset X$, $f(U)$ is open in Y.

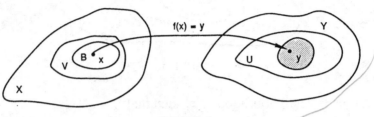

f(x) = y

FIGURE 1

Let $y \in Y$ and U represent a neighborhood of y. For some $x \in X$

$$f(x) = y.$$

Since f is continuous, an open set V exists, such that

$$x \in V \in T \quad \text{and} \quad f(V) \subset U.$$

Since (X, T) is locally compact, there is a compact set B such that

$$x \in Int(B) \subset B \subset V.$$

Then

$$y = f(x) \in f(Int\ B) \subset f(B) \subset U.$$

Since f is open, $f(Int\ B)$ is open. Since B is compact and f is continuous, $f(B)$ is compact. Thus, Y is locally compact.

● PROBLEM 16-38

Prove the existence of a Lebesque number described in this theorem.

THEOREM

If (X, d) is a compact metric space and $\{U_\alpha\}$ is an open cover of X, then there is a positive number $r > 0$, such that

$$B(x, r) \subset U_\alpha \quad \text{for some } \alpha \text{ for any } x \in X.$$ ∎

600

That is, number $r > 0$ exists such that the r–neighborhood $B(x, r)$ of any point $x \in X$ is a subset of at least one of U_α. Number r is called a Lebesque number of the cover.

SOLUTION:

Since $\{U_\alpha\}$ is a cover of X, each $x \in X$ belongs to at least one U_α. Each U_α is open, hence for each $x \in X$

$$x \in B(x, r_x) \subset U_\alpha. \tag{1}$$

The collection of sets

$$\left\{ B\left(x, \frac{r_x}{2}\right) : x \in X \right\} \tag{2}$$

forms an open cover of X and since X is compact, a finite subcover of (2) exists

$$\left\{ B\left(x_1, \frac{r_{x_1}}{2}\right), \ldots, B\left(x_n, \frac{r_{x_n}}{2}\right) \right\}. \tag{3}$$

The Lebesque number is defined by

$$r = \min\left(\frac{r_{x_1}}{2}, \ldots, \frac{r_{x_n}}{2}\right). \tag{4}$$

Indeed, let $x \in X$ then

$$x \in B\left(x_k, \frac{r_{x_k}}{2}\right)$$

for some $1 \le k \le n$. Let

$$y \in B(x, r) \tag{5}$$

then

$$d(y, x_k) \le d(y, x) + d(x, x_k) < r + \frac{r_{x_k}}{2} \le r_{x_k}. \tag{6}$$

Thus

$$B(x, r) \subset B(x_k, r_{x_k}) \subset U_\alpha \tag{7}$$

for some U_α.

● PROBLEM 16–39

Show that the function $f : R \to R$

$$f(x) = 2x \tag{1}$$

is uniformly continuous.

601

SOLUTION:

The function

$$f : X \to Y \qquad (2)$$

from a metric space (X, d) into a metric space (Y, d') is said to be uniformly continuous if for every $\varepsilon > 0$, there is $\delta > 0$ such that for every $x \in X$

$$f(B(x, \delta)) \subset B(f(x), \varepsilon). \qquad (3)$$

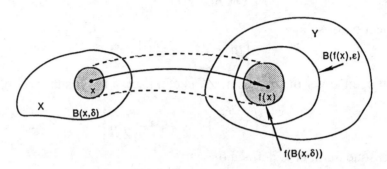

FIGURE 1

Let us choose any $\varepsilon > 0$ and set $\delta = \varepsilon/2$, then for any $x \in X$

$$f(B(x, \delta)) \subset B(f(x), \varepsilon). \qquad (4)$$

Indeed let

$$y \in B(x, \delta)$$

then

$$|x - y| < \varepsilon/2$$

and

$$|f(x) - f(y)| = |2x - 2y| < 2 \cdot \varepsilon/2 = \varepsilon. \qquad (5)$$

Thus, f is uniformly continuous.

● **PROBLEM 16–40**

Let $f : X \to Y$ represent a continuous function from a compact metric space (X, d) into a metric space (Y, d'). Show that f is uniformly continuous.

SOLUTION:

Let ε represent any positive number, $\varepsilon > 0$. We can define an open cover of Y as

$$\{B(y, {}^{\varepsilon}\!/_2) : y \in Y\}. \tag{1}$$

Since f is a continuous function, each $f^{-1}(B(y, {}^{\varepsilon}\!/_2))$ is an open set in X and

$$\{f^{-1}(B(y, {}^{\varepsilon}\!/_2))\} \tag{2}$$

is an open cover of X. But X is a compact metric space. Hence, by Problem 16–38, a Lebesque number r exists for cover (2).

Then for each $x \in X$,

$$f(B(x, q)) \subset B(f(x), \varepsilon). \tag{3}$$

Indeed, from the definition of the Lebesque number $(r = q)$

$$B(x, r) \subset f^{-1}(B(y, {}^{\varepsilon}\!/_2)) \tag{4}$$

where $f(x) = y$

● PROBLEM 16–41

Let (R^2, d) represent R^2 space with the Pythagorean distance. Show that the subset $B(x, 1) \subset R^2$, where $x = (1, 1)$, is totally bounded.

SOLUTION:

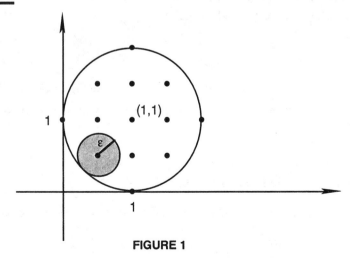

FIGURE 1

Let A represent a subset of a metric space (X, d) and let $\varepsilon > 0$. An ε–net for A is a finite set of points in X

$$N = \{x_1, x_2, \ldots, x_n\}$$

such that for every $x \in A$, $x_k \in N$ exists, such that

$$d(x_1, x_k) < \varepsilon.$$

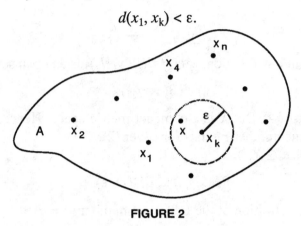

FIGURE 2

DEFINITION

A subset A of a metric space (X, d) is totally bounded if A has an ε-net for every $\varepsilon > 0$.

■

Let $\varepsilon > 0$. As an ε-net for a circle $B(x, 1)$ with the center at $x = (1, 1)$ and radius 1, we can choose all points inside the circle $B(x, 1)$ with coordinates

$$(1, 1), (1, 1 \pm \varepsilon), (1, 1 \pm 2\varepsilon), \ldots , (1 \pm \varepsilon, 1), \ (1 \pm 2\varepsilon, 1), \ \ldots,$$

$$\ldots, (1 \pm k\varepsilon, 1 \pm n\varepsilon)$$

where n and k are positive integers. Note that we are not looking for the smallest number of elements to form an ε-net.

● **PROBLEM 16–42**

Prove:

THEOREM

Totally bounded sets are bounded.

■

SOLUTION:

The diameter of A, $\delta(A)$, is defined by

$$\delta(A) = \sup \{d(x, x') : x, x' \in A\}. \tag{1}$$

Set A is bounded if

$$\delta(A) < \infty \qquad\qquad \qquad (2)$$

Suppose A is a totally bounded set and let

$$N = \{x_1, x_2, \ldots, x_n\} \qquad\qquad (3)$$

be ε-net for A.

Let $x, x' \in A$ be any elements of A. Then two elements of N exist, x_l and x_m, such that

$$d(x_1, x_l) < \varepsilon \quad d(x', x_m) < \varepsilon. \qquad (4)$$

Also for any $x_l, x_m \in N$

$$d(x_l, x_m) < p < \infty \qquad\qquad (5)$$

where p is some positive finite number.

We have

$$d(x, x') \le d(x, x_l) + d(x_l, x') \le$$

$$\le d(x, x_l) + d(x', x_m) + d(x_m, x_l) <$$

$$< \varepsilon + \varepsilon + p < \infty. \qquad\qquad (6)$$

Hence

$$\delta(A) = \sup \{d(x, x') : x, x' \in A\} < \infty \qquad (7)$$

and A is bounded.

● **PROBLEM 16-43**

Give an example of a bounded set which is not totally bounded.

SOLUTION:

We shall first define Hilbert space H. By R^∞ we denote the class of all infinite sequences

$$(a_1, a_2, \ldots) \qquad\qquad (1)$$

such that

$$\sum_{n=1}^{\infty} a_n^2 < \infty. \qquad\qquad (2)$$

Let $a, b \in R^\infty$ and

$$a = (a_1, a_2, \ldots) \quad b = (b_1, b_2, \ldots) \tag{3}$$

then the l_2–metric on R^∞ is defined by

$$d(a,b) = \sqrt{\sum_{n=1}^{\infty} |a_n - b_n|^2}. \tag{4}$$

The metric space (R^∞, d) with the l_2–metric is called Hilbert space, or l_2–space, and denoted by H. Let A represent a subset of H consisting of elements

$$e_1 = (1, 0, 0, \ldots)$$

$$e_2 = (0, 1, 0, \ldots)$$

$$e_3 = (0, 0, 1, \ldots)$$

Then

$$d(e_j, e_k) = \sqrt{1 + 1} = \sqrt{2} \tag{5}$$

and A is bounded because

$$\delta(A) = \sup\{d(e_j, e_k) : e_j, e_k \in A\} = \sqrt{2} < \infty. \tag{6}$$

But A is not totally bounded. Let $\varepsilon = {}^1/_2$ then, ε–net of A consists of all elements of A. The infinite set A cannot be separated into a finite number of subsets, each with diameter less than $\varepsilon = {}^1/_2$.

● PROBLEM 16-44

Show that sequentially compact subsets of a metric space are totally bounded.

SOLUTION:

Let A represent a subset of a metric space (X, d). We shall prove that if A is not totally bounded, then A is not sequentially compact. Suppose A is not totally bounded, then $\varepsilon > 0$ exists, such that no finite ε-net of A exists. Let $a_1 \in A$. See Figure 1.

Then $a_2 \in A$ exists, such that

$$d(a_1, a_2) \geq \varepsilon.$$

Otherwise $\{a_1\}$ would be ε-net. A point $a_3 \in A$ exists, such that

$$d(a_1, a_3) \geq \varepsilon$$

$$d(a_1, a_2) \geq \varepsilon$$

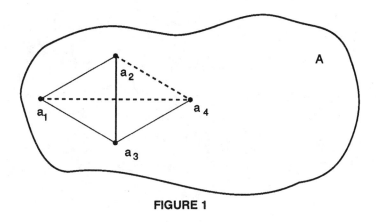

FIGURE 1

Otherwise $\{a_1, a_2\}$ would be ε-net. Following this procedure, we obtain a sequence of points

$$(a_1, a_2, a_3, \ldots),$$

such that

$$d(a_l, a_k) \geq \varepsilon \quad \text{for } l \neq k.$$

This sequence $(a_1, a_2, \ldots)$ does not have any convergent subsequence. Thus, A is not sequentially compact.

● **PROBLEM 16-45**

Show that if A is a countably compact subset of a metric space (X, d), then A is also sequentially compact.

SOLUTION:

Let

$$(a_1, a_2, \ldots) \tag{1}$$

represent a sequence in A. We shall show that a subsequence of (1) exists which converges to a point in A. If the set

$$\{a_1, a_2, \ldots\} \tag{2}$$

is finite then one of the elements, say a_l, of (2) appears infinitely many times in (1). Hence the subsequence

$$(a_l, a_l, a_l, \ldots) \tag{3}$$

of (1) converges to $a_l \in A$.

Suppose (2) is infinite. By hypothesis, A is countably compact. Hence, the infinite subset of A

$$\{a_1, a_2, a_3, ...\}$$

has an accumulation point a in A. Since (X, d) is a metric space, we can choose a subsequence

$$(a_{n_1}, a_{n_2}, a_{n_3}, ...)\tag{4}$$

of (1), which converges to $a \in A$. Thus, A is sequentially compact.

● **PROBLEM 16–46**

Prove this theorem:

THEOREM

If (X, d) is a metric space, then the following statements are equivalent:

1. X is compact.

2. X is countably compact

3. X is sequentially compact.

SOLUTION:

In Problem 16–28, we proved that any compact space is countably compact. Hence

$$1 \Rightarrow 2$$

for any metric space.

In Problem 16–45, we proved that any countably compact metric space is sequentially compact. Hence

$$2 \Rightarrow 3.$$

We shall prove that if a subset A of a metric space (X, d) is sequentially compact, then it is compact, i.e., $3 \Rightarrow 1$.

Let A be sequentially compact and let $\{U_\alpha\}$ be an open cover of A. Set $\{U_\alpha\}$ has a Lebesque number $r > 0$, since every open cover of a sequentially compact subset of a metric space has a Lebesque number. Since A is totally bounded, there is a decomposition of A into a finite number of subsets

$$B_1, B_2, \ldots, B_k$$

such that $d(B_1), d(B_2), \ldots, d(B_k) < r$. But r is a Lebesque number of A, hence open sets exist such that

$$B_1 \subset U_{\alpha_1}, \ldots, B_k \subset U_{\alpha_k}.$$

Thus

$$A \subset B_1 \cup \ldots \cup B_k \subset U_{\alpha_1} \cup \ldots \cup U_{\alpha_k}$$

and $\{U_{\alpha_1}, \ldots, U_{\alpha_k}\}$ is a finite subcover of $\{U_\alpha\}$. A is compact and

$$1 \Leftrightarrow 2 \Leftrightarrow 3.$$

● **PROBLEM 16-47**

Show that any countably compact metric space (X, d) is separable.

SOLUTION:

Let s represent any positive number. For any s, there is a maximal subset A_s of X, such that for any $a, b \in A_s$

$$d(a, b) \geq s. \tag{1}$$

Suppose A_s is infinite for some $s > 0$, then A_s has an accumulation point y. Then $B(y, s/2)$ contains infinitely many points of A_s. But any of these points are closer than s. Thus, a contradiction and A_s is finite for each s. But then for any $x \in X$

$$A_s \cap B(x, s) \neq \phi. \tag{2}$$

Otherwise we would obtain a contradiction to the maximality of A_s. For each positive integer n we find $A_{\frac{1}{n}}$. Then

$$\bigcup_{n \in N} A_{\frac{1}{n}}$$

is a countable dense subset of X. Hence X is separable.

● **PROBLEM 16-48**

Find an example of a compactification of the open interval $(0, 1)$ with the absolute value topology.

SOLUTION:

DEFINITION OF COMPACTIFICATION

Let (X, T) represent any topological space. A compactification of X is a compact space (Y, T'), such that X is homeomorphic to a dense subspace of Y.

■

Note that by compactification of X, we understand a space Y and a homeomorphism from X onto a dense subspace of Y.

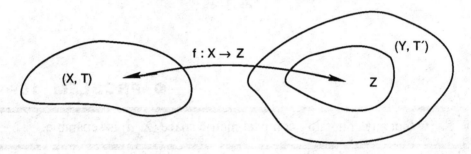

f is homeomorphism
Z is dense in Y

FIGURE 1

Thus, for a given X and Y, it is possible to find more than one compactification (possibly infinitely many).

Two possible compactifications of the open interval $(0, 1)$ are the circle and the closed interval $[0, 1]$.

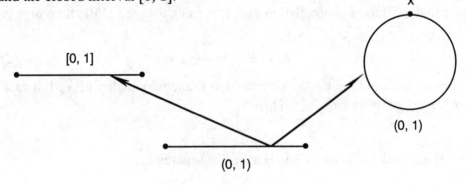

FIGURE 2

Obviously $(0, 1)$ is homeomorphic to a dense subset of $[0, 1]$. Also $(0, 1)$ is homeomorphic to a circle with one point removed.

Let P denote the (x, y) plane in the R^3 space with the Euclidean topology. Let S be the sphere of radius 1 and center at $(0, 0, 1)$.

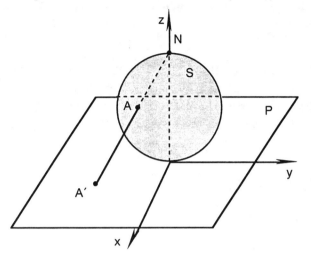

FIGURE 1

Show that S is a compactification of P.

SOLUTION:

Consider a line passing through the point $N = (0, 0, 2)$ on the sphere S and any point A on the sphere as shown in the figure.

Take any point A' on the plane, $A' \in P$. Then a line from A' to N determines the image of A' on the sphere.

$$f : P \to S$$

is defined by

$$f(A') = A.$$

f is one-to-one and onto and both f and f^{-1} are continuous. Hence, f is a homeomorphism from the plane P onto the subset $S - \{N\}$ of S. Obviously P is not compact, while S is compact.

The set $S - \{N\}$ is dense in S. Thus, S is a compactification of P.

● **PROBLEM 16-50**

Define the Alexandroff (sometimes called one-point) compactification of a T_2–space.

SOLUTION:

Let (X, T) represent any T_2–space. If X is compact then we define the Alexandroff compactification of X to be X itself.

Now suppose X is not compact. Let A be any point which does not belong to X. We define

$$Y = X \cup \{A\}$$

and the topology on Y as follows: U is open in Y if U is open in X, i.e., $U \in T$; or if $A \in U$, then $Y - U$ is a compact subset of X.

The Alexandroff, or one-point, compactification of X is defined as the set Y with the topology defined above.

Point A is called an ideal point. It can be shown that

1. Y is a topological space

2. Y is compact

3. X is dense in Y.

It can be also proved that:

THEOREM

The Alexandroff compactification Y of any topological T_2–space, (X, T) is a topological space and it is a compactification of X in the sense of the definition of Problem 16–48.

● **PROBLEM 16–51**

Prove the following:

THEOREM

The Alexandroff compactification of a space (X, T) is T_2 if and only if X is T_2 and locally compact. ∎

SOLUTION:

Suppose (X, T) is compact. Then the Alexandroff compactification of X is X itself.

Then if X is T_2, X is T_2 and also locally compact. Also if X is T_2 and locally compact, then X is T_2.

Suppose X is not compact and Y is its Alexandroff compactification. If Y

is T_2, then X is T_2, since X is a subspace of Y.

Since Y is T_2 and compact, Y is also locally compact. X is locally compact as an open subspace of Y.

Now suppose X is T_2 and locally compact. Let x and y represent elements of Y, $x \neq y$. If $x, y \in X$, then open sets U and V exist such that

$$x \in U \quad \text{and} \quad y \in V$$

$$U \cap V = \phi$$

because X is T_2.

Suppose $x = A$ (A is an ideal point). Then $y \in X$ and a compact subset B of X exists, such that

$$y \in \text{Int } B \subset B.$$

Hence $Y - B$ is a neighborhood of $x = A$ and $\text{Int } B$ is a neighborhood of y and

$$(Y - B) \cap \text{Int } B = \phi.$$

Thus Y is T_2.

CHAPTER 17

CONNECTEDNESS

Show that any set X, consisting of two or more points, with discrete topology is disconnected.

SOLUTION:

DEFINITION OF MUTUAL SEPARATION

In the space (X, T), two nonempty, disjoint sets $U \subset X$, $V \subset X$ are said to be mutually separated if neither U, nor V contains a boundary point of the other.

■

DEFINITION OF CONNECTED SPACES

A space (X, T) is said to be disconnected if the set X can be expressed as the union of multiple, mutually separated, nonempty subsets of X. The set X is connected if it is not disconnected, (see Figure 1).

■

$$X = U \cup V$$

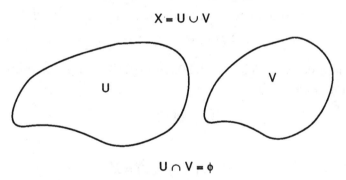

$$U \cap V = \phi$$

FIGURE 1

Space (X, T) with the discrete topology consisting of more than one point is disconnected. Indeed take any

$$x \in X \tag{1}$$

and consider the sets

$$\{x\} \quad \text{and} \quad X - \{x\}. \tag{2}$$

Both sets (2) are open, nonempty, and disjoint. Their union is X

$$\{x\} \cup [X - \{x\}] = X. \tag{3}$$

Any set with the trivial topology is connected since the only nonempty open subset of X is X itself.

Show that connectedness is a topological property.

SOLUTION:

We shall prove that if $f : (X, T) \to (Y, T')$ is a continuous function onto Y and X is connected, then Y is connected.

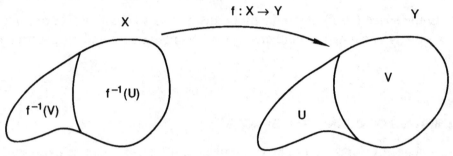

FIGURE 1

Suppose Y is not connected. Then there are open sets U and V such that

$$Y = U \cup V, \quad U \cap V = \phi. \tag{1}$$

Since f is continuous, then both $f^{-1}(U)$, and $f^{-1}(V)$ are open subsets of X. Since f is a bijection then

$$f^{-1}(U) \cap f^{-1}(V) = \phi \tag{2}$$

and

$$f^{-1}(U) \cap f^{-1}(V) = X. \tag{3}$$

Hence X is disconnected which is a contradiction. Therefore, Y is connected.

For example, if X is a connected space and R is an equivalence relation on R, then the identification space $X/_R$ is connected.

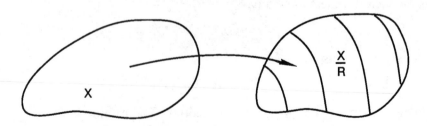

FIGURE 2

The identification mapping $f: X \to X/_R$ is continuous.

Show that if (X, T) is any topological space, then the following statements are equivalent:

1. X is connected.

2. X cannot be expressed as the union of two nonempty, disjoint, closed subsets.

SOLUTION:

1. $\Rightarrow$ 2.

Suppose X is connected and two sets F and G exist such that

$$X = F \cup G \text{ and } F \cap G = \phi \qquad (1)$$

and F, G are nonempty and closed in X. Then

$$G = X - F$$

$$F = X - G \qquad (2)$$

as complements of closed sets are open. Hence,

$$X = F \cup G$$

is the union of disjoint, nonempty, open subsets of X. Thus, X is not connected.

2. $\Rightarrow$ 1.

Suppose X is disconnected. Then open sets U, V exist such that

$$X = U \cup V, \quad U \cap V = \phi \qquad (3)$$

where U, V are nonempty. Then

$$U = X - V \quad \text{and} \quad V = X - U.$$

Hence, U and V are closed, a contradiction with 2.

Let (X, T) be any topological space. Show that the following statements are equivalent:

1. X is connected.

2. The only subsets of X which are open and closed are X and ϕ.

3. No continuous

$$f : X \to Y$$

is onto, where $Y = \{0, 1\}$ with the discrete topology.

SOLUTION:

1. $\Rightarrow$ 2.
   Suppose $A \subset X$ is both open and closed, then

$$A \text{ and } X - A$$

are both open and

$$A \cup (X - A) = X, \quad A \cap (X - A) = \phi.$$

Hence, X is disconnected, a contradiction.

2. $\Rightarrow$ 3.
   Suppose

$$f : X \to \{0, 1\}$$

is a continuous and onto. Then

$$f^{-1}(0) \ne \phi, \ f^{-1}(0) \ne X.$$

Since the subset $\{0\}$ of $\{0, 1\}$ is open and closed,

$$f^{-1}(0)$$

is open and closed in X. Contradiction.

3. $\Rightarrow$ 1.
   Let U and V be disjoint, nonempty, open sets, such that

$$X = U \cup V.$$

The U and V are also closed. The characteristic function

$$\chi_A : X \to \{0, 1\}$$

is continuous and onto.

Suppose (X, T) is a space such that

$$X = U \cup V$$

$$U \cap V = \phi \tag{1}$$

where U and V are open, nonempty subsets of X. Show that if A is any connected subset of X, then either

$$A \subset U \quad \text{or} \quad A \subset V. \tag{2}$$

SOLUTION:

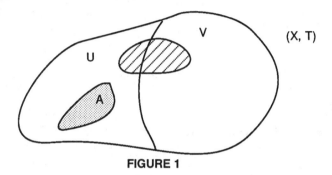

FIGURE 1

Suppose

$$A \cap U \neq \phi \quad \text{and} \quad A \cap V \neq \phi. \tag{3}$$

Then sets $A \cap U$ and $A \cap V$ are nonempty, disjoint subsets of A which are open in A. Then

$$A = (A \cap U) \cup (A \cap V). \tag{4}$$

Hence, A is not connected.

Therefore, either $A \cap U = \phi$ or $A \cap V = \phi$. Hence, we have

$$A \subset V \quad \text{or} \quad A \subset U.$$

Note that a subset A of X is said to be connected if the subspace A is connected.

● **PROBLEM** 17-6

Let A and B be connected sets which are not mutually separated. Show that

$$A \cup B$$

is connected.

SOLUTION:

For working convenience write the definition of mutually separated sets, as given in Problem 17–1, in terms of set theory.

DEFINITION OF MUTUALLY SEPARATED SETS

In the topological space (X, T), $A \subset X$ and $B \subset X$ are said to be mutually separated if

$$A \cap B = \phi$$

$$\overline{A} \cap B = \phi$$

$$A \cap \overline{B} = \phi$$

■

Suppose $A \cup B$ is disconnected. Then

$$A \cup B = U \cup V.$$

Since A is a connected subset of $A \cup B$, we have either

$$A \subset U \quad \text{or} \quad A \subset V$$

by Problem 17–5.
  Also, either

$$B \subset U \quad \text{or} \quad B \subset V.$$

If $A \subset U$ and $B \subset V$ (or $A \subset V$ and $B \subset U$), then

$$A = (A \cup B) \cap U$$

$$B = (A \cup B) \cap V$$

are separated sets, a contradiction with the hypothesis. Thus, either

$$A \cup B \subset U \quad \text{or} \quad A \cup B \subset V.$$

Therefore, $U \cup V$ is not a disconnection of $A \cup B$ and $A \cup B$ is connected.

Prove the following:

THEOREM

The union of any family of connected subsets of any space (X, T) having at least one point in common is also connected.

∎

SOLUTION:

Let $\{A_\alpha\}$ be a family of connected subsets of X such that for each α

$$x_0 \in A_\alpha$$

$$A = \bigcup_\alpha A_\alpha$$

$$x_0 \in \bigcap_\alpha A_\alpha$$

Let (Y, T') be the space $Y = \{0, 1\}$ with discrete topology. Let

$$f : A \to Y$$

be continuous. Since each A_α is connected and each $f \mid A_\alpha$ is continuous, no $f \mid A_\alpha$ is onto. Because for each α

$$x_0 \in A_\alpha$$

we obtain

$$f(x) = f(x_0)$$

for each $x \in A_\alpha$ for all α. Hence, f is not onto and $\bigcup_\alpha A_\alpha$ is connected.

Note that the intersection of two connected sets need not be connected.

● PROBLEM 17-8

Let A be a connected subset of (X, T) and let

$$A \subset B \subset \overline{A} \qquad (1)$$

Show that B is connected.

SOLUTION:

Suppose B is disconnected, then
$$B = U \cup V.$$

Since A is a connected subset of B, we have either
$$A \cap U = \phi \quad \text{or} \quad A \cap V = \phi.$$

Suppose $A \cap U = \phi$. Then $X - U$ is closed and

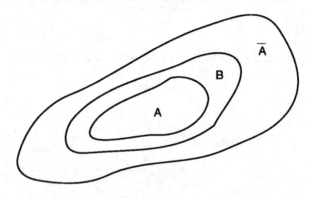

FIGURE 1

$$A \subset B \subset \overline{A} \subset X - U.$$

Thus,

$$B \cap U = \phi.$$

This is in contradiction with $B = U \cup V$. Hence, B is connected.

● **PROBLEM 17-9**

Show that the only connected subsets of R with the absolute value topology, having more than one point, are R and the intervals (open, closed, half-open).

SOLUTION:

We shall show that
$$(Y \text{ connected subset of } R) \Leftrightarrow (Y \text{ is an interval}).$$

$\Rightarrow$ Suppose Y is connected and is not an interval. Then there are
$$a, b \in Y \quad \text{and} \quad c \notin Y \tag{1}$$

622

such that $a < c < b$. Then

$$Y \cap \{x : x > c\}$$

$$Y \cap \{x : x < c\} \tag{2}$$

is a decomposition of Y, since both sets (2) are open, disjoint, and nonempty.

$\Leftarrow$ Suppose Y is an interval and is not connected. Then disjoint, nonempty, open sets A, B exist such that

$$Y = A \cup B \tag{3}$$

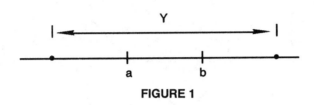

FIGURE 1

we can find $a \in A$, $b \in B$, such that

$$a < b. \tag{4}$$

Let us define

$$\omega = \sup \{x : [a, x) \subset A\} \tag{5}$$

and since Y is an interval,

$$\omega \in Y \quad \omega \le b. \tag{6}$$

We have

$$\omega \in \overline{A}_Y \tag{7}$$

and since

$$A = Y - B$$

is closed in Y, we have

$$\omega \in A. \tag{8}$$

But, A is also open in Y, hence, $\varepsilon > 0$ exists such that

$$(\omega - \varepsilon, \omega + \varepsilon) \subset A \tag{9}$$

which contradicts the definition of ω.

Prove the following variation of the fixed-point theorem:

THEOREM 1

If $f: [0, 1] \rightarrow [0, 1]$ is continuous, then $x_0 \in [0, 1]$ exists such that
$$f(x_0) = x_0.$$

■

SOLUTION:

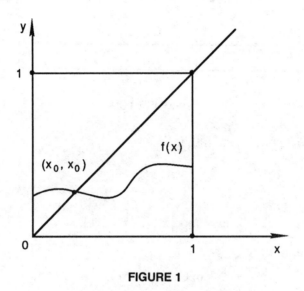

FIGURE 1

In Problem 17–2, we proved that a continuous image of a connected set is connected. The only connected subsets (consisting of more than one point) of R are intervals. Hence, the conclusion in the form of a theorem.

THEOREM 2

If $f: X \rightarrow R$ is a continuous function defined on a connected set X, then f assumes all values between any two of its values.

■

Theorem 1 is a conclusion of Theorem 2. Also note that the graph of the continuous function
$$f: [0, 1] \rightarrow [0, 1]$$
is located in the square
$$[0, 1] \times [0, 1].$$

Hence, it must intersect the diagonal joining (0, 0) with (1, 1). Therefore, point $x_0 \in [0, 1]$ exists such that

$$f(x_0) = x_0.$$

Consider Theorem 1 of Problem 17–10. Prove this theorem using the sets

$$U = \{(x, y) : x < y\}$$

$$V = \{(x, y) : x > y\}. \tag{1}$$

Observe that the graph of f is connected.

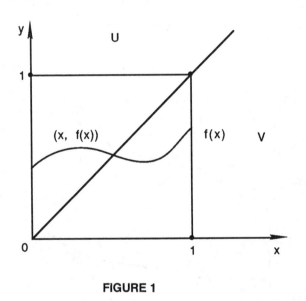

FIGURE 1

SOLUTION:

If $f(0) = 0$ or $f(1) = 1$, the theorem is proved. Otherwise

$$f(0) > 0 \quad \text{and} \quad f(1) < 1 \tag{2}$$

then

$$(0, f(0)) \in U \quad (1, f(1)) \in V. \tag{3}$$

Suppose the graph of f

$$F = \{(x, y) : x \in [0, 1], y = f(x)\} \tag{4}$$

does not contain any points of the diagonal

$$D = \{(x, y) : x = y\}. \tag{5}$$

Then $U \cup V$ is a disconnection of F. But F is connected as a continuous image of a connected set. Contradiction. Thus, F contains a point (x_0, x_0) of the diagonal, hence

$$f(x_0) = x_0. \tag{6}$$

● **PROBLEM 17–12**

Let (X, T) be a space such that for any two elements

$$x, y \in X \tag{1}$$

a connected set $A \subset X$ exists such that

$$x, y \in A. \tag{2}$$

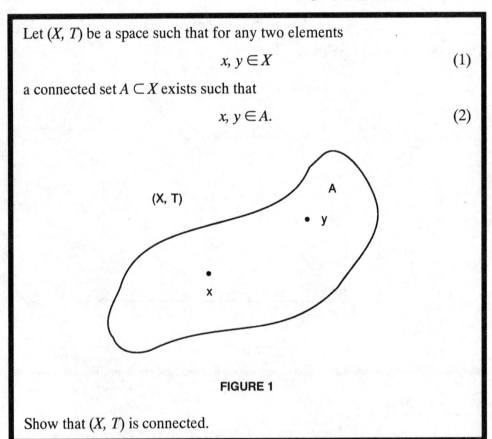

FIGURE 1

Show that (X, T) is connected.

SOLUTION:

Let x be a fixed element of X. Then, for any $y \in X$ there exists a connected subspace $A(x, y)$ of X such that

$$x, y \in A(x, y).$$

Consider the family of sets

$$\{A(x, y) : y \in X\} \qquad (3)$$

which consists of connected subspaces of X and

$$X = \bigcup_{y \in X} A(x,y). \qquad (4)$$

The intersection of (3) is nonempty

$$X \in \bigcap_{y \in X} A(x,y). \qquad (5)$$

By Problem 17–7, we conclude that (X, T) is connected.

● **PROBLEM 17-13**

Apply Problem 17–12 to show that the Euclidean n-space R^n $(n > 1)$ with one point S removed

$$R^n - \{S\}$$

is connected.

SOLUTION:

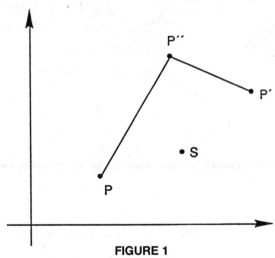

FIGURE 1

Suppose P and P' are any two points of $R^n - \{S\}$. Any closed line segment in the Euclidean space R^n is homeomorphic to the closed interval $[0, 1]$. Hence, it is a connected subspace of R^n. We choose point $P'' \in R^n - \{S\}$ such that

$$S \notin \overline{PP''}$$

and

627

$$S \notin \overline{P'' P'}.$$

The sets $\overline{PP''}$ and $\overline{P'' P'}$ are connected and their intersection is nonempty

$$\overline{PP''} \cap \overline{P''P'} = \{P''\} \neq \phi.$$

Hence,

$$\overline{PP''} \cup \overline{P''P'}$$

is connected.

Points P and P' belong to the connected subspace of $R^n - \{S\}$. Hence, $R^n - \{S\}$ is connected. Note that this is not true for $n = 1$.

● **PROBLEM 17–14**

Which of the subsets of R^2 are polygonally connected?

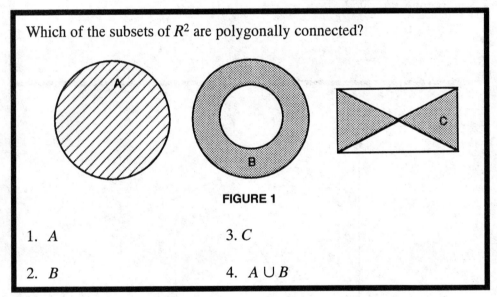

FIGURE 1

1. A

2. B

3. C

4. $A \cup B$

SOLUTION:

DEFINITION OF A PATH AND PATH-CONNECTED SPACES

A subspace Y of any space (X, T) is said to be a path in X if there is a continuous function from $[0, 1]$ (with the absolute value topology)

$$f : [0, 1] \rightarrow Y$$

onto Y.

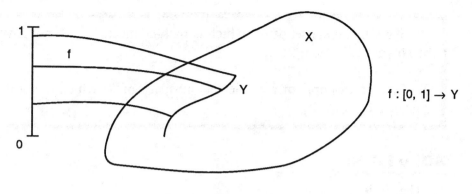

FIGURE 2

X is said to be path-connected if any two points x, y of X belong to some path in X.

∎

Consider the Euclidean space R^n. A subset $A \subset R^n$ is said to be polygonally connected if for any x, $y \in A$ there are points

$$x = x_0, x_1, \ldots, x_{k-1}, x_k = y$$

such that

$$\bigcup_{i=0}^{k-1} \overline{x_i x_{i+1}} \subset A$$

where $\overline{x_i x_{i+1}}$ is a closed segment joining x_i and x_{i+1}.

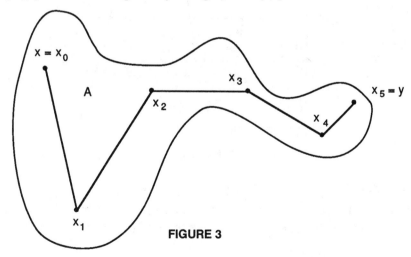

FIGURE 3

Hence, A, B, and C are polygonally connected, while $A \cup B$ is not.

629

1. Show that a subset of R^n, which is path-connected, does not have to be polygonally connected.

2. Give an example of a connected subspace of R^2 which is not path-connected.

SOLUTION:

1. The circle

$$\{(x, y) : x^2 + y^2 = 1\} \quad R^2$$

is path-connected, but is not polygonally connected.

2. Consider the set

$$A = \{(x, y) : y = \sin \frac{1}{x}, x > 0\} \cup \{0, 0\}$$

which is a subset of R^2.

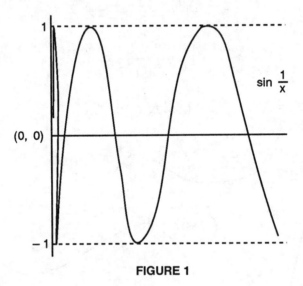

FIGURE 1

From Problem 17–8 we see that A is connected.

But A is not path-connected. There is no path in A which contains $(0, 0)$ and any point $P \neq (0, 0)$ of A. Otherwise the function $f(x) = \sin \frac{1}{x}, x > 0$ would be continuous.

Prove that any polygonally connected subspace of R^n is path-connected.

SOLUTION:

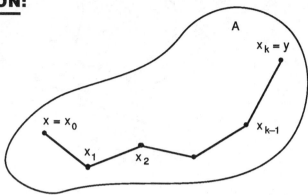

FIGURE 1

Let $A \subset R^n$ be polygonally connected and let x, y be any points in A.

Since A is polygonally connected, there are points $x = x_0, x_1, \ldots, x_k = y$ such that

$$\bigcup_{i=0}^{k-1} \overline{x_i x_{i+1}} \subset A$$

Set $\bigcup_{i=0}^{k-1} \overline{x_i x_{i+1}}$ is a path. A continuous function exists

$$f : [0,1] \to \bigcup_{i=0}^{k-1} \overline{x_i x_{i+1}}.$$

Hence, A is path-connected.

Let

$$f : [0, 1] \to X, f(0) = a, f(1) = b$$

be a path from $a \in X$ to $b \in X$ and

$$g : [0, 1] \to X, g(0) = b, g(1) = c$$

be a path from $b \in X$ to $c \in X$.

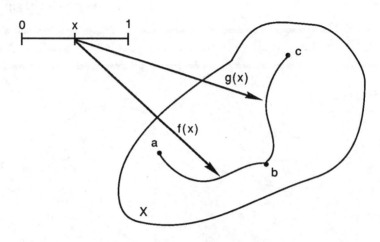

FIGURE 1

Show that the juxtaposition of the two paths f and g is a path from a to c.

SOLUTION:

The juxtaposition of the two paths f and g, denoted by $f \circ g$ is defined as follows:

$$f \circ g : [0, 1] \to X$$

$$(f \circ g)(t) = \begin{cases} f(2t) & \text{for} \quad 0 \le t \le \frac{1}{2} \\ g(2t - 1) & \text{for} \quad \frac{1}{2} \le t \le 1 \end{cases}.$$

Note that $\qquad (f \circ g)(0) = f(0) = a$

$$(f \circ g)(^1/_2) = f(1) = b = g(0)$$

and $\qquad (f \circ g)(1) = g(1) = c.$

Thus, $f \circ g$ is a path from a to c.

● **PROBLEM 17–18**

1. Show that if (X, T) is a path-connected space, then X is connected.

2. Explain why any convex subset of R^n is connected.

SOLUTION:

1. In Problem 17–7 we proved that if (X, T) is a space and
$$X = \bigcup_\alpha A_\alpha$$
where $\{A_\alpha\}$ is a collection of connected subspaces of X and
$$\bigcap_\alpha A_\alpha \neq \phi$$
contained in some connected subspace of X, then X is connected.

   Suppose X is path-connected and $x \in X$. Then for each $y \in X$, $P(x, y)$ denotes a path from x to y.

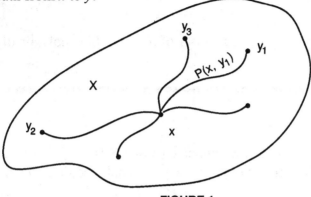

FIGURE 1

Then
$$X = \bigcup_{y \in X} P(x,y)$$

and
$$X \in \bigcap_{y \in X} P(x,y) \neq \phi.$$

Since each $P(x, y)$ is connected (as a continuous image of a connected set $[0, 1]$), X is connected.

2.

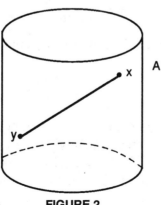

FIGURE 2

A subset A of R^n is convex if, for any points x and y of A, the closed segment $\overline{xy}$ is a subset of A.

Any convex subset of R^n is polygonally connected and, hence, by Problem 17–16, is path-connected and, hence, is connected.

● **PROBLEM 17–19**

Prove the following theorem concerning the Euclidean R^n spaces.

THEOREM

If U is an open connected subset of R^n, then U is polygonally connected.

∎

SOLUTION:

Suppose $U \subset R^n$ is open and connected and $u \in U$.

Let X be all points of U which can be polygonally connected with u

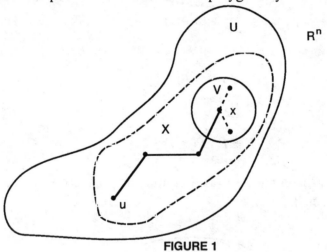

FIGURE 1

$$X = \{x \in U : x \text{ can be polygonally connected to } u \in U\}. \qquad (1)$$

Let

$$Y = U - X. \qquad (2)$$

Since U is open for any $x \in X$, there is an open neighborhood V of x such that u can be polygonally connected with any point of V; hence,

$$V \subset X. \qquad (3)$$

Therefore, X is open.

634

The set $Y = U - X$ consists of elements which cannot be polygonally connected to u. By the same argument Y is open. Thus,

$$U = X \cup Y, X \cap Y = \phi \tag{4}$$

where X and Y are disjoint open subsets of U. Since U is connected, we have two options:

either $\qquad\qquad\qquad X = \phi, \qquad Y = U$

or $\qquad\qquad\qquad X = U, \qquad Y = \phi. \tag{5}$

Since $\qquad\qquad\qquad u \in X, X \neq \phi$

and $\qquad\qquad\qquad X = U, \quad Y = \phi$

and U is polygonally connected.

● PROBLEM 17-20

Show that for any $n > 1$, R^n is not homeomorphic to R.

SOLUTION:

Consider the space R (with the absolute value topology) with one point removed $R - \{a\}$.

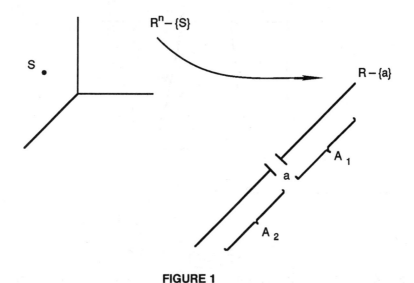

FIGURE 1

635

Then R is disconnected into two disjoint open sets

$$A_1 = \{x \in R : x > a\}$$

$$A_2 = \{x \in R : x < a\}$$

Hence, $R - \{a\}$ is disconnected. Suppose R^n $(n > 1)$ is homeomorphic to R, then for any point $S \in R^n$, $R^n - \{S\}$ is not connected.

In Problem 17–13, we showed that $R^n - \{S\}$ is connected. Therefore, R^n is not homeomorphic to R.

● **PROBLEM 17–21**

In Problem 17–18 we proved that

$$((X, T) \text{ is path-connected}) \Rightarrow ((X, T) \text{ is connected}).$$

Show that the converse of this theorem is not true.

SOLUTION:

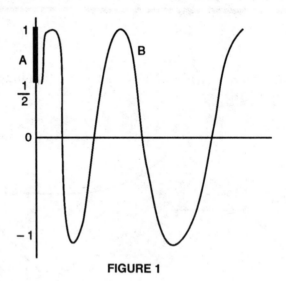

FIGURE 1

Let A and B be the subsets of R^2

$$A = \{(0, y) : {}^1\!/_2 \le y \le 1\}$$

$$B = \{(x, \sin {}^1\!/_x) : 0 < x \le 1\}.$$

Set A is a closed interval and B is a continuous image of a closed interval; hence, both are connected. Each point of A is an accumulation point of B, hence A and B are not separated. Thus,

636

$$A \cup B$$

is connected.

But $A \cup B$ is not path-connected. There is no path connecting any point of A with any point of B.

The converse implication is not true.

● **PROBLEM 17-22**

Show that the continuous image of a path-connected set is path-connected.

SOLUTION:

Let $A \subset X$ be path-connected and let $f : X \rightarrow Y$ be continuous. We shall show that $f(A)$ is path-connected. Let $a, b \in f(A)$, then $a', b' \in A$ (not necessarily uniquely defined) exist that

$$f(a') = a \quad \text{and} \quad f(b') = b.$$

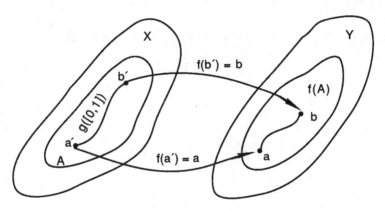

FIGURE 1

Since A is path-connected, there is a continuous function

$$g : [0, 1] \rightarrow A$$

such that $g(0) = a'$, $g(1) = b'$.

Since both f and g are continuous,

$$f \circ g : [0, 1] \rightarrow f(A)$$

is continuous and

$$f(g(0)) = f(a') = a$$

637

$$f(g(1)) = f(b') = b.$$

Thus, $f(A)$ is path-connected.

1. Show that an open disc in the Euclidean R^2 plane is path-connected (do not use results of Problem 17–19).

2. Let P be a class of path-connected subsets of X with a nonempty intersection

$$P = \{A_\alpha : \alpha \in A\}.$$

Show that

$$A = \bigcup_\alpha A_\alpha$$

is path-connected.

SOLUTION:

1. Let p and q be any points of an open disc in the R^2 plane.

$$p = (x_1, y_1)$$

$$q = (x_2, y_2)$$

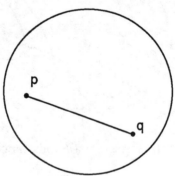

FIGURE 1

A path from p to q may be defined as

$$f : [0, 1] \rightarrow R^2$$

$$f(t) = (x_1 + t(x_2 - x_1), y_1 + t(y_2 - y_1))$$

for $t \in [0, 1]$. Then $f(t)$ is the line segment between p and q. For all points p and q in the open disc, $f(t)$ is a path contained completely in the disc. Therefore the open disc in R^2 is path-connected.

2. Let $x, y \in A$. Path-connected sets exist such that

$$x \in A_x, \quad y \in A_y$$

$$A_x, A_y \in P.$$

Nonempty intersection $\bigcap_\alpha A_\alpha$ contains at least one point, say q

$$q \in \bigcap_\alpha A_\alpha.$$

Then $x, q \in A_x$ and there is a path

$$f : [0, 1] \to A_x$$

$$f(0) = x, f(1) = q.$$

Similarly, there is a path

$$g : [0, 1] \to A_y$$

$$g(0) = q, \quad q(1) = y.$$

The juxtaposition of f and g is a path from x to y (see Problem 17–17). Hence, A is path-connected.

● **PROBLEM 17-24**

Let X be connected and let $P = \{A_\alpha\}$ be an open cover of X. Show that any pair of points of X can be joined by a simple chain consisting of elements of P.

SOLUTION:

DEFINITION OF A SIMPLE CHAIN

Subsets $A_1, A_2, \ldots, A_K$ of X are said to form a simple chain joining points x and y in X if

1. only A_1 contains x

2. only A_K contains y

3. $A_i \cap A_j = \phi$ iff $|i-j| > 1$

∎

Let x be any point in X and let B be the set consisting of all points of X which can be joined to x by a simple chain consisting of elements of P.

Since $x \in B$, $B \neq \phi$. We shall prove that B is both open and closed, i.e., $B = X$ because X is connected.

B is open.

Let $b \in B$, then there is a simple chain $A_1, A_2, ..., A_l$ from x to b.

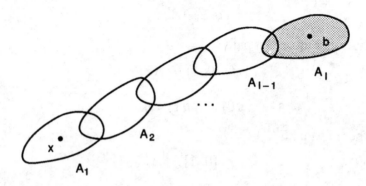

FIGURE 1

Note that all elements of A_l belong to B (chain $A_1, ..., A_l$ connects any element of A_l with x). Hence,

$$b \in A_l \subset B.$$

Since A_l is open, B is open.

B is closed.

Suppose

$$y \notin B.$$

Since P is a cover of X, there is an open $A_t \in P$ such that

$$y \in A_t$$

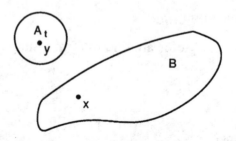

FIGURE 2

640

Suppose

$$A_t \cap B \neq \phi$$

then there is some chain from y to x and $y \in B$, contradiction.

Thus,

$$A_t \cap B = \phi$$

and

$$y \in A_t \subset X - B.$$

Hence, $X - B$ is open and B is closed. We conclude

$$X = B.$$

● **PROBLEM 17-25**

Let C_1 and C_2 be subsets of R^2

$$C_1 = \{(x, y) : x^2 + y^2 < 1\}$$

$$C_2 = \{(x, y) : (x - 2)^2 + y^2 < 1\}$$

Are C_1 and C_2 mutually separated?

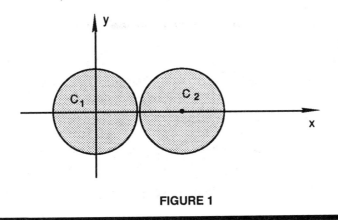

FIGURE 1

SOLUTION:

DEFINITION OF MUTUALLY SEPARATED SETS

Two sets A_1 and A_2 of a space (X, T) are said to be mutually separated if

$$\overline{A_1} \cap A_2 = A_1 \cap \overline{A_2} = \phi.$$

Where $\overline{A_1}$ indicates the closure of A_1. Hence,

$$\overline{C_1} = \{(x, y) : x^2 + y^2 \leq 1\}$$

$$\overline{C_2} = \{(x, y) : (x-2)^2 + y^2 \leq 1\}$$

and

$$\overline{C_1} \cap C_2 = C_1 \cap \overline{C_2} = \phi.$$

The sets C_1 and C_2 are mutually separated.

● **PROBLEM 17-26**

Not every subspace of a connected space is connected. The following theorem enables us to determine whether or not a subspace of a given space is connected.

THEOREM

A subspace A of (X, T) is connected iff A cannot be expressed as the union $C \cup D$, where C and D are nonempty, mutually separated subsets of X.

∎

Prove it.

SOLUTION:

If A is not connected, then

$$A = C \cup D \quad \text{and} \quad C \cap D = \phi \qquad (1)$$

where C and D are nonempty subsets of A, which are open and closed in A. Suppose

$$x \in C \cap \overline{D} \quad \text{then} \quad x \in C \text{ and } x \in \overline{D}. \qquad (2)$$

Since

$$C \subset A$$

$$x \in A \cap \overline{D} = \overline{D}_A = D. \qquad (3)$$

Then

$$x \in C \cap D = \phi \qquad (4)$$

a contradiction.

642

Hence,

$$C \cap \overline{D} = \phi. \qquad (5)$$

Similarly, we show that

$$D \cap \overline{C} = \phi. \qquad (6)$$

Now, suppose

$$A = C \cup D$$

$$C \cap \overline{D} = D \cap \overline{C} = \phi \quad \text{and} \quad C, D \neq \phi. \qquad (7)$$

Then

$$\overline{C_A} = \overline{C} \cap A = (C \cup D) \cap \overline{C} =$$

$$= (C \cap \overline{C}) \cup (D \cap \overline{C}) = C \cup \phi = C. \qquad (8)$$

Similarly,

$$\overline{D_A} = D \qquad (9)$$

where $\overline{D_A}$ indicates the closure of D in A, i.e., $\overline{D_A} = \overline{D} \cap A$.
Hence, C and D are closed in A, and A is disconnected.

● **PROBLEM 17-27**

1. Show that if a space (X, T) contains a connected dense subspace, then X is connected.

2. Show that if (X, T) is connected then any compactification Y of X is connected.

SOLUTION:

1. Let $A \subset X$ be a connected dense subspace of X.

$$\overline{A} = X.$$

In Problem 17–8 we proved that if A is a connected subset of (X, T) and

$$A \subset B \subset \overline{A}$$

then B is connected.
   Thus,

$$A \subset X \subset \overline{A}$$

and since A is connected, X is also connected.

2. Remember that a compactification of (X, T) is a compact space (Y, T') such that X is homeomorphic to a dense subspace of Y.

In other words, X is a dense connected subspace of Y; hence, Y is connected.

● **PROBLEM 17-28**

Let

$$X = \{(x, y) : 0 < x, y = \sin \frac{1}{x}\} \cup \{(0, 0)\} \subset R^2.$$

Apply Problem 17–8 to show that X is connected. (See following figure.)

SOLUTION:

Consider the set

$$A = X - \{(0, 0)\}.$$

We define a function

$$f : R \rightarrow R^2$$

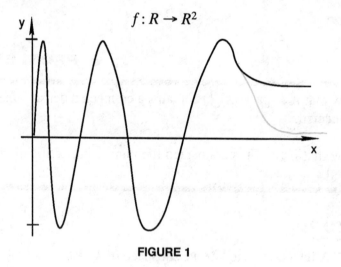

FIGURE 1

by

$$R \supset \{x : x > 0\} \ni x \rightarrow f(x) = (x, \sin \frac{1}{x}) \in R^2.$$

Note that f is continuous. Hence, since $\{x : x > 0\}$ is connected, $A = X - \{(0, 0)\}$ is connected as a continuous image of a connected set.

Since every neighborhood of $(0, 0)$ contains an infinite amount of points of A,

$$(0, 0) \in \bar{A}.$$

We have

$$A \subset X \subset \overline{A}.$$

Hence, since A is connected, by Problem 17–8, we conclude that X is connected.

● **PROBLEM 17-29**

Show that a component A of a space (X, T) is closed.

SOLUTION:

A disconnected space has connected subspaces.

DEFINITION OF COMPONENT

A component A of (X, T) is a maximal connected subset of X.

■

We see that A is connected and A is not a proper subset of any connected subset of X. By Problem 17–8 since

$$A \subset \overline{A} \subset \overline{A}$$

$\overline{A}$ is connected.
But

$$A \subset \overline{A}$$

and A is a maximal connected subspace of X; hence,

$$A = \overline{A}.$$

Therefore, A is closed.

● **PROBLEM 17-30**

1. Let R be the space of real numbers with the absolute value topology and let

$$A_+ = \{x \in R : x > a\}$$

$$A_- = \{x \in R : a > x\}$$

where $a \in R$. Find the components of $R - \{a\}$.

2. Let R^2 be the plane

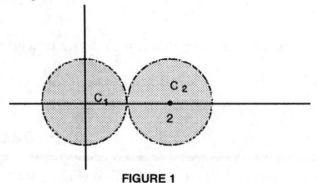

FIGURE 1

with the Pythagorean topology and

$$C_1 = \{(x, y) : x^2 + y^2 < 1\}$$

$$C_2 = \{(x, y) : (x - 2)^2 + y^2 < 1\}$$

Find the components of the space $C_1 \cup C_2 \subset R^2$.

SOLUTION:

1. The components of

$$R - \{a\} = A_+ \cup A_-$$

are A_+ and A_-. Each is the maximal connected subspace of $R - \{a\}$.

2. Here the components of $C_1 \cup C_2$ are C_1 and C_2.

● **PROBLEM 17–31**

Show that each element of (X, T) is contained in a component of X.

SOLUTION:

Obviously each element x of X is contained in at least one connected subspace of X, namely $\{x\} \subset X$. We will show that there is a maximal connected subspace of X which contains x.

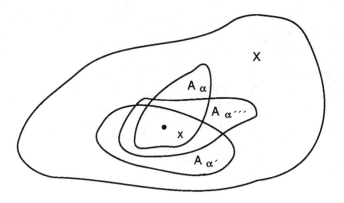

FIGURE 1

Let

$$\{A_\alpha : \alpha \in A\}$$

be the family of all connected subspaces of X which contains x,

$$x \in A_\alpha \quad X$$

and A_α is connected for each $\alpha \in A$. Then

$$\bigcup_\alpha A_\alpha$$

is a connected subspace of X which contains x. Indeed, since all A_α are connected and

$$x \in \bigcap_\alpha A_\alpha \neq \phi$$

then by Problem 17–7, $\bigcup_\alpha A_\alpha$ is connected.

$\bigcup_\alpha A_\alpha$ is a maximal connected subspace of X which contains x and is unique.

● **PROBLEM 17–32**

Let $\{A_\alpha\}$, $\alpha \in A$ be the set of components of a space (X, T). Show that the components of X form a partition of X, i.e., (see Figure 1 on following page)

$$X = \bigcup_\alpha A_\alpha$$

if $\alpha \neq \beta$, then $A_\alpha \cap A_\beta = \phi$.

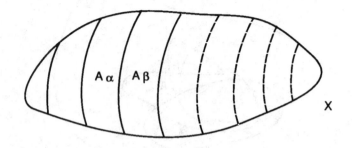

FIGURE 1

SOLUTION:

Any element x of X belongs to at least one connected subspace of X (for example, $x \in \{x\} \subset X$). Thus, a maximal connected subspace of X contains x. We have

$$X = \bigcup_{\alpha} A_{\alpha}.$$

We will show that if $\alpha \neq \beta$, then $A_{\alpha} \cap A_{\beta} = 0$. Suppose $\alpha \neq \beta$ and $A_{\alpha} \cap A_{\beta} \neq 0$, then $A_{\alpha} \cup A_{\beta}$ is a connected subspace of X which contains both A_{α} and A_{β}. A contradiction arises because A_{α} and A_{β} are maximal.

● PROBLEM 17–33

Prove that if a space has only a finite number of components, then each component is open.

SOLUTION:

Let

$$A_1, A_2, \ldots, A_K$$

be components of X.

648

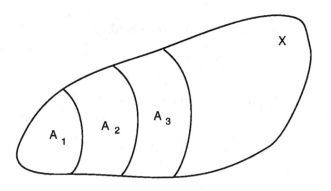

FIGURE 1

Then

$$X = A_1 \cup \dots \cup A_K.$$

Since each pair of components are disjoint sets

$$A_i \cap A_j = \phi \quad \text{for } i \neq j$$

we have

$$A_1 = X - (A_2 \cup \dots \cup A_K).$$

In Problem 17–29 we proved that a component of a space is closed. Therefore, since there are a finite number of components,

$$A_2 \cup \dots \cup A_K$$

is a closed set. Hence, A_1 is an open set. Similarly, we show that all sets $A_1, A_2,$ ..., A_K are open.

● **PROBLEM 17–34**

Show that a finite product of connected spaces is connected.

SOLUTION:

Let X and Y be connected spaces and let

$$p = (x_1, y_1) \in X \times Y$$

$$q = (x_2, y_2) \in X \times Y. \tag{1}$$

The set

$$\{x_1\} \times Y$$

649

is homeomorphic to Y and since Y is connected, so is $\{x_1\} \times Y$. Similarly, we show that

$$X \times \{y_2\}$$

is connected since it is homeomorphic to X.

Therefore,

$$\{x_1\} \times Y \cup X \times \{y_2\}$$

is connected because

$$\{(x_1, y_2)\} = \{x_1\} \times Y \cap X \times \{y_2\} \neq \phi.$$

Also,

$$p, q \in \{x_1\} \times Y \cup X \times \{y_2\}.$$

Hence, p and q belong to the same component. But p and q were chosen arbitrarily; hence, $X \times Y$ has only one component — itself.

We conclude that a finite product of connected spaces is connected.

● **PROBLEM 17–35**

Show that a product of connected spaces is connected. *in The Prod.*

SPACE

SOLUTION:

Let

$$\{X_\alpha : \alpha \in A\}$$

be a collection of connected spaces and let

$$X = \underset{\alpha \in A}{\times} X_\alpha$$

be the product space.

Let

$$x \in X$$

$$x = (X_\alpha : \alpha \in A)$$

and let $P \quad X$ be the component to which x belongs, $x \in P$.

We will show that an arbitrary point $y \in X$ belongs to P.

Let

$$F = X\{X_\alpha : \alpha \neq \alpha_1, \ldots, \alpha_m\} \times U_{\alpha_1} \times \ldots \times U_{\alpha_m}$$

650

be any basic open set containing $y \in X$. The set

$$G = X\{\{X_\alpha\} : \alpha \neq \alpha_1, \ldots, \alpha_m\} \times X_{\alpha_1} \times \ldots \times X_{\alpha_m}$$

is homeomorphic to

$$X_{\alpha_1} \times \ldots \times X_{\alpha_m}$$

hence, it is connected.

Since $x \in G$

$$G \subset P$$

where P is the component of x.

The set

$$F \cap G \neq \phi$$

and

$$P \cap P \neq \phi.$$

Thus,

$$y \in P = P.$$

X has one component; hence, it is connected.

● **PROBLEM 17-36**

Show that the space of real numbers R with the topology T generated by the closed-open intervals $[a, b) \subset R$ is totally disconnected.

SOLUTION:

DEFINITION OF TOTALLY DISCONNECTED SPACES

A topological space (X, T) is said to be totally disconnected if, for each

$$x, y \in X$$

open nonempty sets exist, U and V such that

$$X = U \cup V$$

$$U \cap V = \phi$$

and

$$x \in U, \quad y \in V.$$

■

Let $x, y \in R$ and $x < y$. We define

$$U = (-\infty, y)$$

$$V = [y, +\infty).$$

Sets U and V are disjoint, open, nonempty sets and $R = U \cup V$. Also,

$$x \in U \quad \text{and} \quad y \in V.$$

Thus, R with the topology generated by the closed-open intervals is totally disconnected.

● **PROBLEM 17-37**

1. Show that the set Q of rational numbers with the absolute value topology is totally disconnected.

2. Show that a totally disconnected space is T_2 (i.e., Hausdorff).

SOLUTION:

1. Let $x, y \in Q$ and

$$x < y.$$

An irrational number a exists such that

$$x < a < y.$$

Define sets U and V

$$U = \{z \in Q : z < a\}$$

$$V = \{z \in Q : z > a\}.$$

Thus, $U \cup V$ is a disconnection of Q and

$$x \in U, y \in V.$$

Q is totally disconnected.

2. Let (X, T) be totally disconnected and let

$$x, y \in X.$$

Open, nonempty, disjoint sets U, V exist such that

$$x \in U \quad \text{and} \quad y \in V$$

$$U \cap V = \phi.$$

Hence, X is a Hausdorff space.

● **PROBLEM 17-38**

Prove that the components of a totally disconnected space are the singleton sets.

That statement leads to an equivalent definition of totally disconnected spaces. A space (X, T) is said to be totally disconnected if X is not connected and the only connected subspaces of X are the singleton subsets of X and ϕ.

SOLUTION:

Suppose $A \subset X$ is a component of X and

$$x, y \in A, \quad x \neq y.$$

Since X is totally disconnected, there are sets U and V such that

$$X = U \cup V, \quad U \cap V = \phi$$

$$x \in U, \quad y \in V.$$

U and V are open.

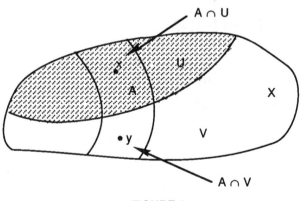

FIGURE 1

Hence, the sets

$$A \cap U \quad \text{and} \quad A \cap V$$

are nonempty, and $U \cup V$ is a disconnection of A. But as a component, A has to be connected.

Contradiction. Thus, A consists of one point.

Show that a space X with the discrete topology is totally disconnected and locally connected.

SOLUTION:

The only connected subsets of a discrete space are singleton sets and ϕ. Hence, X is totally disconnected.

DEFINITION OF LOCALLY CONNECTED SPACE

A space (X, T) is locally connected, if for any $x \in X$ and any neighborhood U of x, there is a connected neighborhood V of x such that

$$V \subset U.$$

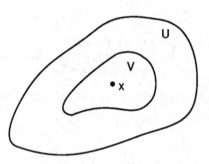

FIGURE 1

We can use this equivalent definition:

DEFINITION

Space (X, T) is said to be locally connected if there is an open neighborhood system for T such that for each $x \in X$, N_x consists of connected subspaces.

Let (X, T) be the space with discrete topology. For each

$$x \in X$$

we define

$$N_x = \{\{x\}\}$$

which establishes an open neighborhood system. Each subspace $\{x\}$ is connected.

Consider the space

$$X = \{(x, y) : 0 < x, y = \sin \frac{1}{x}\} \cup \{(0, 0)\}.$$

Show that X is not locally connected. Use this space to show that local connectedness is not preserved by a continuous function.

SOLUTION:

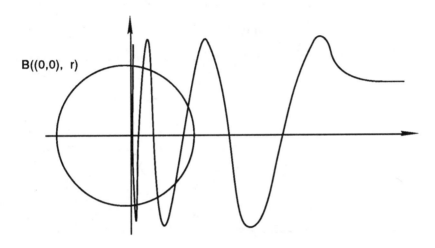

B((0,0), r)

FIGURE 1

For any $0 < r < 1$, $B((0, 0), r)$ has an infinite number of components. X is connected but not locally connected.

On the other hand, X is a continuous image of a locally connected space.

Hence, the local connectedness is not generally perserved by continuous functions.

● PROBLEM 17-41

1. Let X and Y be locally connected. Show that $X \times Y$ is locally connected.

2. Show that a component of a locally connected space X is open.

SOLUTION:

1. Space X is locally connected if and only if X has a base B consisting of connected sets. Also, Y has a base B' consisting of connected sets. Since $X \times Y$ is a finite product, a base for $X \times Y$ is defined by

$$\{U \times V : U \in B, V \in B'\}.$$

Each of the sets $U \times V$ is connected, since both U and V are connected. The product $X \times Y$ has a base consisting of connected sets; hence, $X \times Y$ is locally connected.

2. Let A be a component of a locally connected space X and

$$a \in A.$$

Since X is locally connected, a belongs to at least one open connected set $U_a \subset A$.

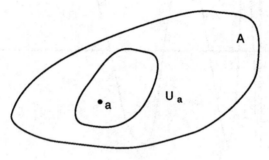

FIGURE 1

But A is a component of a; hence,

$$a \in U_a \subset A$$

and

$$A = \bigcup_{a \in A} U_a.$$

Thus A is open, as the union of open sets is.

● **PROBLEM 17-42**

Show that a compact, locally connected space (X, T) has a finite number of components.

SOLUTION:

Let

$$(A_\alpha : \alpha \in A\}$$

be the family of components of X. Hence, $\{A_\alpha\}$ forms an open cover of X.

Since X is compact, a finite subcover exists, say

$$A_{\alpha_1}, A_{\alpha_2}, \dots, A_{\alpha_k}.$$

On the other hand, since $\{A_\alpha\}$ is the family of components, for $\alpha \neq \beta$ and α and $\beta \in A$, we have

$$A_\alpha \cap A_\beta = \phi.$$

Therefore, no A_α can be omitted from $\{A_\alpha\}$ and the remaining components still form a cover of X.

Hence,

$$A_{\alpha_1}, A_{\alpha_2}, \dots, A_{\alpha_k}$$

are all of the components of X.

● **PROBLEM 17–43**

Continuous functions do not preserve local connectedness. But continuous open functions do. Prove:

THEOREM

If f is a continuous open function from a locally connected space (X, T) onto (Y, T').

$$f : X \to Y$$

then Y is locally connected.

∎

SOLUTION:

f is a continuous open and onto function

$$f : X \to Y.$$

Let U be any neighborhood of $y \in Y$.

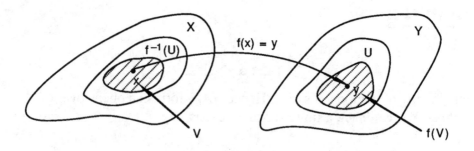

FIGURE 1

Since f is onto, $x \in X$ exists such that

$$f(x) = y.$$

f is continuous; hence, $f^{-1}(U)$ is an open set $x \in f^{-1}(U)$. But X is locally connected; thus, there is a connected neighborhood V of x such that

$$x \in V \subset f^{-1}(U).$$

Since f is continuous and open, $f(V)$ is a connected open set and

$$y \in f(V) \subset U.$$

Hence, Y is locally connected.

● **PROBLEM 17-44**

Prove the following:

THEOREM

If $\{X_\alpha : \alpha \in A\}$ is a family of connected locally connected spaces, then $\underset{\alpha}{\times} X_\alpha$ is locally connected. ∎

SOLUTION:

Let

$$x = (x_\alpha) \in \underset{\alpha}{\times} X_\alpha$$

and let $U \subset \underset{\alpha}{\times} X_\alpha$ be the neighborhood of x. A member of the defining base exists

$$B = U_{\alpha_1} \times \dots \times U_{\alpha_k} \times (\underset{\alpha \neq \alpha_1, \dots, \alpha_k}{x} X_\alpha)$$

such that

$$x \in B \subset U$$

and

$$x_{\alpha_l} \in U_{\alpha_l}.$$

Each coordinate space is locally connected; hence. connected open subsets V_{α_l} exist such that

$$V_{\alpha_l} \subset X_{\alpha_l}$$

and

$$x_{\alpha_l} \in V_{\alpha_l} \subset U_{\alpha_l}$$

for $l = 1, 2, \dots, k$.

Let

$$V = V_{\alpha_1} \times \dots \times V_{\alpha_k} \times (\underset{\alpha \neq \alpha_1, \dots, \alpha_k}{x} X_\alpha).$$

V is connected because each X_α is connected and each V_{α_l} is connected. Since V is open and

$$x \in V \subset B \subset U,$$

X is locally connected.

● **PROBLEM 17-45**

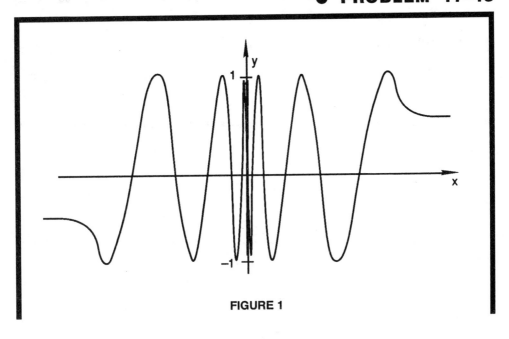

FIGURE 1

Which of the following sets is a continuum:

1. closed interval

2. closed n-dimensional cube

3. the set P of points defined by

$$y = \sin {}^1/_x \text{ for } 0 < |x| \leq 1$$

$$|y| \leq 1 \quad \text{for} \quad x = 0$$

SOLUTION:

A compact connected T_2–space (X, T) is called a continuum.

∎

Note that any closed, bounded, and connected subset of R^n for any n is a continuum. This is due to the fact that the closure of any bounded subset of R^n is compact.

Also, any path in a T_2–space is a continuum.

We see that each of the spaces described in 1, 2, and 3 is a continuum.

● **PROBLEM 17–46**

Prove this:

THEOREM

The union of two continuua which have a common point is a continuum.

∎

SOLUTION:

Let (X, T) and (Y, T') be compact connected spaces. We proved that a topological space which is the union of two compact sets is compact.

Hence, $X \cup Y$ is compact. We also proved the following:

If the sets A and B are connected and are not separated, then their union is connected.

Hence, $X \cup Y$ is connected.

Therefore, $X \cup Y$ is a continuum.

Let X and Y be continua such that

$$Y \land X$$

and let $X - Y$ be the union of two disjoint open sets A and B. Show that

$$Y \cup A \quad \text{and} \quad Y \cup B$$

are continua.

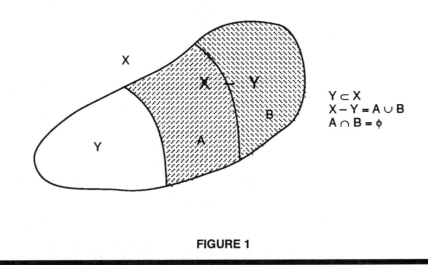

$$Y \subset X$$
$$X - Y = A \cup B$$
$$A \cap B = \phi$$

FIGURE 1

SOLUTION:

We shall apply the following:

THEOREM

If Y is a connected subset of the connected space X and

$$X - Y = A \cup B$$

where A and B are separated, then the sets

$$Y \cup A \quad \text{and} \quad Y \cup B$$

are connected.

Note that sets A and B are separated, if

$$\overline{A} \cap B = \phi \quad \text{and} \quad A \cap \overline{B} = \phi.$$

We conclude that $Y \cup A$ and $Y \cup B$ are continua.

Prove the following:

THEOREM

If $f : X \to Y$ is a homeomorphism and X is a continuum, then Y is a continuum.

■

SOLUTION:

We shall apply two theorems.

THEOREM

The continuous image of a compact set is compact.

■

THEOREM

The image under a continuous mapping of a connected space is a connected space.

■

Also, if $f : X \to Y$ is a homeomorphism and X is T_2, then Y is T_2.

Hence, if the spaces are homeomorphic and one is a continuum, then the other one is also a continuum.

The above proved theorem is a generalization of the theorem from analysis. Suppose X is a continuum and f is a homeomorphic real-valued function, then $f(X)$ is either a single point or a closed interval.

CHAPTER 18

METRIZABLE SPACES

Show that if X is metrizable, then $X \times X$ is also metrizable.

SOLUTION:

Suppose (X, T) is metrizable.

DEFINITION OF METRIZABLE SPACE

A space (X, T) is said to be metrizable if a metric d can be defined on X such that the topology induced by d is T. Otherwise, the space X is called non-metrizable.

■

Metric d exists such that d induces T. We define topology on $X \times X$ in the usual way. It can be proved that this topology is induced by the metric

$$d'((x_1, x_2), (y_1, y_2)) = d(x_1, y_1) + d(x_2, y_2).$$

Hence, if X is metrizable, so is $X \times X$.

R is the space of real numbers with the topology T generated by the basis consisting of open intervals. Show that (R, T) is metrizable.

SOLUTION:

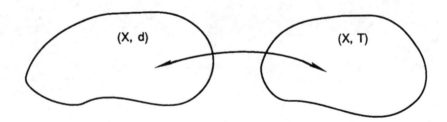

FIGURE 1

Suppose a metric space (X, d) is given. The metric d induces a certain topology T_d on X.

$$(X, d) \to (X, T_d)$$

In the metric space (X, d) a d-ball of radius r and center a is defined by

$$B(a, r) = \{x \in X : d(x, a) < r\}.$$

The family

$$\{B(a, r) : a \in X, r > 0\}$$

of all d-balls in X can serve as the basis for topology T.

Hence, we obtain a topology T on X determined (or induced) by the metric d.

Let R be the space of real numbers with the absolute value metric. Then this metric induces a topology with the open intervals as the basis. Hence, R with the topology generated by the basis consisting of open intervals is metrizable. The metric we are looking for is the absolute value metric.

● **PROBLEM 18-3**

Let $\{(X_n, d_n) : n \in N\}$ be a countable family of second countable metric spaces.

1. Show that the product space

$$\underset{n \in N}{\times} X_n$$

is second countable.

2. Show that

$$d(x, y) = \sum_{n \in N} \frac{\min(d_n(x_n, y_n), 1)}{2^n}.$$

is a metric on xX_n, where

$$x = (x_1, x_2, \ldots, x_n, \ldots); \; x_n, y_n \in X_n$$

$$y = (y_1, y_2, \ldots, y_n, \ldots)$$

SOLUTION:

1. Since each X_n is second countable, $\times X_n$ is second countable. We proved earlier that the product space of a countable family of nonempty spaces is second countable, if and only if each component space is second countable.

2. We have

$$d(x, y) = \sum_n \frac{\min(d_n(x_n, y_n), 1)}{2^n} \leq \sum_n \frac{1}{2^n}.$$

Hence, $d(x, y)$ is defined for all

$$x, y \in \underset{n}{\times} X_n.$$

Also,

$$d(x, y) \geq 0$$

$d(x, y) = 0$ iff $x_n = y_n$ for all $n \in N$, hence, $x = y$.

$$d(x, y) = d(y, x)$$

$$d(x, y) = \sum_n \frac{\min(d_n(x_n, y_n), 1)}{2^n} \leq 1$$

$$\leq \sum_n \left[\frac{\min(d_n(x_n, z_n), 1)}{2^n} + \frac{\min(d_n(z_n, y_n), 1)}{2^n} \right] =$$

$$= d(x, z) + d(z, y).$$

Hence, d is a metric on $\underset{n}{\times} X_n$.

● **PROBLEM 18–4**

Using the metric defined in Problem 18–3, show that the space $\underset{n \in N}{\times} X_n$ is metrizable, where $\{(X_n, d_n) : n \in N\}$ is a countable family of second countable metric spaces.

SOLUTION:

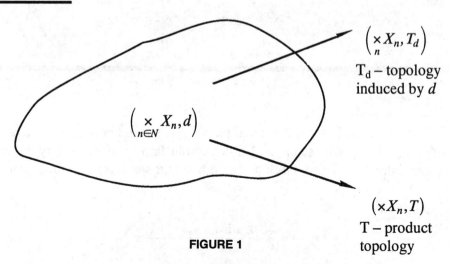

$\left(\underset{n}{\times} X_n, T_d \right)$

T_d – topology induced by d

$\left(\underset{n \in N}{\times} X_n, d \right)$

$\left(\times X_n, T \right)$

T – product topology

FIGURE 1

666

We have to show that

$$T_d = T.$$

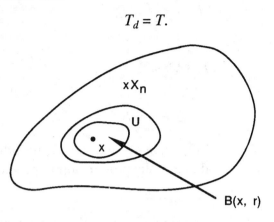

FIGURE 2

Let $x \in xX_n$ and let U be the basic neighborhood of x in the product topology. Then

$$U = \underset{n \in N}{\times} V_n$$

where V_n is open in X_n and for all but at most finitely many n, say

$$n_1, \ldots, n_k$$

$$V_n = X_n$$

The sets $V_{n_1}, \ldots, V_{n_k}$ are open in $X_{n_1}, \ldots, X_{n_k}$ respectively. Hence, positive numbers $r_1, \ldots, r_k$ exist such that

$$x_{n_1} \in B(x_{n_1}, r_1) \subset V_{n_1}$$

$$\vdots \qquad \vdots \qquad \vdots$$

$$x_{n_k} \in B(x_{n_k}, r_k) \subset V_{n_k}$$

Choose

$$r = \min(r_1, \ldots, r_k).$$

Then if $y \in B(x, r)$

$$y_n \in V_n$$

thus

$$y \in U \quad \text{and} \quad B(x, r) \subset U.$$

To complete the proof, we have to show that for any $B(x, r)$ there is U such that $U \subset B(x, r)$. Let $B(x, r)$ be a d-neighborhood of x. Choose $m \in N$ such that

667

$$\sum_{n=m}^{\infty} \frac{1}{2^n} < \frac{r}{2}$$

and let $r_1, \ldots, r_{m-1}$ be such that

$$r_1 + \ldots + r_{m-1} < {}^r/_2.$$

We define

$$U = \underset{n \in N}{x} \; B_n$$

where $B_n = B(x_n, r_n)$ for $n = 1, 2, \ldots, m-1$ and $B_n = X_n$ for other n's.
    U is a basic neighborhood of x in the product topology and

$$U \subset B(x, r).$$

Hence, the product and metric topologies on xX_n are equivalent and $\underset{N}{\times} X_n$ is metrizable.

● PROBLEM 18-5

Explain why the Hilbert cube $\underset{n \in N}{\times} [0,1]$ is second countable and metrizable.

SOLUTION:

Since the closed interval $[0, 1]$ is second countable, the Hilbert cube $\underset{N}{\times} [0,1]$ is second countable.

The closed interval $[0, 1]$ with the absolute value topology is metrizable. By Problem 18–4 the product of the countable family of second countable metric spaces is metrizable.

● PROBLEM 18-6

Show that the sequence

$$a_n = \frac{1}{n}, \; n \in N$$

in the space of real numbers R with the absolute value metric is a Cauchy sequence.

SOLUTION:

We shall define Cauchy sequences.

DEFINITION OF CAUCHY SEQUENCES

A sequence (a_n) in (X, d) is a Cauchy sequence iff

$\forall \, \varepsilon > 0, \, \exists \, N_0 \in N$ such that

$$n, \, m > N_0 \Rightarrow d(a_n, a_m) < \varepsilon.$$

■

We take an arbitrary fixed $\varepsilon > 0$. The solution of the problem lies in finding $N_0 = N^0(\varepsilon)$ such that for every $n, \, m > N_0$

$$\left| \frac{1}{n} - \frac{1}{m} \right| < \varepsilon.$$

Let N_0 be such that

$$\frac{1}{N_0} < \frac{\varepsilon}{2}.$$

Note that we do not have to find the smallest possible N_0.

For any $n, \, m > N_0$

$$\frac{1}{n} < \frac{1}{N_0} \quad \text{and} \quad \frac{1}{m} < \frac{1}{N_0}$$

and

$$\left| \frac{1}{n} - \frac{1}{m} \right| < \frac{1}{n} + \frac{1}{m} < \frac{1}{N_0} + \frac{1}{N_0} <$$

$$< \frac{\varepsilon}{2} + \frac{\varepsilon}{2} = \varepsilon.$$

Hence, $(1/_n)$ is a Cauchy sequence in $(R, 1 \cdot 1)$.

● **PROBLEM 18-7**

Show that every convergent sequence in a metric space (X, d) is a Cauchy sequence.

SOLUTION:

Let (a_n) be a convergent sequence in (X, d)

$$a_n \underset{n \to \infty}{\to} a.$$

In other words,

669

$$d(a_n, a) \to 0.$$

Then for every $\varepsilon > 0$ there exists N_0 such that

$$n > N_0 \Rightarrow d(a_n, a) < \varepsilon/2.$$

Hence, by the triangle inequality

$$n, m > N_0 \Rightarrow d(a_n, a_m) \leq d(a_n, a) + d(a_m, a) <$$

$$< \frac{\varepsilon}{2} + \frac{\varepsilon}{2} = \varepsilon.$$

Thus, (a_n) is a Cauchy sequence.

● **PROBLEM 18–8**

Find an example which illustrates that the converse of the statement in Problem 18–7 is not true; that is, not every Cauchy sequence is convergent.

SOLUTION:

Consider the space Q of rational numbers with the absolute value metric $(Q, 1 \cdot 1)$. The sequence is given as follows:

$$a_1 = 1$$

$$a_2 = 1.4$$

$$a_3 = 1.41$$

$$a_4 = 1.414$$

$$\vdots$$

Sequence (a_n) described above is the decimal expansion of $\sqrt{2}$. It is easy to verify that (a_n) is a Cauchy sequence. But (a_n) does not converge in Q (it converges in R). Hence, (a_n) is a sequence which is Cauchy but not convergent.

● **PROBLEM 18–9**

Show that every Cauchy sequence (a_n) in a metric space (X, d) is bounded and totally bounded.

SOLUTION:

Let (a_n) be a Cauchy sequence in (X, d). We shall prove that (a_n) is totally bounded (and thus bounded).

A subset A of a metric space (X, d) is totally bounded if A has an ε-net for every $\varepsilon > 0$. An ε-net for A is a finite set of points $\{a_1, \ldots, a_n\}$, such that for every $a \in A$ there is $a_k \in \{a_1, \ldots, a_n\}$ such that

$$d(a, a_k) < \varepsilon.$$

Let $\varepsilon > 0$. Since (a_n) is a Cauchy sequence,

$\exists N_0 \in N$ such that for every

$$n, m > N_0, d(a_n, a_m) < \varepsilon.$$

Hence, the diameter of the set

$$\{a_{N_0+1}, a_{N_0+2}, \ldots\}$$

is at most ε.

The ε-net for the sequence (a_n) is

$$\{a_1, a_2, \ldots, a_{N_0}, a_{N_0+1}\}.$$

● **PROBLEM 18-10**

Let (X, d) be a space with the trivial metric

$$d(x,y) = \begin{cases} 0 & \text{if } x = y \\ 1 & \text{if } x \neq y \end{cases}.$$

Find a general formula for a Cauchy sequence in this space.

SOLUTION:

Suppose (a_n) is a Cauchy sequence in the trivial space.
Let $\varepsilon = 1/2$. Then there exists N_0 such that

$$n, m > N_0 \Rightarrow d(a_n, a_m) < 1/2 \text{ that is}$$

$$a_n = a_m.$$

Hence, a Cauchy sequence in the space with the trivial metric is of the form

$$(a_1, a_2, \ldots, a_k, a, a, a \ldots)$$

that is constant from a certain term on.

671

Let (a_n) be a Cauchy sequence in (X, d) and let (a_{K_n}) be its subsequence. Show that

$$d(a_n, a_{K_n}) \xrightarrow[n \to \infty]{} 0.$$

SOLUTION:

Let $\varepsilon > 0$ be an arbitrary positive number. Since (a_n) is a Cauchy sequence $\exists\, N_0 \in N$ such that,

$$n, m > N_0 - 1 \Rightarrow d(a_n, a_m) < \varepsilon.$$

Hence,

$$K_{N_0} \geq N_0 > N_0 - 1 \quad \text{and} \quad d(a_{N_0}, a_{K_{N_0}}) < \varepsilon.$$

That is

$$\lim_{n \to \infty} d(a_n, a_{K_n}) = 0.$$

Let (a_n) be a sequence in a metric space (X, d). Let

$$A_1 = \{a_1, a_2, \ldots\}$$

$$A_2 = \{a_2, a_3, \ldots\}$$

$$A_3 = \{a_3, a_4, \ldots\}$$

$$\vdots$$

Show that

$$((a_n) \text{ is a Cauchy sequence}) \Leftrightarrow (\lim_{n \to \infty} \delta(A_n) = 0)$$

where $\delta(A_n)$ denotes the diameter of A_n.

SOLUTION:

$\Rightarrow$ Suppose (a_n) is a Cauchy sequence. For $\varepsilon > 0$ there exists $N_0 \in N$ such that for $n, m > N_0$

$$d(a_n, a_m) < \varepsilon.$$

Thus, for $n > N_0$

$$\delta(A_n) < \varepsilon.$$

Since ε was chosen arbitrarily,

$$\lim_{n \to \infty} \delta(A_n) = 0.$$

$\Leftarrow$

$$\lim_{n \to \infty} \delta(A_n) = 0.$$

For any $\varepsilon > 0$ there exists N_0 such that for any $n > N_0$

$$\delta(A_n) < \varepsilon.$$

In other words, for any $n, m > N_0$

$$d(a_n, a_m) < \varepsilon$$

and (a_n) there is a Cauchy sequence.

● **PROBLEM 18-13**

Let (a_n) be a Cauchy sequence in (X, d) and (b_n) be a sequence in (X, d) such that

$$d(a_n, b_n) < \frac{1}{n}$$

for every $n \in N$. Show that:

1. (b_n) is also a Cauchy sequence.

2. (b_n) converges to $a \in X$ iff
    (a_n) converges to $a \in X$.

SOLUTION:

1. Applying the triangle inequality,

$$d(b_m, b_n) \le d(b_m, a_m) + d(a_m, b_n) \le$$

$$\le d(b_m, a_m) + d(b_n, a_n) + d(a_n, a_m).$$

Let $\varepsilon > 0$. We can find $N_1 \in N$ such that for

$$n, m > N_1$$

we have

$$d(b_m, b_n) < \frac{\varepsilon}{3} + \frac{\varepsilon}{3} + d(a_m, a_n).$$

Since (a_n) is a Cauchy sequence,

$\exists\, N_2 \in N$ such that

$$n, m > N_2 \Rightarrow d(a_m, a_n) \le \frac{\varepsilon}{3}.$$

Setting

$$N_3 = \max\{N_1, N_2\}$$

we get

$$n, m > N_3 \Rightarrow d(b_m, b_n) < \varepsilon.$$

Hence, (b_n) is a Cauchy sequence.

2. Similarly,

$$d(a_n, a) \le d(a_n, b_n) + d(b_n, a).$$

Hence,

$$\lim_{n \to \infty} d(a_n, a) \le \lim_{n \to \infty} d(a_n, b_n) + \lim_{n \to \infty} d(b_n, a).$$

But

$$\lim d(a_n, b_n) = 0.$$

If $b_n \to a$, then

$$\lim_{n \to \infty} d(a_n, a) \le \lim_{n \to \infty} d(b_n, a) = 0$$

and

$$\lim_{n \to \infty} a_n = a$$

Similarly, if $\quad a_n \underset{n \to \infty}{\to} a \quad$ then $\quad b_n \underset{n \to \infty}{\to} a.$

● **PROBLEM 18-14**

Show that if (x_n) is a Cauchy sequence in Euclidean m-space R^m

$$x_1 \;=\; (x_1^{(1)}, x_1^{(2)}, \ldots, x_1^{(m)})$$

$$\vdots$$

$$x_n \;=\; (x_n^{(1)}, x_n^{(2)}, \ldots, x_n^{(m)})$$

$$\vdots$$

674

then each of the sequences

$$\left(x_K^{(1)}\right), \left(x_K^{(2)}\right), \ldots, \left(x_K^{(m)}\right)$$

$k = 1, \ldots, n$ is a Cauchy sequence in R.

SOLUTION:

Let $\varepsilon > 0$. Since (x_n) is a Cauchy sequence,

$\exists\, N_0 \in N$ such that for $j, l > N_0$

$$d(x_j, x_l) = \sqrt{\left|x_j^{(1)} - x_l^{(1)}\right|^2 + \ldots + \left|x_j^{(m)} - x_l^{(m)}\right|^2} < \varepsilon.$$

Thus, for each term we obtain

$$\left|x_j^{(1)} - x_l^{(1)}\right| < \varepsilon$$

$$\vdots$$

$$\left|x_j^{(m)} - x_l^{(m)}\right| < \varepsilon.$$

Each of the m sequences

$$\left(x_k^{(1)}\right), \ldots, \left(x_k^{(m)}\right)$$

is a Cauchy sequence.

● PROBLEM 18-15

1. Show that the real line R with the usual metric is complete.

2. Give an example of a subspace of R which is not complete.

SOLUTION:

DEFINITION OF COMPLETE METRIC SPACE

A metric space (X, d) is complete if every Cauchy sequence (a_n) in X converges to a point $a \in X$.

$$\lim_{n \to \infty} a_n = a \in X.$$

The real line R is complete. This is the result of the Cauchy theorem: Every Cauchy sequence of real numbers converges to a real number.

2. Consider the open interval]0, 1[with the absolute metric. The sequence $(^1/_2, ^1/_3, ^1/_4, ^1/_5, \dots)$ is a Cauchy sequence which converges, in the limit as

$$\lim_{n \to \infty} \frac{1}{n} = 0.$$

However, 0 does not belong to]0, 1[, hence,]0, 1[is not complete.

● **PROBLEM 18-16**

1. Is completeness a topological property?

2. Find the diameters of A, B, and C

$$A = \{(x, y) : x^2 + y^2 \le 1\} \subset R^2$$

$$B = \{(x, y) : x^2 + y^2 = 1\} \subset R^2$$

$$C = \{(x, y) : |x| < 1, |y| \le 2\} \subset R^2$$

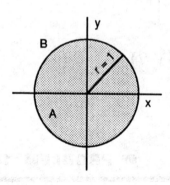
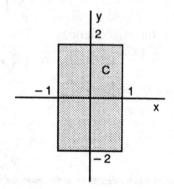

FIGURE 1

SOLUTION:

1. The real line R with the absolute metric is complete. As demonstrated in Problem 18-15, the open interval]0, 1[with the absolute metric is not complete. But R is homeomorphic to]0, 1[, then completeness is not a topological property.

2. The diameter of a subset A of a metric space (X, d) is defined to be the least upper bound of

$$\{d(x, y) : x, y \in A\}.$$

We denote the diameter of A by $\delta(A)$ or $d(A)$.

The diameter of A, B, and C are

$$\delta(A) = 1$$

$$\delta(B) = 1$$
$$\delta(C) = \sqrt{20}.$$

● **PROBLEM 18-17**

Show that for any subset A of a metric space (X, d)

$$\delta(A) = \delta(\overline{A}). \qquad (1)$$

SOLUTION:

We shall prove that for any positive number $r > 0$

$$\delta(\overline{A}) < \delta(A) + r \qquad (2)$$

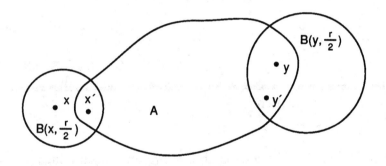

FIGURE 1

Suppose

$$x, y \in \overline{A}. \qquad (3)$$

Then

$$B(x, {}^r\!/_2) \cap A \neq \phi$$

$$B(y, {}^r\!/_2) \cap A \neq \phi. \qquad (4)$$

We choose x' and y' such that

$$x' \in B(x, {}^r\!/_2) \cap A$$

$$y' \in B(y, {}^r\!/_2) \cap A \qquad (5)$$

677

Then

$$d(x, y) \le d(x, x') + d(y, x') \le$$

$$\le d(x, x') + d(y, y') + d(x', y') < {}^r/_2 + {}^r/_2 + \delta(A) =$$

$$= r + \delta(A). \tag{6}$$

Therefore,

$$\delta(\overline{A}) < \delta(A) + r \tag{7}$$

and

$$\delta(\overline{A}) \le \delta(A). \tag{8}$$

Since $A \subset \overline{A}$,

$$\delta(A) \le d(\overline{A}). \tag{9}$$

Thus,

$$\delta(A) = \delta(\overline{A}). \tag{10}$$

● **PROBLEM 18-18**

Prove the Cantor theorem.

CANTOR THEOREM

Let $\{A_n\}$, $n \in N$ be a countable family of closed nonempty subsets of complete space X such that

$$A_1 \supset A_2 \supset A_3 \ldots \quad \text{and} \quad \delta(A_n) \xrightarrow[n \to \infty]{} 0 \tag{1}$$

Then

$$\bigcap_{n \in N} A_n \ne \phi. \tag{2}$$

∎

SOLUTION:

For every $n \in N$ choose $a_n \in A_n$. We shall show that (a_n) is a Cauchy sequence. Let $\varepsilon > 0$. There exists

$$N_0 \in N$$

such that for $n > N_0$

$$\delta(A_n) < {}^\varepsilon/_2. \tag{3}$$

For $k, l > N_0$, a_k, $a_l \in A_{N_0 + 1}$.

Hence,

$$d(a_k, a_l) < \varepsilon. \tag{4}$$

Thus, (a_n) is a Cauchy sequence in (X, d). Since, by hypothesis, X is complete

$$\lim_{n \to \infty} a_n = a \in X. \tag{5}$$

Note that for any n

$$(a_n, a_{n+1}, \ldots)$$

is also a sequence which converges to a.

$$\{a_n, a_{n+1}, \ldots\} \subset A_n. \tag{6}$$

Since for each n, A_n is closed,

$$a \in \overline{A_n} = A_n. \tag{7}$$

Thus,

$$a \in \bigcap_{n \in N} A_n \quad \text{and} \quad \bigcap_{n \in N} A_n \neq \phi. \tag{8}$$

● **PROBLEM 18-19**

Prove the following:

THEOREM

A metric space (X, d) is complete if and only if every nested sequence (i.e., $A_1 \supset A_2 \supset A_3 \ldots$) of nonempty closed sets such that $\delta(A_n) \to 0$ has a nonempty intersection, $\bigcap_n A_n \neq \phi$. ∎

SOLUTION:

We have to show that

$$((X, d) \text{ complete}) \Leftrightarrow \left(\bigcap_n A_n \neq \phi \right).$$

For any countable family $\{A_n\}$ of closed, nonempty subsets of X such that

$$A_1 \supset A_2 \supset A_3 \ldots$$

and

$$\delta(A_n) \to 0.$$

In Problem 18–18 we proved $\Rightarrow$.

Now we shall show $\Leftarrow$.

Let $A_1 \supset A_2 \supset A_3 \ldots$ be any sequence of closed, nonempty subsets of X such that

$$\delta(A_n) \to 0, \cap A_n \neq \phi.$$

Let $\{X_n\}$, $n \in N$ be a Cauchy sequence in X. We define

$$B_1 = \{x_1, x_2, x_3, \ldots\}, A_1 = \overline{B}_1$$

$$B_2 = \{x_2, x_3, \ldots\}, A_2 = \overline{B}_2$$

$$B_3 = \{x_3, x_4, \ldots\}, A_3 = \overline{B}_3$$

$$\vdots$$

Then $A_1 \supset A_2 \supset A_3 \supset \ldots$ and $\delta(A_n) \to 0$ and $\{A_n\}$ are closed, nonempty subsets of X. Hence,

$$\bigcap_{n \in N} A_n \neq \phi.$$

Let

$$x \in \bigcap_n A_n.$$

We show that

$$x_n \to x \in X.$$

Let $\varepsilon > 0$. Then

$\exists\, N_0 \in N$ such that for $n > N_0$

$$\delta(B_n) < \varepsilon.$$

Hence (by Problem 18–17),

$$\delta(B_n) = \delta(A_n) < \varepsilon.$$

For all $n > N_0$

$$d(x_n, x) < \varepsilon.$$

Hence,

$$x_n \to x \in X.$$

> Show that any compact metric space (X, d) is complete.

SOLUTION:

A space X is said to be compact if given any open cover $\{U_i\}$, $i \in I$, of X, there is a finite subcover of $\{U_i\}$, $i \in I$.

Suppose (A_n), $n \in N$, are closed, nonempty subsets of X such that

$$A_1 \supset A_2 \supset A_3 \supset \dots \tag{1}$$

We shall apply the following: X is compact if and only if for any family $\{F_i\}$, $i \in I$, of closed subsets of X such that the intersection of any finite number of the F_i is nonempty,

$$\bigcap_{i \in I} F_i \neq \phi. \tag{2}$$

Then

$$\bigcap_{n \in N} A_n \neq \phi. \tag{3}$$

Hence, by Problem 18–19, the space (X, d) is complete.

> 1. Show that a subspace of a complete metric space can be incomplete.
>
> 2. Show that any separable metric space (X, d) is homeomorphic to a dense subspace of a complete metric space.

SOLUTION:

1. The set of rational numbers forms an incomplete subspace of the complete set of real numbers.

2. Space X is said to be separable if X contains a countable dense subset.

Any separable metric space (X, d) has a metrizable compactification, say (Y, d'). If a metric space (Y, d') is compact, then (Y, d') is complete.

Show that any closed subspace A of a complete metric space (X, d) is a complete metric space.

SOLUTION:

Suppose

$$(a_n), n \in N$$

is a Cauchy sequence in A. Then (a_n) is also a Cauchy sequence in X. Since X is complete

$$a_n \to a \in X.$$

But

$$a \in \overline{A} = A.$$

Hence, (a_n) converges in A and A is complete.

● PROBLEM 18-23

Show that

$$((x X_n, d) \text{ is complete}) \Leftrightarrow (\text{each } (X_n, d_n) \text{ is complete}).$$

SOLUTION:

Let us define d as follows:

$$d(x,y) = \sum_{n \in N} \frac{\min(d_n(x_n, y_n), 1)}{2^n}.$$

where
$$x = (x_1, x_2, x_3, \ldots)$$

$$y = (y_1, y_2, y_3, \ldots).$$

It can be shown that d is a metric in $\underset{n}{\times} X_n$.

$\Leftarrow$ Suppose each (X_n, d_n) is complete. Let

$$(a_k), k \in N$$

be a Cauchy sequence in $\underset{n}{\times} X_n$. The l-th coordinate of a_k is denoted by

$$a_k(l).$$

The sequence $(a_k(l))$, $k \in N$ is a Cauchy sequence in X_l. Hence, $(a_k(l))$ converges in X_l for each $l \in N$. Thus, (a_k) converges in $\underset{n}{x} X_n$.

$\Rightarrow$ Suppose (X_1, d_1) is not complete. There is a Cauchy sequence

$$(a_k(1)), k \in N$$

in X_1, which does not converge. Choose from each X_n, $n > 1$ a point

$$x_n \in X_n.$$

We define a sequence as

$$(a_K(l)) = \begin{cases} a_K(1) & \text{for} \quad l = 1 \\ \\ x_l & \text{for} \quad l \neq 1 \end{cases}$$

which is a Cauchy sequence but does not converge in $(\underset{n}{x} X_n, d)$.
   Contradiction.

● **PROBLEM 18-24**

1. Show that

$$f : R^2 \to R^2$$

$$f(x) = \frac{x}{2}$$

defined on Euclidean R^2 space is a contracting mapping.

2. Show that a contracting mapping is continuous.

SOLUTION:

DEFINITION OF CONTRACTING MAPPING

   A function $f : X \to X$, defined as (X, d) is called a contracting mapping if a real number $a \in R$, $0 \le a \le 1$ exists there, such that for every $x, y \in X$

$$d(f(x), f(y)) \le a \, d(x, y) < d(x, y).$$

∎

1.
$$d(f(x), f(y)) = \|f(x) - f(y)\| = \left\| \frac{x}{2} - \frac{y}{2} \right\| =$$

$$= \frac{1}{2} \|x - y\| = \frac{1}{2} d(x, y).$$

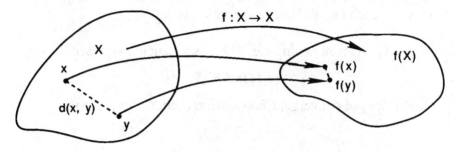

FIGURE 1

2. Let $f : X \to X$ be a contracting mapping on a metric space (X, d). A real number $0 \le a < 1$ exists such that for every $x, y \in X$

$$d(f(x), f(y)) \le a \, d(x, y).$$

Let $x_1 \in X$ and $\varepsilon > 0$.
   We have

$$d(x, x_1) < \varepsilon \Rightarrow d(f(x), f(x_1)) \le a \, d(x, x_1) \le \varepsilon \, a < \varepsilon.$$

Hence, f is continuous.

● **PROBLEM 18–25**

Show that if $f : X \to X$ is a contracting mapping on a complete metric space (X, d) then there exists one and only one point $x_0 \in X$ such that

$$f(x_0) = x_0. \tag{1}$$

SOLUTION:

Let $a_0 \in X$. We define

$$f(a_0) = a_1, f(a_1) = a_2 = f^2(a_0), \dots$$

$$f(a_{n-1}) = a_n = f^n(a_0), \dots \tag{2}$$

and show that

$$a_0, a_1, a_2, \dots$$

is a Cauchy sequence.

$$d(f^{k+l}(a_0), f^l(a_0)) \le a \, d(f^{k+l-1}(a_0), f^{l-1}(a_0)) \le \dots$$

$$\dots \le a^l \, d(f^k(a_0), a_0) \le$$

684

$$\leq a^l[d(a_0, f(a_0)) + d(f(a_0), f^2(a_0)) + \ldots + d(f^{k-1}(a_0), f^k(a_0))]. \quad (3)$$

But

$$d(f^{m+1}(a_0), f^m(a_0)) \leq a^m d(f(a_0), a_0). \quad (4)$$

Hence, from (3) and (4) we obtain

$$d(f^{k+l}(a_0), f^l(a_0)) \leq a^l d(f(a_0), a_0) [1 + a + a^2 + \ldots + a^{k-1}]. \quad (5)$$

Since $a < 1$,

$$1 + a + a^2 + \ldots + a^{k-1} \leq \frac{1}{1-a}. \quad (6)$$

From (5) and (6) we find

$$d(f^{k+l}(a_0), f^l(a_0)) \leq \frac{a^l}{1-a} d(f(a_0), a_0). \quad (7)$$

For $\varepsilon > 0$, we define

$$\lambda = \begin{cases} \varepsilon(1-a) & \text{if } d(f(a_0), a_0) = 0 \\ \dfrac{\varepsilon(1-a)}{d(f(a_0), a_0)} & \text{if } d(f(a_0), a_0) \neq 0 \end{cases}. \quad (8)$$

Since $a < 1$, $M \in N$ exists such that

$$a^M < \lambda. \quad (9)$$

For $k \geq l > M$,

$$d(a_k, a_l) \leq a^l \cdot \frac{1}{1-a} \cdot d(f(a_0), a_0) < \frac{\lambda}{1-a} \cdot d(f(a_0), a_0) \leq \varepsilon. \quad (10)$$

Thus, $(a_1, a_2, a_3, \ldots)$ is a Cauchy sequence and since X is complete $x_0 \in X$ exists such that

$$\lim_{n \to \infty} a_n = x_0. \quad (11)$$

Function f is continuous, thus

$$f(x_0) = f(\lim a_n) = \lim f(a_n) = \lim a_n = x_0. \quad (12)$$

Point x_0 is unique. For, suppose $f(x_0) = x_0$ and $f(y_0) = y_0$, then

$$d(x_0, y_0) = d(f(x_0), f(y_0)) \leq a \, d(x_0, y_0). \quad (13)$$

But $a < 1$. Therefore,

$$d(x_0, y_0) = 0$$

and $x_0 = y_0$.

Show that the set of real numbers R is a completion of the set of rational numbers Q.

SOLUTION:

DEFINITION OF COMPLETION OF METRIC SPACE

A metric space X^* is called a completion of a metric space X if X^* is complete and X is isometric to a dense subset of X^*.

∎

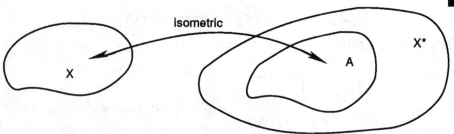

X^* – complete.
X and A are isometric.
$\overline{A} = X^*$.

FIGURE 1

Consider the set Q of rational numbers. The closure of Q is the entire set R of real numbers, i.e.,

$$\overline{Q} = R.$$

The set R is complete. Hence, the set R of real numbers is a completion of the set Q of rational numbers.

● **PROBLEM 18-27**

Let $C[X]$ denote the collection of all Cauchy sequences on X. The relation ~ on $C[X]$ is defined by

$$((a_n) \sim (b_n)) \Leftrightarrow (\lim_{n \to \infty} d(a_n, b_n) = 0). \qquad (1)$$

Show that ~ is an equivalence relation on $C[X]$.

SOLUTION:

$$(a_n) \sim (a_n)$$

is obvious, since $d(x, x) = 0$

$$((a_n) \sim (b_n)) \Rightarrow ((b_n) \sim (a_n))$$

because

$$d(x, y) = d(y, x).$$

Suppose $(a_n) \sim (b_n)$ and $(b_n) \sim (c_n)$. Then

$$\lim d(a_n, b_n) = \lim d(b_n, c_n) = 0.$$

From the definition of a metric,

$$d(a_n, c_n) \leq d(a_n, b_n) + d(b_n, c_n).$$

Taking the limit

$$\lim d(a_n, c_n) \leq \lim d(a_n, b_n) + \lim d(b_n, c_n) = 0.$$

Thus,

$$(a_n) \sim (c_n).$$

The relation $\sim$ is an equivalence relation on $C[X]$.

● **PROBLEM 18–28**

Show that

$$((a_n) \sim (b_n)) \Leftrightarrow \begin{pmatrix} (a_n) \text{ and } (b_n) \text{ are subsequences} \\ \text{of some Cauchy sequence } (c_n) \end{pmatrix}.$$

SOLUTION:

$\Leftarrow$ A Cauchy sequence (c_n) exists such that

$$(a_n) = (c_{k_n}) \quad \text{and} \quad (b_n) = (c_{l_n}).$$

Then

$$\lim_{n \to \infty} d(a_n, b_n) = \lim_{n \to \infty} d(c_{k_n}, c_{l_n}) = 0$$

since (c_n) is a Cauchy sequence.

$\Rightarrow$ Suppose

$$(a_n) \sim (b_n)$$

that is

$$\lim_{n \to \infty} d(a_n, b_n) = 0.$$

We define a sequence

$$(c_n) = (a_1, b_1, a_2, b_2, a_3, b_3, \ldots)$$

that is

$$(c_n) = \begin{cases} a_{\frac{n+1}{2}} & \text{if } n \text{ is odd} \\ b_{\frac{n}{2}} & \text{if } n \text{ is even} \end{cases}.$$

and show that (c_n) is a Cauchy sequence. Let $\varepsilon > 0$, then $\exists\, n_1 \in N$ such that

$$n > n_1 \Rightarrow d(a_n, b_n) < \varepsilon/2$$

$\exists\, n_2 \in N$ such that

$$m, n > n_2 \Rightarrow d(a_m, a_n) < \varepsilon/2$$

$\exists\, n_3 \in N$ such that

$$m, n > n_3 \Rightarrow d(b_m, b_n) < \varepsilon/2$$

Set $n_0 = \max(n_1, n_2, n_3)$. Then for $m > 2n_0$

$$\frac{1}{2} m > n_1, \frac{1}{2} m > n_3$$

$$\frac{m+1}{2} > n_1, \frac{m+1}{2} > n_2.$$

We shall show that

$$m, n > 2n_0 \Rightarrow d(c_m, c_n) < \varepsilon.$$

We have

$$m, n \text{ even} \Rightarrow c_m = b_{\frac{m}{2}}, c_n = b_{\frac{n}{2}}$$

then

$$d(c_m, c_n) < \varepsilon/2 < \varepsilon$$

$$m, n \text{ odd} \Rightarrow c_m = a_{\frac{m+1}{2}}, c_n = a_{\frac{n+1}{2}}$$

then

$$d(c_m, c_n) < \varepsilon/2 < \varepsilon$$

m,n one odd and one even $\Rightarrow c_m = \dfrac{b_m}{2}$, $c_n = \dfrac{a_{n+1}}{2}$

then

$$d(c_m,c_n) \le d(\frac{a_m}{2},\frac{b_m}{2}) + d(\frac{a_m}{2},\frac{a_{n+1}}{2}) < \tfrac{\varepsilon}{2} + \tfrac{\varepsilon}{2} = \varepsilon.$$

Hence, (c_n) is a Cauchy sequence.

● PROBLEM 18-29

Describe the construction of a completion X^* of a metric space (X, d).

SOLUTION:

A metric space (X, d) is given. By $C[X]$ we denote the collection of all Cauchy sequences in X. Among Cauchy sequences the relation ~ is introduced as follows:

$$(a_n) \sim (b_n) \text{ iff } \lim_{n \to \infty} d(a_n, b_n) = 0.$$

In Problem 18–27 we showed that ~ is an equivalence relation. We define X^*

$$X^* = \frac{C[X]}{\sim}$$

as the quotient space. Elements of X^* are the equivalence classes $[(a_n)]$ of Cauchy sequences (a_n) of $C[X]$. Thus, if $(a_n) \in [(a_n)]$ and $(b_n) \in [(a_n)]$ we have $(a_n) \sim (b_n)$ and $\lim_{n \to \infty} d(a_n, b_n) = 0$.

Later we shall show that the function δ defined by

$$\delta([(a_n)], [(b_n)]) = \lim_{n \to \infty} d(a_n, b_n)$$

is a metric on X^*.

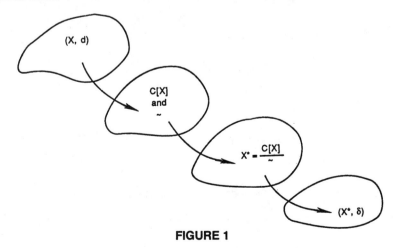

FIGURE 1

689

Show that the function δ defined in Problem 18–29 by

$$\delta([(a_n)], [(b_n)]) = \lim_{n \to \infty} d(a_n, b_n)$$

is well-defined, i.e., that

$$\begin{pmatrix} (a'_n) \sim (a_n) \\ (b'_n) \sim (b_n) \end{pmatrix} \Rightarrow (\lim\ d(a'_n, b'_n) = \lim\ d(a_n, b_n)).$$

SOLUTION:

Suppose $(a'_n) \sim (a_n)$ and $(b'_n) \sim (b_n)$ then

$$\lim_{n \to \infty} d(a'_n, a_n) = 0$$

and

$$\lim_{n \to \infty} d(b'_n, b_n) = 0.$$

Set

$$a = \lim d(a_n, b_n)$$

$$a' = \lim d(a'_n, b'_n).$$

We have

$$d(a_n, b_n) \le d(a_n, a'_n) + d(b_n, b'_n) + d(a'_n, b'_n).$$

Let $\varepsilon > 0$, then

$\exists\ n_1 \in N$ such that

$$n > n_1 \Rightarrow d(a_n, a'_n) < {}^{\varepsilon}\!/_3$$

$\exists\ n_2 \in N$ such that

$$n > n_2, \Rightarrow d(b_n, b'_n) < {}^{\varepsilon}\!/_3$$

$\exists\ n_3 \in N$ such that

$$n > n_3 \Rightarrow d(a'_n, b'_n) - a' < {}^{\varepsilon}\!/_3$$

For

$$n > \max(n_1, n_2, n_3)$$

we get

$$d(a_n, b_n) < \frac{\varepsilon}{3} + \frac{\varepsilon}{3} + \frac{\varepsilon}{3} + a' = a' + \varepsilon$$

and

$$\lim_{n \to \infty} d(a_n, b_n) = a \leq a' + \varepsilon.$$

Hence, since $\varepsilon > 0$ is arbitrary

$$a \leq a'.$$

In the same way we show that

$$a' \leq a$$

and conclude that

$$a' = a.$$

● PROBLEM 18-31

Show that

$$\delta([(a_n)], [(b_n)]) = \lim_{n \to \infty} d(a_n, b_n)$$

is metric on

$$X^* = \frac{C[X]}{\sim}$$

(compare Problem 18–29).

SOLUTION:

We have already shown that δ is well-defined (Problem 18–30).

1. $$\delta([a_n)], [(b_n)] \geq 0$$

because $d(a_n, b_n) \geq 0$.

$$\delta([a_n)], [(a_n)] = 0$$

because $d(a_n, b_n) = 0$.

2. $$\delta([(a_n)], [(b_n)]) = \delta([(b_n)], [(a_n)])$$

because $d(a_n, b_n) = d(b_n, a_n)$.

3. $$\delta([(a_n)], [(b_n)]) = \lim_{n \to \infty} d(a_n, b_n) \leq$$

691

$$\leq \lim_{n \to \infty} d(a_n, c_n) + \lim_{n \to \infty} d(b_n, c_n) =$$

$$= \delta([(a_n)], [(c_n)]) + \delta([(b_n)], [(c_n)]).$$

4. If $[(a_n)] \neq [(b_n)]$ then

$$\delta([(a_n)], [(b_n)]) > 0.$$

Indeed

$$\delta([(a_n)], [(b_n)]) = \lim_{n \to \infty} d(a_n, b_n) > 0$$

because

$$(a_n) \text{ is not in relation } \sim \text{ with } (b_n) \Rightarrow \lim d(a_n, b_n) > 0.$$

Hence,

$$\delta([(a_n)], [(b_n)]) > 0.$$

● **PROBLEM 18-32**

Suppose $a \in X$ then $(a, a, a, ...)$ is a Cauchy sequence

$$(a, a, a, ...) \in C[X].$$

We define space $\hat{X}$ as follows:

$$\hat{a} \in \hat{X} = \{\hat{a} : a \in X\}$$

$\hat{a} = [(a, a, a, ...)]$. Space $\hat{X}$ is a subspace of $X *$.
  Now, suppose $(a_n) \in C[X]$. Show that

$$a = [(a_n)] \in X^*$$

is the limit of the sequence

$$(\hat{a}_1, \hat{a}_2, \hat{a}_3, ...)$$

in $\hat{X}$.

SOLUTION:

Sequence (a_n) is a Cauchy sequence in X. We have

$$\lim_{n \to \infty} \delta(a, \hat{a}_n) = \lim_{n \to \infty} (\lim_{n \to \infty} d(a_m, a_n)) =$$

$$= \lim_{\substack{n \to \infty \\ m \to \infty}} d(a_m, a_n) = 0.$$

Thus,

$$(\hat{a}_n) \to a.$$

1. Show that X is isometric to space $\hat{X}$ defined in Problem 18–32.

2. Show that $\hat{X}$ is dense in X^*.

DENSE

SOLUTION:

1. Suppose $a, b \in X$. Then

$$\hat{a} = [(a, a, a, ...)] \in \hat{X}$$

$$\hat{b} = [(b, b, b, ...)] \in \hat{X}$$

and

$$\delta(\hat{a}, \hat{b}) = \lim_{n \to \infty} d(a, b) = d(a, b).$$

Hence, X is isometric to $\hat{X}$.

2. We shall show that every point in X^* is the limit of a sequence in $\hat{X}$. Let

$$a = [(a_1, a_2, a_3, ...)] \in X^*.$$

Then $(a_1, a_2, a_3, ...)$ is a Cauchy sequence in X. Hence, a is the limit of the sequence

$$(\hat{a}_1, \hat{a}_2, \hat{a}_3, ...)$$

in $\hat{X}$. Thus, $\hat{X}$ is dense in X^*.

DENSE

● **PROBLEM 18–34**

Show that (X^*, d) is a completion of X, i.e., that every Cauchy sequence in (X^*, δ) converges.

SOLUTION:

Let

$$(a_1, a_2, a_3, \ldots)$$

be a Cauchy sequence in (X^*, δ). In Problem 18–33 we proved that

$$\hat{X} \text{ is dense in } X^*.$$

Hence,

$$\forall \, n \in N \quad \exists \, \hat{a}_n \in \hat{X} \text{ such that}$$

$$\delta(\hat{a}_n, a_n) < \frac{1}{n}.$$

Hence, $(\hat{a}_1, \hat{a}_2, \hat{a}_3, \ldots)$ is also a Cauchy sequence.

By Problem 18–33 $(\hat{a}_1, \hat{a}_2, \hat{a}_3, \ldots)$ converges to

$$a = [(a_1, a_2, \ldots)] \in X^*.$$

Thus, (a_n) also converges to a. We conclude that (X^*, δ) is complete.

● **PROBLEM 18–35**

Show that

$$(Y^* \text{ is a completion of } X) \Rightarrow (Y^* \text{ is isometric to } X^*).$$

SOLUTION:

X is a subspace of Y^*. Thus for every

$$y \in Y^*$$

there exists a Cauchy sequence $(y_1, y_2, y_3, \ldots)$ in X converging to y. Define a mapping

$$f : Y^* \to X^*$$

by

$$f(y) = [(y_1, y_2, \ldots)].$$

We shall show that f is an isometry between Y^* and X^*. Mapping f is well-defined. Indeed, suppose

$$(y'_1, y'_2, y'_3, \ldots) \to y$$

where (y'_n) is a sequence in X.

Then

$$\lim_{n \to \infty} d(y_n, y'_n) = 0$$

and

$$[(y_n)] = [(y'_n)].$$

Also f is subjective. Suppose

$$[(x_1, x_2, \ldots)] \in X^*.$$

Then

$$(x_1, x_2, \ldots)$$

is a Cauchy sequence in $X \subset Y^*$. But Y^* is complete. Hence, (x_n) converges to $x \in Y^*$ and

$$f(x) = [(x_n)].$$

Now suppose $x, y \in Y^*$. Then there are sequences (x_n) and (y_n) in X such that

$$x_n \to x$$

$$y_n \to y.$$

We have

$$\delta(f(x), f(y)) = \delta([(x_n)], [(y_n)]) =$$

$$= \lim_{n \to \infty} d(x_n, y_n) = d(x, y).$$

Hence, f is an isometry between Y^* and X^*.

● PROBLEM 18-36

1. Show that the set Z of integers is a nowhere dense subset of the real line R.

2. Show that the set Q of rational numbers is not nowhere dense in R.

SOLUTION:

1. A subset A of a topological space (X, T) is nowhere dense in X iff the interior of the closure of A is empty

$$int(\overline{A}) = \phi.$$

The set Z is closed and its interior is empty. Hence,

$$int(\overline{Z}) = int(Z) = \phi.$$

2. The closure of Q is R

$$\overline{Q} = R$$

and

$$int(\overline{Q}) = int(R) = R \neq \phi.$$

Thus, the set Q of rational numbers is not nowhere dense in R.

● **PROBLEM 18-37**

DENSE   Let U be an open subset of the metric space (X, d) and let M be nowhere dense in X. Show that $x \in X$ exists and $\varepsilon > 0$ such that

$$B(x, \varepsilon) \subset U$$

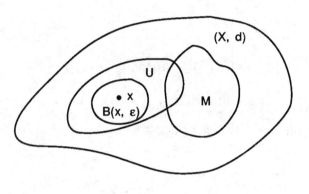

FIGURE 1

SOLUTION:

Let us denote

$$U \cap \overline{M}^c = V.$$

Then

$$V \subset U \quad \text{and} \quad V \cap M = \phi.$$

DENSE   Set V is nonempty because U is open and $\overline{M}^c$ is dense in X.

$$V \neq \phi.$$

Hence,

$$x \in V.$$

696

V is open because U is open and $\overline{M}^c$ is open.

Thus, $\varepsilon > 0$ exists such that

$$B(x, \varepsilon) \subset V$$

and

$$B(x, \varepsilon) \cap M = \phi.$$

● PROBLEM 18-38

Show that if M is a nowhere dense subset of X then $\overline{M}^c$ is dense in X. (We denote $M^c = X - M$.)

DENSE

SOLUTION:

Suppose on the contrary that $\overline{M}^c$ is not dense in X. Then $x \in X$ exists and an open set U such that

$$x \in U$$

$$U \cap \overline{M}^c = \phi.$$

Hence,

$$x \in U \subset \overline{M}$$

and

$$x \in int(\overline{M}).$$

It is a contradiction, because M is nowhere dense in X, $int(\overline{M}) = \phi$. Thus, $\overline{M}^c$ is dense in X.

● PROBLEM 18-39

Prove Baire's Category Theorem.

THEOREM

Every complete metric space (X, d) is of second category.

■

SOLUTION:

DEFINITION OF FIRST CATEGORY SPACES

A topological space (X, T) is said to be of first category if X is the count-
~~DENSE~~ able union of nowhere dense subsets of X. All other spaces are of second category.

■

Let $A \subset X$ be of first category. We will show that $A \neq X$. Since A is of first
category, by definition

$$A = M_1 \cup M_2 \cup \dots \tag{1}$$

~~DENSE~~ where each M_n is a nowhere dense subset of X.

Since M_1 is nowhere dense in X, there exists $x_1 \in X$ and $\varepsilon_1 > 0$ such that

$$B(x_1, \varepsilon_1) \cap M_1 = \phi. \tag{2}$$

Then

$$\overline{B(x_1, \frac{\varepsilon_1}{2})} \cap M_1 = \phi. \tag{3}$$

Since $B(x_1, \frac{\varepsilon_1}{2})$ is open and M_2 is nowhere dense in X, by Problem 18–38 we
conclude that

$\exists\, x_2 \in X,\, \exists\, \varepsilon_2 > 0$ such that

$$B(x_2, \varepsilon_2) \subset B(x_1, \frac{\varepsilon_1}{2}) \subset \overline{B(x_1, \frac{\varepsilon_1}{2})} \tag{4}$$

and

$$B(x_2, \varepsilon_2) \cap M_2 = \phi. \tag{5}$$

Set

$$\varepsilon_2 \leq \frac{\varepsilon_1}{2}. \tag{6}$$

Then

$$\overline{B(x_2, \frac{\varepsilon_2}{2})} \subset \overline{B(x_1, \frac{\varepsilon_1}{2})} \tag{7}$$

and

$$\overline{B(x_2, \frac{\varepsilon_2}{2})} \cap M_2 = \phi. \tag{8}$$

In this way we get a nested sequence of closed sets

$$\overline{B(x_1, \frac{\varepsilon_1}{2})} \supset \overline{B(x_2, \frac{\varepsilon_2}{2})} \supset \overline{B(x_3, \frac{\varepsilon_3}{2})} \supset \dots \tag{9}$$

such that for every $n \in N$

$$\overline{B(x_n, \frac{\varepsilon_n}{2})} \cap M_n = \phi \tag{10}$$

698

and

$$\frac{\varepsilon_n}{2} \le \frac{\varepsilon_1}{2^n}.$$

Hence,

$$\lim_{n \to \infty} \varepsilon_n \le \lim_{n \to \infty} \frac{\varepsilon_1}{2^{n-1}} = 0$$

and $x \in X$ exists such that

$$x \in \bigcap_{n-1}^{\infty} \overline{B(x_n, \tfrac{\varepsilon_n}{2})}.$$

(11)

Also, $x \notin M_n$ for every $n \in N$ and

$$x \notin A.$$

Thus, (X, d) is of second category.

ELEMENTS OF HOMOTOPY THEORY

Show that the two continuous functions,

$$f : X \to R$$

$$g : X \to R$$

where R is the space of real numbers with standard topology, are always homotopic.

SOLUTION:

DEFINITION OF HOMOTOPIC FUNCTIONS

Two continuous functions

$$f : X \to Y$$

$$g : X \to Y$$

are given. Functions f and g are said to be homotopic, if a continuous function h

$$h : X \times [0, 1] \to Y$$

exists, such that

$$h(x, t) \in Y$$

$$h(x, 0) = f(x) \quad \text{and} \quad h(x, 1) = g(x).$$

■

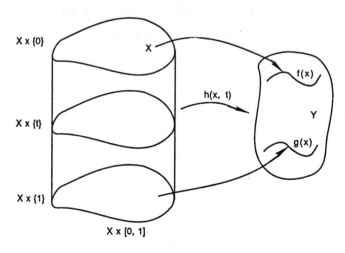

FIGURE 1

701

If $Y = R$ (or more generally, $Y = R^n$), then the functions f and g are always homotopic. Indeed,

$$h(x, t) = f(x) + t[g(x) - f(x)]$$

and $h(x, t)$ is continuous.

Show that the identity function on the unit disk is homotopic to a constant function which maps the whole disk into $(0, 0)$.

SOLUTION:

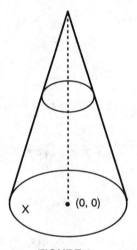

FIGURE 1

Suppose X and Y are unit disks and f is the identity function

$$f(x) = x \quad \text{for} \quad x \in X.$$

Let $g(x)$ denote a constant function defined on the unit disk

$$g(x) = (0, 0) \quad \text{for} \quad x \in X.$$

We shall define function $h(x, t)$.

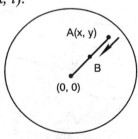

FIGURE 2

702

Let $A = (x, y)$ represent a point in the disk. Point $h(A, t) = B$ is the point on $\overline{(0, 0)\,(x, y)}$ that is t times the distance from (x, y) to $(0, 0)$. Hence

$$h(A, 0) = A$$

$$h(A, 1) = (0, 0) = \text{const.}$$

● **PROBLEM 19-3**

Let Y^X denote the set of all continuous functions from X into Y, and let $f \sim g$ dneote that f is homotopic to g. Show that if $f \in Y^X$, then

$$f \sim f.$$

SOLUTION:

The homotopy

$$h : X \times [0, 1] \rightarrow Y$$

between f and g is given by

$$h(x, t) = f(x)$$

for all $x \in X$ and all $t \in [0, 1]$.

We must show that $h(x, t)$ is continuous and that

$$h(x, 0) = h(x, 1) = f(x).$$

Suppose U is an open subset of Y. Then

$$h^{-1}(U) = f^{-1}(U) \times [0, 1].$$

Since f is continuous $f^{-1}(U)$ is an open subset of X and $f^{-1}(U) \times [0, 1]$ is an open subset of $X \times [0, 1]$. Hence, $h(x, t)$ is continuous.

● **PROBLEM 19-4**

Show that if $f \sim g$, then $g \sim f$.

SOLUTION:

Since $f \sim g$, there is a homotopy

$$h : X \times [0, 1] \rightarrow Y$$

703

such that

$$h(x, 1) = f \quad \text{and} \quad h(x, 0) = g.$$

Let us define H by

$$H : X \times [0, 1] \rightarrow Y$$

$$H(x, t) = h(x, 1 - t)$$

for all $x \in X$ and all $t \in [0, 1]$.

We have

$$H(x, 0) = h(x, 1) = f$$

$$H(x, 1) = h(x, 0) = g.$$

Since $h(x, t)$ is continuous, so is $H(x, t)$.

Hence, $H(x, t)$ is a homotopy between g and f.

● **PROBLEM 19–5**

Show that the relation, $f \sim g$

f is homotopic to g

is an equivalence relation.

SOLUTION

In Problem 19–3, we have shown that

$$f \sim f$$

and in Problem 19–4, we have shown that

if $f \sim g$, then $g \sim f$.

It remains to be shown that, if $f \sim g$ and $g \sim k$, then $f \sim k$. Since $f \sim g$, there is a homotopy

$$h_1 : X \times [0, 1] \rightarrow Y$$

such that

$$h_1 (x, 0) = g(x) \quad \text{and}$$

$$h_1 (x, 1) = f(x) \text{ for all } x \in X.$$

Similarly, since $g \sim k$, there is a homotopy

$$h_2 : X \times [0, 1] \to Y$$

such that

$$h_2(x, 0) = k(x) \quad \text{and}$$

$$h_2(x, 1) = g(x) \text{ for all } x \in X.$$

Let us define mapping

$$h : X \times [0, 1] \to Y$$

by

$$h(x,t) = \begin{cases} h_2(x, 2t) & \text{for} \quad 0 \leq t \leq \frac{1}{2} \\ h_1(x, 2t - 1) & \text{for} \quad \frac{1}{2} \leq t \leq 1 \end{cases}$$

for all $x \in X$.

Note that

$$h(x, 0) = h_2(x, 0) = k(x)$$

$$h(x, 1) = h_1(x, 1) = f(x)$$

$$h(x, \tfrac{1}{2}) = h_1(x, 0) = h_2(x, 1) = g(x).$$

Mapping $h(x, t)$ is well-defined and since both h_1 and h_2 are continuous, so is $h(x, t)$.

Thus, $h(x, t)$ is a homotopy between f and k.

The relation $\sim$ is an equivalence relation on Y^X.

● **PROBLEM 19–6**

1. Why do the homotopy classes of Y^X form a partition of Y^X?

2. Can you find any relationship between the extension of functions and functions being homotopic?

SOLUTION:

DEFINITION OF A HOMOTOPY CLASS

The family of all continuous functions from X into Y, which are homotopic to a continuous function f, is called the homotopy class of f. ■

The homotopy classes of Y^X form a partition of Y^X because ~ is an equivalence relation on Y^X (See Problem 19–5).

2. Suppose two continuous functions f and g are given, $f, g \in Y^X$. We define function h as follows:

$$h : (X \times \{0\}) \cup (X \times \{1\}) \to Y$$

$$h(x, 0) = f(x) \quad \text{and} \quad h(x, 1) = g(x)$$

for all $x \in X$.

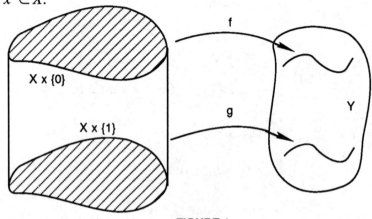

FIGURE 1

It is easy to see that f and g are homotopic, if and only if function h can be extended to a continuous function H

$$H : X \times [0, 1] \to Y$$

such that

$$H \,|\, (X \times \{0\}) \cup (X \times \{1\}) = h.$$

● PROBLEM 19–7

Why are any two paths in R^2 homotopic?

SOLUTION:

A continuous function from the interval $[0, 1]$ into any space (X, T) is called a path in X. (See Figure 1.)
   The space

$$[0, 1] \times [0, 1]$$

with the product topology is a normal space and its subset

$$A = \{(x, y) : y = 1\} \cup \{(x, y) : y = 0\}$$

is a closed set.

If f is any continuous function from A into R^2, then, by Tietze's Extension Theorem, f has a continuous extension F.

Thus, any two paths in R^2 are homotopic.

13-33

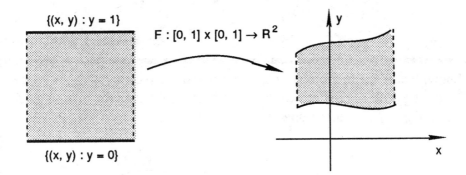

● **PROBLEM 19-8**

In Problem 19–2, we have shown that the identity function $f_1 = i$ on the unit disk Y is homotopic to the function f_0, which maps the whole disk into $(0, 0)$. Show that $f_1 = i$ is homotopic to the function

$$g : Y \rightarrow Y$$

defined by

$$g(x, y) = (0, {}^1\!/_2)$$

for all $(x, y) \in Y$.

SOLUTION:

Let h denote the homotopy between $f_1 = i$ and f_0. We define

$$H((x,y),t) = \begin{cases} h((x,y),1-2t) & \text{for } 0 \le t \le \tfrac{1}{2} \\ (0, t - \tfrac{1}{2}) & \text{for } \tfrac{1}{2} \le t \le 1 \end{cases}.$$

Note, that $H((x, y), t)$ on the segment $[0, {}^1\!/_2]$ contracts Y to $(0, 0)$ and then on the segment $[{}^1\!/_2, 1]$, slides it from $(0, 0)$ to $(0, {}^1\!/_2)$ along the straight line between these two points.

707

Since $H | (Y \times [0, \frac{1}{2}])$ and $H | (Y \times [\frac{1}{2}, 1])$ are continuous and

$$H((x, y), \frac{1}{2}) = h((x, y), 0) = (0, 0)$$

is well-defined, H is continuous. Thus H is a homotopy, since

$$H | (Y \times \{0\}) = g$$

$$H | (Y \times \{1\}) = f_1 = i.$$

● PROBLEM 19-9

Show, that if the homotopic mappings f_0 and f_1 have values in the space Y and if g_0 and g_1 are defined on Y and homotopic, then the mappings g_0 o f_0 and g_1 o f_1 are homotopic, i.e., prove that

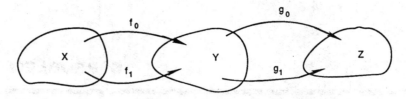

FIGURE 1

$$\begin{pmatrix} f_0 \sim f_1 \\ g_0 \sim g_1 \end{pmatrix} \Rightarrow (g_0 \, o \, f_0 \sim g_1 \, o \, f_1).$$

SOLUTION:

Since $f_0 \sim f_1$, a continuous function

$$h_1(x, t) \in Y$$

exists, such that

$$h_1(x, 0) = f_0(x) \, ; h_1(x, 1) = f_1(x).$$

Since $g_0 \sim g_1$, a continuous function

$$h_2(y, t) \in Z$$

exists, such that

$$h_2(y, 0) = g_0(y) \, ; h_2(y, 1) = g_1(y).$$

To prove that

$$g_0 \, o \, f_0 \sim g_1 \, o \, f_1$$

708

we have to show that a continuous function

$$H(x, t) \in Z$$

exists, such that

$$H(x, 0) = g_0 \circ f_0 \quad \text{and} \quad H(x, 1) = g_1 \circ f_1.$$

Let us define

$$H(x, t) = h_2(h_1(x, t), t)$$

then

$$H(x, 0) = h_2(h_1(x, 0), 0) = h_2 (f_0(x), 0) =$$

$$= g_0(f_0(x)) = g_0 \circ f_0(x)$$

and

$$H(x, 1) = h_2(h_1(x, 1), 1) = h_2 (f_1(x), 1) =$$

$$= g_1(f_1(x)) = g_1 \circ f_1(x).$$

Since h_1 and h_2 are continuous, H is also continuous. Thus, H is a homotopy between $g_0 \circ f_0$ and $g_1 \circ f_1$.

● PROBLEM 19-10

A homeomorphism of $[0, 1]$ into any space (X, T) is called an arc. A continuous function of $[0, 1]$ into any space is called a path. In Problem 19–7, we showed that any two paths in R^2 are homotopic. Thus, we can state that any two arcs in R^2 are homotopic. Let A represent any point in R^2. Show that two arcs in $R^2 - \{A\}$ are not necessarily homotopic.

SOLUTION:

Let a_1 and a_2 denote distinct arcs in $R^2 - \{A\}$, such that

$$a_1(0) = a_2(0) \quad \text{and} \quad a_1(1) = a_2(1).$$

Obviously a_1 and a_2 are homotopic in R^2. We choose a_1 and a_2 in such a way, that A is a point in the area bounded by a_1 and a_2. (See Figure 1.)

Then a_1 and a_2 are not homotopic in $R^2 - \{A\}$ because a_1 cannot be continuously transformed into a_2.

709

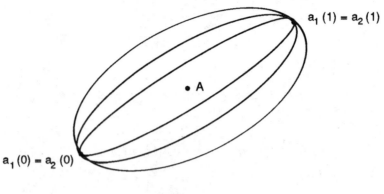

$$a_1(0) = a_2(0)$$

$$a_1(1) = a_2(1)$$

• A

FIGURE 1

When are two loops a_0 and a_1 in X with base point x_0 homotopic relative to x_0?

SOLUTION:

A continuous function a

$$a : [0, 1] \rightarrow X$$

such that

$$a(0) = a(1) = x_0$$

is called a loop in X with base point y_0.

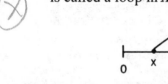

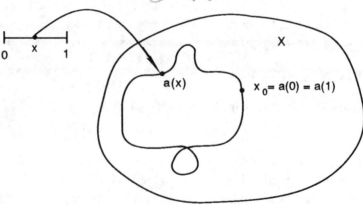

0   x   1

X

$a(x)$

$x_0 = a(0) = a(1)$

FIGURE 1

Two loops a_0 and a_1 in X with base point x_0 are said to be homotopic

710

relative to x_0, if a homotopy H between a_0 and a_1 exists, i.e.,

$$H(x, t) \in X; \, x, t \in [0, 1]$$

$$H(x, 0) = a_0 \,; \; H(x, 1) = a_1$$

such that, for each $t \in [0, 1]$

$$H(0, t) = H(1, t) = x_0.$$

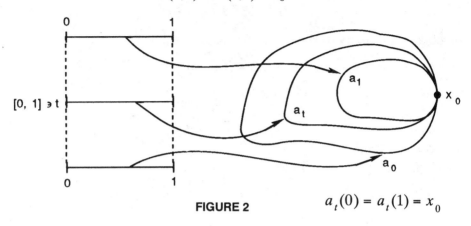

FIGURE 2

$$a_t(0) = a_t(1) = x_0$$

That is, for each $t \in [0, 1]$

$$H \,|\, [0, 1] \times \{t\}$$

is a loop in X with base point x_0. See Figure 2.

● **PROBLEM 19-12**

Let $L(X, x_0)$ denote the set of all loops in X with base point x_0. Show that the relation "homotopic relative to x_0 to" (denoted by $\sim$) defined on $L(X, x_0)$, is an equivalence relation.

SOLUTION:

In Problem 19–5, we showed that "f, homotopic to g", is an equivalence relation on Y^X.

For any

$$a \in L(X, x_0)$$

a is homotopic relative to x_0 to a

$$a \sim a.$$

711

Let

$$a \sim b.$$

Then, the homotopy H exists between a and b, such that

$$H(0, t) = H(1, t) = x_0.$$

Setting

$$H'(x, t) = H(x, 1 - t)$$

we obtain homotopy between b and a, such that

$$H'(0, t) = H'(1, t) = x_0.$$

Hence,

$$b \sim a.$$

Suppose $a \sim b$ and $b \sim c$.
    Homotopies exist, such that

$$h_1(0, t) = h_1(1, t) = x_0$$

$$h_1(0, t) = h_2(1, t) = x_0.$$

Then

$$h(x,t) = \begin{cases} h_2(x, 2t) & \text{for} \quad 0 \le t \le \frac{1}{2} \\ h_1(x, 2t - 1) & \text{for} \quad \frac{1}{2} \le t \le 1 \end{cases}$$

is a homotopy between a and c, such that

$$h(0, t) = h(1, t) = x_0.$$

● PROBLEM 19-13

Let $L(X, x_0)$ denote the set of all loops in X with base point x_0. For any a_1, $a_2 \in L(X, x_0)$, we define an operation $\oplus$ as follows:

$$(a_1 \oplus a_2)(t) = \begin{cases} a_1(2t), & 0 \le t \le \frac{1}{2} \\ a_2(2t - 1), & \frac{1}{2} \le t \le 1 \end{cases}.$$

Show that

$$a_1 \oplus a_2 \in L(X, x_0).$$

See Figure 1.

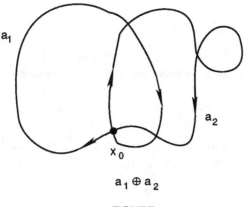

$a_1 \oplus a_2$

FIGURE 1

SOLUTION:

Loop $a_1 \oplus a_2$ is formed by moving arround a_1 once and then around a_2 once.

$$a_1 \oplus a_2 : [0, 1] \to X.$$

$a_1 \oplus a_2$ is continuous on $[0, \frac{1}{2}]$ and on $[\frac{1}{2}, 1]$ because both a_1 and a_2 are continuous and at $t = \frac{1}{2}$

$$(a_1 \oplus a_2)\,(\tfrac{1}{2}) = a_1(1) = x_0 = a_2(0).$$

Also

$$(a_1 \oplus a_2)\,(0) = a_1(0) = x_0$$

$$(a_1 \oplus a_2)\,(1) = a_2(1) = x_0.$$

Hence,

$$a_1 \oplus a_2 \in L(X, x_0).$$

● **PROBLEM 19-14**

We proved that ~ (is homotopic relative y_0 to) is an equivalence relation on $L(X, x_0)$. We denote

$$\frac{L(X,x_0)}{\sim} = \pi(X,x_0)$$

the set of homotopy relative to x_0 classes of $L(X,x_0)$. If $a \in L(X, x_0)$, then by

$$[a] \in \pi(X, x_0)$$

we denote the equivalence class of a.

713

Justify the following definition:

$$[a_1] \oplus [a_2] = [a_1 \oplus a_2].$$

SOLUTION:

We define $[a_1] \oplus [a_2]$ to be the equivalence class of the loop $a_1 \oplus a_2$. In Problem 19–13, we proved that $a_1 \oplus a_2$ is a loop.

$$a_1 \oplus a_2 \in L(X, x_0).$$

Taking its homotopy equivalence class $[a_1 \oplus a_2]$, we obtain an element of $\pi(X, x_0)$.

To make sure that $[a_1] \oplus [a_2] = [a_1 \oplus a_2]$ is a well-defined operation on $\pi(X, x_0)$, it must be shown that this definition is independent of the representatives a_1 and a_2 that we choose. This will be done in the next problem.

● PROBLEM 19–15

Let

$$a_1, a_2, a_3 \in L(X, x_0).$$

Show, that if $a_1 \sim a_3$, then

$$a_1 \oplus a_2 \sim a_3 \oplus a_2.$$

SOLUTION:

Since $a_1 \sim a_3$, a homotopy relative to x_0 exists between a_1 and a_3, such that

$$h : [0, 1] \times [0, 1] \to X.$$

We define

$$H : [0, 1] \times [0, 1] \to X$$

by

$$H(r,t) = \begin{cases} h(2r,t), & 0 \le r \le \frac{1}{2} \\ a_2(2r - 1), & \frac{1}{2} \le r \le 1 \end{cases}$$

Thus,

714

$$H(r,0) = \begin{cases} h(2r,0) = a_1(2r), & 0 \le r \le \frac{1}{2} \\ a_2(2r-1), & \frac{1}{2} \le r \le 1 \end{cases}$$

and

$$H(r, 0) = (a_1 \oplus a_2)(r) \text{ for } r \in [0, 1].$$

Also

$$H(r, 1) = (a_3 \oplus a_2)(r) \text{ for } r \in [0, 1].$$

We have

$$H(0, t) = h(0, t) = x_0$$

$$H(1, t) = a_2(1) = x_0.$$

Continuity of H can easily be shown. Hence,

$$a_1 \oplus a_2 \sim a_3 \oplus a_2.$$

● **PROBLEM 19-16**

Show that the operation
$$[a_1] \oplus [a_2] = [a_1 \oplus a_2] \qquad (1)$$
defined in Problem 19–14 is well-defined.

SOLUTION:

Suppose $a_1 \sim a_3$ and $a_2 \sim a_4$. In Problem 19–15, we proved that

$$a_1 \oplus a_2 \sim a_3 \oplus a_2. \qquad (2)$$

Similary, we can show that

$$a_1 \oplus a_2 \sim a_1 \oplus a_4. \qquad (3)$$

Thus,

$$[a_1 \oplus a_2] = [a_1] \oplus [a_2] = [a_3 \oplus a_2] =$$
$$= [a_3] \oplus [a_2] \qquad (4)$$

and

$$[a_1] \oplus [a_2] = [a_1] \oplus [a_4]. \qquad (5)$$

715

We have

$$a_1 \oplus a_2 \sim a_3 \oplus a_2 \sim a_3 \oplus a_4. \tag{6}$$

Hence, if $[a_1] = [a_3]$ and $[a_2] = [a_4]$, then

$$[a_1] \oplus [a_2] = [a_3] \oplus [a_4]. \tag{7}$$

● PROBLEM 19-17

Draw the homotopy $H(r, t)$ described in Problem 19–15. Explain why $H(r, t)$ is continuous on $[0, 1] \times [0, 1]$.

SOLUTION:

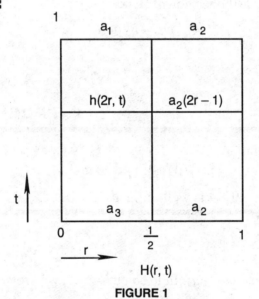

FIGURE 1

On $[0, \frac{1}{2}] \times [0, 1]$, we continuously deform a_1 into a_3, while a_2 remains fixed on $[\frac{1}{2}, 1] \times [0, 1]$.

For $r = \frac{1}{2}$, we have

$$H(\tfrac{1}{2}, t) = h(1, t) = a_2(1) = x_0$$

$$H(\tfrac{1}{2}, t) = a_2(0) = x_0.$$

The homotopy $H(r, t)$ is continuous on $[0, \frac{1}{2}] \times [0, 1]$ and on $[\frac{1}{2}, 1] \times [0, 1]$, hence, it is continuous on $[0, 1] \times [0, 1]$.

Show that operation $\oplus$, defined on $\pi(X, x_0)$, is an associative operation; that is, show that

$$([a_1] \oplus [a_2]) \oplus [a_3] = [a_1] + ([a_2] \oplus [a_3]).$$

SOLUTION:

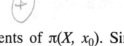

Let $[a_1]$, $[a_2]$, $[a_3]$ represent any elements of $\pi(X, x_0)$. Since operation $\oplus$ defined on $\pi(X, x_0)$ does not depend on the choice of representatives, it suffices to show, that

$$(a_1 \oplus a_2) \oplus a_3 \sim a_1 \oplus (a_2 \oplus a_3).$$

By definition

$$((a_1 \oplus a_2) \oplus a_3)(r) = \begin{cases} (a_1 \oplus a_2)(2r); & 0 \le r \le \frac{1}{2} \\ a_3(2r - 1); & \frac{1}{2} \le r \le 1 \end{cases}.$$

Thus,

$$((a_1 \oplus a_2) \oplus a_3)(r) = \begin{cases} a_1(4r); & 0 \le r \le \frac{1}{4} \\ a_2(4r - 1); & \frac{1}{4} \le r \le \frac{1}{2} \\ a_3(2r - 1); & \frac{1}{2} \le r \le 1 \end{cases}.$$

Similarly,

$$(a_1 \oplus (a_2 \oplus a_3))(r) = \begin{cases} a_1(2r); & 0 \le r \le \frac{1}{2} \\ (a_2 \oplus a_3)(2r - 1); & \frac{1}{2} \le r \le 1 \end{cases} =$$

$$= \begin{cases} a_1(2r); & 0 \le r \le \frac{1}{2} \\ a_2(4r - 2); & \frac{1}{2} \le r \le \frac{3}{4} \\ a_3(4r - 3); & \frac{3}{4} \le r \le 1 \end{cases}.$$

The homotopy between $(a_1 \oplus a_2) \oplus a_3$ and $a_1 \oplus (a_2 \oplus a_3)$ is defined as follows:

$$H(r,t) = \begin{cases} a_1(\frac{4r}{1+t}); & 0 \le r \le \frac{1}{4}(1+t) \\ a_2(4r-1-t); & \frac{1}{4}(1+t) \le r \le \frac{1}{4}(2+t). \\ a_3(1-\frac{4(1-r)}{2-t}); & \frac{1}{4}(2+t) \le r \le 1 \end{cases}$$

One can directly compute that $H(r, t)$ is indeed the homotopy between $(a_1 \oplus a_2) \oplus a_3$ and $a_1 \oplus (a_2 \oplus a_3)$.

● **PROBLEM 19-19**

Draw the homotopy defined in Problem 19–18. Explain why it is a suitable homotopy.

SOLUTION:

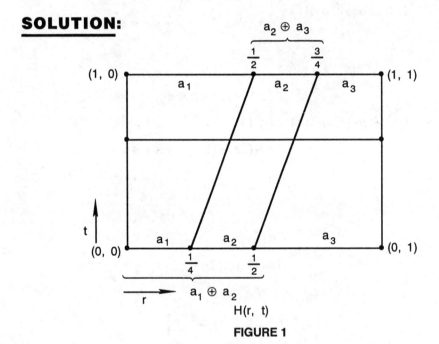

FIGURE 1

The homotopy $H(r, t)$ was defined as follows:

$$H(r,t) = \begin{cases} a_1(\frac{4r}{1+t}); & 0 \le r \le \frac{1}{4}(1+t) \\ a_2(4r-1-t); & \frac{1}{4}(1+t) \le r \le \frac{1}{4}(2+t). \\ a_3(1-\frac{4(1-r)}{2-t}); & \frac{1}{4}(2+t) \le r \le 1 \end{cases}$$

Note that for $t = 0$

$$H(r,0) = \begin{cases} a_1(4r); & 0 \leq r \leq \frac{1}{4} \\ a_2(4r - 1); & \frac{1}{4} \leq r \leq \frac{1}{2} \\ a_3(2r - 1); & \frac{1}{2} \leq r \leq 1 \end{cases}$$

which is $((a_1 \oplus a_2) \oplus a_3) (r)$.

Similarly, for $t = 1$, we have

$$H(r,1) = \begin{cases} a_1(2r); & 0 \leq r \leq \frac{1}{2} \\ a_2(4r - 2); & \frac{1}{2} \leq r \leq \frac{3}{4} \\ a_3(4r - 3); & \frac{3}{4} \leq r \leq 1 \end{cases}$$

which is $(a_1 \oplus (a_2 \oplus a_3)) (r)$. It is evident that $H(r, t)$ is continuous.

● **PROBLEM 19-20**

Let

$$f : [0, 1] \to x_0 \in X$$

denote a constant function that maps $[0, 1]$ onto $x_0 \in X$. Show that $[f]$ is an identity for $\pi(X, x_0)$ with respect to $\oplus$.

SOLUTION:

We shall prove, that for any $[a] \in \pi(X, x_0)$,

$$[a] \oplus [f] = [f] \oplus [a] = [a].$$

By definition

$$[a] \oplus [f] = [a \oplus f]$$

and

$$(a \oplus f)(r) = \begin{cases} a(2r); & 0 \leq r \leq \frac{1}{2} \\ f(2r - 1); & \frac{1}{2} \leq r \leq 1 \end{cases}.$$

719

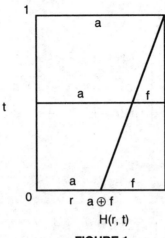

H(r, t)

FIGURE 1

The figure illustrates the homotopy between $a \oplus f$ and a. Hence,

$$[a] \oplus [f] = [a].$$

Similarly, we show that

$$[f] \oplus [a] = [a].$$

● **PROBLEM 19-21**

Let a denote a loop in X with base point x_0, $a \in L(X, x_0)$. Define an inverse loop a^{-1}.

SOLUTION:

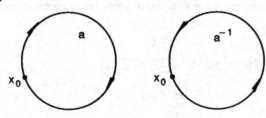

FIGURE 1

We define a^{-1} by

$$a^{-1}(r) = a(1 - r)$$

for each $r \in [0, 1]$.
Then

$$a^{-1}(0) = a(1) = x_0$$

720

and

$$a^{-1}(1) = a(0) = x_0.$$

Also, since a is continuous, so is a^{-1}, hence,

$$a^{-1} \in L(X, x_0).$$

Note that a^{-1} has the shape of a, but is drawn in the opposite direction.

● PROBLEM 19-22

Define the fundamental group of the space X, given that $x_0 \in X$.

SOLUTION:

A topological space (X, T) is given and $x_0 \in X$. $L(X, x_0)$ is the set of all loops in X with base point x_0.

On the set of homotopy, the classes of $L(X, x_0)$ denoted by $\pi(X, x_0)$, we defined operation $\oplus$, which is associative and has an identity $[f]$ with respect to x.

It is easy to prove, that

$$a \oplus a^{-1} \sim a^{-1} \oplus a \sim f$$

by using the homotopy between $a \oplus a^{-1}$ and f

$$H(r,t) = \begin{cases} a(2r(1-t)); & 0 \le r \le \frac{1}{2} \\ a(2(1-r)(1-t)); & \frac{1}{2} \le r \le 1 \end{cases}.$$

Similarly, we can define the homotopy between $a^{-1} \oplus a$ and f. Thus, the inverse of $[a] \in \pi(X, x_0)$ is $[a^{-1}]$.

The space $\pi(X, x_0)$ with operation $\oplus$ is a group with identity $[f]$. Each $[a] \in \pi(X, x_0)$ has its inverse $[a^{-1}] \in \pi(X, x_0)$.

DEFINITION

$(\pi(X, x_0), \oplus)$ is called the fundamental group of the space X based on $x_0 \in X$.

■

Consider the space

$$X = \{(x, y) : x^2 + y^2 = 1\} \subset R^2.$$

Let x_0 represent any point of X and a the loop $a \subset X$, which goes once around the circle in the clockwise direction. For each integer $n \in Z$ and each loop $a \in L(X, x_0)$, we define

$$na = \begin{cases} \overbrace{a \oplus a \oplus \ldots \oplus a}^{n \text{ times}} & \text{for } n \text{ positive} \\ f & \text{for } n = 0 \\ \underbrace{a^{-1} \oplus a^{-1} \oplus \ldots a^{-1}}_{-n \text{ times}} & \text{for } n \text{ negative} \end{cases}$$

where

$$f : [0, 1] \to x_0.$$

Find the fundamental group of X based on x_0.

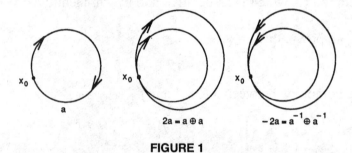

$2a = a \oplus a$ 　　$-2a = a^{-1} \oplus a^{-1}$

FIGURE 1

SOLUTION:

Note, that

$$m\,a \sim n\,a$$

if and only if $m = n$. For every loop $a_1 \in L(X, x_0)$, there is $n \in Z$, such that

$$a_1 \in [n\,a].$$

Thus, there is an isomorphism h between

$$(\pi(X, x_0), \oplus) \text{ and } (Z, +)$$

given by

$$h([n\,a]) = n.$$

Group $(\pi(X, x_0), \oplus)$ is generated by one element $[a]$.

Find the fundamental group based on $\vec{x_0}$ of the space X

$$X = \{(x, y, z) : x^2 + y^2 + z^2 = 1\} \subset R^3$$

$\vec{x_0} \in X$.

SOLUTION:

Let A represent any point of X. Then $X - \{A\}$ is homeomorphic to R^2 and therefore, a contractible space. The plane λ is tangent to X at A'. A' is the point of X antipodal to A.

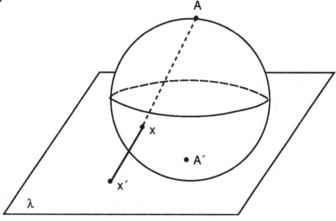

FIGURE 1

The projection of a point $x \in X - \{A\}$ is a point $x' \in \lambda$ determined by the intersection of a straight line $\overline{Ax}$ with the plane λ. Thus, $X - \{A\}$ and λ are homeomorphic.

If a is any loop in X and A is any point in $X - a$, then a is a loop in R^2 (remember R^2 and $X - \{A\}$ are homeomorphic).

Thus, a is homotopic to the identity function f of $\pi(X, x_0)$. Hence, any loop in X based on $x_0 \in X$ is homotopic to f.

Therefore, $\pi(X, x_0)$ consists only of $[f]$.

GLOSSARY OF SYMBOLS AND ABBREVIATIONS

Symbol	Meaning	See Problem where symbol appears
$\alpha, \beta, \gamma, \delta, \ldots$	sentences	1-1
0,1	logical values	1-1
$\equiv$	equivalent	1-1
$\vee$	or	1-1
$\wedge$	and	1-1
$'$	not	1-1
$\Rightarrow$	implies	1-5
$\Leftrightarrow$	if and only if	1-5
iff	if and only if	1-5
$\div$	symmetric difference	1-11
■	end of a proof, theorem, or definition	1-6
$a \in A$	a is an element of A	2-1
$a \notin A$	a is not an element of A	2-1
$\{a, b, c, \ldots\}$	set consisting of elements $a, b, c, \ldots$	2-1
$\{x : p(x)\}$	set of all x such that $p(x) \equiv 1$	2-1
N	set of all natural numbers	
Z	set of all integers	
Q	set of all rational numbers	
R	set of all real numbers	
C	set of all imaginary numbers	
$A \subset B$	A is a subset of B	2-3
$A \cup B$	union of A and B	2-4
$A \cap B$	intersection of A and B	2-4
$A - B$	difference of A and B	2-4
ϕ	empty set	2-7
$P(A)$ (or 2^A)	power set of A	2-8
A^C (or $X - A$)	complement of A	2-12

724

INDEX

Numbers on this page refer to <u>PROBLEM NUMBERS</u>, not page numbers.

Accumulation pt 11-31/P.441
13-6(P.504

729

Numbers on this page refer to **PROBLEM NUMBERS**, not page numbers.

13-33

TIETZE'S EXT. THM. 19-7